T0338251

Visualization of Fields and Applications in Engineering

Visualization of Fields and Applications in Engineering

Stephen Tou

A John Wiley and Sons, Ltd., Publication

This edition first published 2011
© 2011 John Wiley & Sons Ltd.

Registered office
John Wiley & Sons Ltd, The Atrium, Southern Gate, Chichester, West Sussex, PO19 8SQ,
United Kingdom

For details of our global editorial offices, for customer services and for information about how
to apply for permission to reuse the copyright material in this book please see our website at
www.wiley.com.

Library of Congress Cataloging-in-Publication Data

Tou, Stephen.
  Visualization of fields and applications in engineering / Stephen Tou.
    p. cm.
  Includes bibliographical references and index.
  ISBN 978-0-470-97397-4 (hardback)
    1. Engineering mathematics.   2. Fluid dynamics–Mathematics.   3. Electromagnetic
fields–Mathematical models.
4. Gravitational waves–Mathematical models.   5. Information visualization.   I. Title.
  TA330.T68 2011
  620.001'51–dc22

                                                                        2010039468

A catalogue record for this book is available from the British Library.

Print ISBN: 978-0-470-97397-4
ePDF ISBN: 978-0-470-97826-9
oBook ISBN: 978-0-470-97825-2
ePub ISBN: 978-0-470-97846-7

Set in 10/12pt Palatino by Aptara Inc., New Delhi, India.

Contents

Preface

Visual representation is an art in science to deliver the beauty of patterns and graphs otherwise difficult to convey in words, if not impossible. The purpose of this manuscript is to present the basic techniques for tensor field visualization and mapping from an engineering approach. A tremendous amount of research and development has been devoted to the scientific visualization of field data in computer science, engineering and data/image processing technology. The manuscript focuses on the fundamental aspects of post processing digital database and applications. An attempt is made to explore existing theories and their integration in tensor field visualization, transformation and analysis. Examples cover a variety of engineering problems with applications beyond and across disciplines.

Evolution and development of visual science driven by computer technology have been so rapid and extensive in arts and physical sciences that it is timely to introduce the subject from an applied approach. The materials are suitable for academic use. The manuscript is also intended to serve as a source of references for those who work in data representation, transformation/analysis or in scientific visualization and applications.

The acknowledgement due on completion of this manuscript is far beyond words. The author would like to dedicate this work to those who made it possible and to express his gratitude for the understanding and support from his family. The author thanks them all from the bottom of his heart.

S.K.W. Tou
2010

1

Introduction

1.1 A General View

Newton's law of universal gravitation describes the interaction between discrete point masses at distances apart. The nature of gravitational forces and the idea of an object interacting on a distant object of finite sizes are easier to visualize when a model of force fields is employed. This idea of a field model came about because of the conceptual problems inherent in Newton's law of universal gravitation interpreted as action at a distance. The idea lends itself to pictorial description and explains how one object knows the existence of another object in a field model. With the introduction of concept of force transmission and concepts of fields and potentials, we think of the space surrounding the earth as permeated by a gravitational field of force created by the earth. Any mass placed in the field experiences a force due to this field. Gravitational field is one of the physical fields in nature that exerts a force at a mass center point in space with a specific magnitude and direction towards the earth center, giving rise to a field line (line of force). Gravitational fields are represented by vector diagrams using a system of line of forces radiating from the earth center like a pencil of rays. It is then said there exists a first order tensor field. Tensors are multivariate data embedded with complex information in time and space. A vivid pictorial representation of tensor field by field line distributions is essential for an understanding and perception of field characteristics and behaviors. It adds significantly to our ability to interpret physical phenomena. An accurate graphical representation of a first order tensor field must convey both magnitudes and directions of the fields in space. In addition, the point or line of action is also essential. Several graphical display methods are in common use. In a

Visualization of Fields and Applications in Engineering, First Edition. Stephen Tou.
© 2011 John Wiley & Sons, Ltd. Published 2011 by John Wiley & Sons, Ltd.

vector diagram, the length or thickness of each vector line segment at selected points is drawn proportional to its magnitude with arrows pointing in the field directions. These discrete geometric vectors making up a field of arrows are regarded as position vectors emanating from the points of interest. They generally cause cluster. In this respect, field line trajectories are commonly used to compliment a vector diagram with continuity. A tangent along a field line trajectory defines the direction of vector fields at a point. Field line trajectories are a graphical representation of lines of action continuously throughout the field domains. They are a system of non-intersecting curves in space except at a singular/critical point. This field trajectory method is generally applicable to any vector fields regardless of their nature, source or domain dimension.

The other method employs a family of flux lines (streamlines) generated from the corresponding vector potential (stream function) when vector fields are steady, two dimensional and solenoidal. Similarly, tangents along a flux line trajectory define the vector field directions. Flux line trajectories are spaced in such a way that the number of lines crossing a unit area placed perpendicular to the field at some points is proportional to the magnitude of the field at that point. When flux lines are displayed at equal contour intervals in a physical plane, the resulting flux density is a measure of field strength since flux bounded between flux lines is invariant in solenoidal fields. Flux lines are crowded closely where the field is strong or conversely. They obviate the cumbersome use of arrow segments having various length or thickness. Flux line trajectories are a contour representation of stream function, which assumes a constant different from one streamline to another. The engagement with stream function gives flux line trajectories the advantage over field line trajectories. Stream function ensures equation of continuity is satisfied everywhere. It is able to determine vector field directions, magnitudes and property transport quantitatively.

In irrotational fields, another set of lines referred to as equipotentials is produced to provide another means of vector field visualization. They are derived from the corresponding scalar potential (or potential function) which describes field transport capability quantitatively in terms of potential levels with respect to a reference datum. Potential gradients indicate vector field magnitudes and directions, which correspond to the tangents along flux lines. Along these directions maximum spatial rate of potential change occurs. In two-dimensional fields, equipotentials are also spaced at constant contour intervals such that equipotential density is a measure of field strength. Equipotential contours are crowded closely where the field is strong or conversely.

For harmonic fields, both equipotentials and streamlines exist and are mapped together to produce a field map in a physical plane as shown in Figure 1.1. These two sets of curves cross each other orthogonally forming a

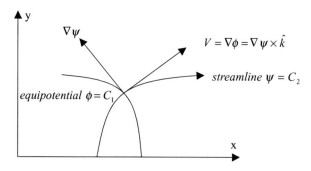

Figure 1.1 Field map.

network of curvilinear squares. In particular, when field maps are produced with a constant ratio of contour intervals equal to curvilinear square aspect ratio of unity, they are an effective means for graphical representation of vector fields. Thus, scalar potentials and vector potentials play a central role in forming the platform for vector field visualization and analysis. In the present study, they are also employed to construct complex potentials paving the way for mapping fields between complex planes.

Graphical representations of second order tensors remain challenging. Traditionally they are visualized through their principal states. In the present study, the tensor rank reduction method is applied to extract tensor vectors on a reference plane or at a point by contraction. The method is extended to include the use of tensor invariants. Last but not least, the hybrid method of displaying evolving tensor ellipse icons along principal axes trajectories is effective for second order tensor field visualization.

1.2 Historical Development and Progress in Visual Science

Graphical representation and visualization of fields and transport phenomena go back centuries. Ancient Babylonians/Egyptians and historians of architecture/artists developed the concept of measurement and descriptive geometry to describe patterns and beauty of shapes. Leonardo da Vinci, a great scientist and artist in the Renaissance period described movement of water, vortices and floating objects and presented sketches of botany and streamline/shapes. Regular polygons/polyhedra were subjects of special study by ancient Greeks. They led to visual conceptual graphics in the field of astronomy reflected in the work of Kepler. In his scientific exploration of the universe, Kepler succeeded in describing the planet systems and orbits by arranging a set of regular polyhedral in three-dimensions. As the arts of visual representation of geometry continue to evolve, an abstract geometry of several dimensions or hypergraphics was then systematically developed. The combination of arts and sciences has led to unprecedented technological advancements and revolutionary developments. From classical concept of action-at-a-distance to the theory of contiguous action and modern concept of fields in field theory, the development has made significant contributions to the pool of knowledge in visual sciences. Historically, R. Norman as a navigational instrument maker about Kepler's time put the idea of a magnetic field forward. Gilbert continued his work to visualize magnetic effects with the birth of field conception coined orbis virtutis. Until then the concept of field lines was originated from Faraday in his study of electrical sciences in an attempt to describe electric field structure. Field lines have become one of the basic techniques of visual representations to interpret field phenomena ever since. They not only lay the foundation ideas of field visualization but also opened the window of applications to a wide range of field phenomena in physical sciences. Field lines are regarded as the structural elements of a field and are represented by straight lines perpendicular to the charge surface. Faraday was convinced on experimental grounds that the physical nature of line of force was action propagating between contiguous particles through a medium. The physical reality to these field lines was then reckoned as material lines consisting of a chain of polarized particles subject to longitudinal strain. This gives rise to an electric phenomenon referred to as the arrangement of electric dipoles. As the ideas took shape, they made significant influence on modern field theory development. Maxwell developed Faraday's ideas into mathematical forms of a theory of contiguous action. He conveyed the importance and physical reality of fields as stress-transmitting media through which electromagnetic waves propagate in specific directions at some finite speeds. The theory integrated forces in nature and exhibited the underlying similarity of field phenomena with a unified view, which remains hidden from direct perception. At about the same time, the concept of continuum and notion of a field theory began to emerge in the work of Euler in hydrodynamics and in the theory of sounds. He laid the foundation for Eulerian description of fluid motions and presented mathematical models to describe the field of motion of inviscid fluids in an inertia frame. The law of motion was applied to point masses in the differential form of conservation of linear momentum for continuous media, bridging up with Newtonian mechanics.

In hydrodynamics, D'Alembert was credited with the introduction of stream function concept for flow visualization. In the study of elasticity, the introduction of Airy stress function and/or strain potential function makes it possible to describe the state of a second order tensor in terms of a single scalar function. Prandtl then developed another version of stress function to obtain stress trajectories produced by torsion. It plays the same role as stream functions introduced by Lagrange and Stokes who laid the foundation work of graphical representation of vector fields in hydrodynamics. Bejan and his co-workers (Kimura and Bejan 1983, 1985; Morega and Bejan 1993; Trevisan and Bejan 1987) then extended streamline technique to transport phenomena involving heat and mass. They are known as heatfunction and massfunction.

Poincare established connections between vector field topology and differential equations. His work laid the foundations for phase-space technique in the study of vector spaces and dynamic systems. This is followed by the development of theory in two-dimensions in the middle of the twentieth century. Mapping of fields has added a new horizon in visual science. Needham (1997) presents a vivid example of electromagnetic fields in television/radio, which is described completely by mapping of electric fields and magnetic fields. Nevertheless, all these pioneer works show the vital importance of tensor field representations and visualizations in a vast range of physical science applications. They have set the stage for modern development and practice in the arts of visual representation.

Scientific visualization developments have gone beyond and across various disciplines. A brief review of current state-of-the-art scientific visualization is a useful starting point to follow its progress, research and directions of development. Graphical representation of data and object has appeared in the work of statistics, botany and earth sciences in the last couple of decades. Since then, computer technology has changed the face of the world and dominated human activities in all trades. Scientific visualization and image processing research have been active among computer scientists and engineers. Computer scientists have keen interests in the visualization of tensor field topological structure and in the development of visual icons ranging from point, line, surface and volume domains at various information levels. Researchers who presented investigations on the topology of simulated tensor fields have suggested the methods of field topology classification and representation based on analysis of critical points (Helman and Hesselink, 1989a, 1989b, 1990, 1991). This data compression/extraction technique was applied to the visualization of higher order critical points.

Computer graphics based iconographic techniques play an important role in the development of visual icons. They are effective visual attributes useful in general applications. In particular, computer graphics techniques make use of color/texture encoding with appealing effects. Elementary vector icons such as arrows (point icon), streamlines (line icon) and stream surfaces/stream ribbons (surface icon) have been in use for sometime. Elementary tensor icons such as Lame stress ellipse and Cauchy stress quadric were introduced in the theory of elasticity at an earlier date. Other areas of development include for instance, ellipsoids, tensor glyphs and probes. These are tensor point icons showing the state of a tensor field at a point. Tensor line icons (lines of principal stress) were introduced formally by Woods (1903), Dickinson (1989, 1991) in elasticity referred to as tensor field trajectories based on principal axes transformation. These replace the point-wise glyphs, which lack data continuity and cause cluster. A series of such trajectories is used to display the continuous distributions of principal directions, which describe load transmissions and/or transport paths in a system. The corresponding principal values are usually presented in contour forms to depict field strength. Hyperstreamlines are then presented by researchers (Delmarcelle and Hesselink, 1992, 1993, 1994) as an extension of tensor field trajectories. Hyperstreamlines sample tensor fields along their principal trajectories and integrate a continuous distribution of point icons. These topological skeletons embed complex information through colors and textures.

Following the development of phase-space technique in data visualization, a number of methods has since then sprung-off. Research work in unsteady field visualization was in progress in a great number of research centers around the globe. Graphical line integral convolution method was introduced to represent the texture of vector fields. This improved rendering technique converted data into images and was free of seeding problems. Development in feature-based visualization includes image based flow visualization. Three dimensional visualization techniques are also in the pipeline of research and development. Three-dimensional icons were then developed to produce three-dimensional texture from two-dimensional domains. Work is also in progress in combining experimental and computational visualization methods in complex environments such as combustions. Comparison studies on vector field visualization methods, classifications and research issues have been presented in literature.

In parallel to these research activities, engineers are also actively engaged in the development of visualization techniques in transport processes. A novel approach to heat transfer visualization based on the concept of heatlines was first introduced by Kimura and Bejan (1983, 1985) in their study of natural convection in enclosure heated from the side. It is an energy analog of streamlines in hydrodynamics. The analogy carried over to the introduction of heatfunction, which plays the same role as stream function. A useful feature of heatlines is the depiction of transport paths, which facilitate heat transfer analysis. The techniques were then applied extensively in a wide spectrum of transport processes in anisotropic medium, in porous medium, in unsteady buoyancy-driven flow, and in turbulent flows. Trevisan and Bejan (1987) then introduced the mass transfer analog of the same idea referred to as masslines. Other researchers presented a unified formulation of the two systems of lines in non-reacting and reacting jets. They also introduce the concept of conserved scalars to extend heatlines and masslines to fields with sources in post-processing visualization.

1.3 Scientific Visualization Philosophy, Techniques and Challenges

Computer technology leads to the generation of an increasingly huge amount of multivariate data of great complexity. The phenomenon of knowledge growth and flow of information is on an incredible scale. The database is not useful unless its information can be conveyed to the users effectively. There is a need to develop scientific visual technology blended with human visual systems. The work on scientific visualizations generally involves data management, data mining, graphical display, transformation, analysis and integration/correlation. The input effort has resulted in the production of various visualization systems; each has its merits or limitations. Scientific visualization technology aims at extracting meaningful representations from multivariate data sources suitable for visual communication and comprehension systems. Data mining methods are used to meet with a variety of goals. It includes searching for patterns, discovering useful data structure and exploring field properties and behaviors. Graphs/icons are an effective means of serving such a purpose to give a visual impression of information at a glance. The general philosophy is to integrate simplicity of graphical representations with continuity of the fields and analysis capability in quantitative terms. Implicit representation of curves/surfaces has played a key role in forging the basic ideas and techniques in field visualization. Implicit function makes it possible to represent relations as the graph of a function and has certain advantages over explicit expressions. This is because an explicit expression from an implicit expression may not exist. Implicit representation offers more flexibility and renders two-dimensional contour curves easily back to three-dimensional surface with the constant assuming the vertical axis. It is also ideal to test graphs/functions for possible symmetries. Visualization and identification of curves/surfaces are expedited in accordance with the value of parameter constant specified by the implicit function. Alternatively, parametric representation is capable of describing complicated curves/surfaces directly by considering a system of equations simultaneously as a function of a third independent variable. Certain graph visualization techniques are effective in displaying the characteristics and behaviors of fields based on the function itself. Plane/line-symmetry, point-symmetry or rotation symmetry, if it exists is useful for displaying field patterns as mirror reflection images in the same or opposite fashion. By observing and taking advantage of symmetry or anti-symmetry of any kind, it is possible to limit the study/visualization of field patterns in certain domains or regions. Similarly, periodicity is to be observed. Asymptotes including dividing streamlines are suitable for shaping the graph of a function in general and to separate the fields into distinct regions of behavior in particular. Asymptotes help to study the behavior of function at infinity. Critical/singular points are important visual elements as they contribute significantly to the nature of fields. Distributions of critical points or singular points provide key information about field behavior since they possess unique topological skeletons or field patterns, which unveil the nature of these points and the fields as well. The displays of field topology in the neighborhood of critical points not only reveal local phenomena but also have connections to global behaviors. The way dividing streamlines split at a critical point sheds light on the interaction between dynamic aspects of fields and boundary geometry/conditions. By extracting salient views and tensor topological structure, they can be blended to provide a comprehensive understanding of the original fields.

The method of transformation is of great value in the analysis and visualization of field behaviors. Mapping of field patterns from a given plane onto another with a great variety of transformation complements each other and produces new information from old. Gradient mapping takes a point in the domain of a scalar field onto a gradient vector pointing in the direction of maximum spatial rate of property change at that point. The gradient system is a vivid display of "runoff" of water flowing downhill towards local minima or away from local maxima. Mapping of singular points to points at infinity in some planes makes it possible to evaluate field properties such as shape factor by graphical means and to interpret the concept of infinity in a geometric setting. Other special points of interest include, but are not limited to, multiple points, higher order critical points, degenerate points of higher multiplicity, inflection points, cuspidal points, points of contact, vertices and so on. These points are invariant in coordinate transformation and help to shape the trajectories in one way or another. Analytic functions of complex variables open the window of exploration for new complex potentials through synthesizing or mapping (composite/differentiation/integration) in the search for target field models. We could study field patterns on different domain boundaries or produce distortion/warping effects on images by conformal transformation. Hodograph method and mapping make it possible to visualize the free boundary of unknown shapes in the physical plane or the inverse images on other complex planes. Hodograph representation of trajectories or field patterns compliments the position hodograph representation and is useful for interpretation of field phenomena. This includes visualization and mapping of trajectories/tangents, straight boundaries (impervious/equipotential/seepage) or radial lines/circles about origin from hodograph plane to physical plane and conversely. Critical points and dividing trajectories are mapped to origin and curves passing through on the hodograph plane. By mapping circles about origin from the hodograph plane, we could determine locus of constant speed or kinetic energy contours on the physical plane. Likewise, by mapping straight lines passing through origin from hodograph plane to physical plane, we could visualize locus of isoclines that is the constant directions of vector fields equal to the slope of the straight lines.

Icons, colors and textures are useful visual attributes to take advantage of human visual systems. As the amount of rendering needs to be economized, there appears to be no need to render every piece of data by a glyph or some other means. It is, however, essential to deliver quantitative and qualitative information at selected points of interest and to display salient features of the fields with continuity. The objectives are to capture field structure in a subset of data and to zoom onto regions of interest through integration and correlation of information from different perspectives. Displays of both velocity arrow icons and acceleration arrow icons at a point are effective at describing the state of motions. By decomposing the acceleration vector into streamline tangential and normal directions through the law of parallelogram, we could determine whether flow is accelerating, decelerating or changing direction. Speed will increase or decrease when the tangential component is in the same or in the opposite velocity direction. Motion is at a constant speed if this component vanishes or the acceleration vector and velocity are mutually perpendicular. Alternatively, we could arrive at the same findings based on the values (positive, negative or zero) of the dot product of the two vectors. The normal component of acceleration vector assumes the role of centripetal acceleration always points towards the center of curvature indicating direction changes. If this component vanishes, it may be a point of inflection or a straight path. The same idea could be used in other vector fields such as mechanical energy flux or heat flux. The displays of mechanical energy flux vector, heat flux vector and velocity at a point allow us to assess the energy transfer processes. To maintain data continuity, we focus on computer graphic techniques of visualization using line icons such as streamlines (flux line trajectories), field line trajectories and potential contours. In this connection, vector potentials including, but are not limited to mechanical energy function, vorticity function and warping flux function, are used to visualize transport trajectories analogous to stream function and streamlines.

Because of the limitations of complex functions in two-dimensions and the complexity of flux (dual) functions in three-dimensions, the present work focuses on a two-dimensional approach. The cross section method is effective to present graphs from multi-dimensions on coordinate planes by holding the spatial variables constant systematically. A pseudo three-dimensional field could be constructed by staggering up a combination of coordinate planes. With regard to second order tensor fields, we apply tensor rank reduction techniques. These include contractions, tensor invariants, tensor icons and principal axes trajectories. The latter is connected to the idea of independently moving graphical icons leading to two-dimensional ellipse-based principal axes trajectories (Kerlick, 1990). The hybrid method is able to maintain data continuity and convey decoding information visually by a sequence of evolving tensor ellipses animated along a transport path. The connections between vector field visualization, mapping and those in tensor fields lay the platform for integration and correlation in mixed field phenomena.

Almost half a century has elapsed since the birth of the first two-dimensional television set. Today a three-dimensional television set remains to be seen. The difficulty and cost of developing three-dimensional visual technology and high order tensor representations present a compelling challenge. Simulation of three-dimensional scenes/images is possible in certain fields but has not yet become widespread. The development in stereoscopic technique, virtual reality, two-dimensional data integration with an elevation model, computer-aided design, optical images, graphical simulation and visual comprehension are encouraging and have attracted a great deal of research interest. Intriguing work has been presented in the art of hyper-graphics for multi-dimension geometry. It appears that the inherent problems of presenting three-dimensional graphs on two-dimensional views/papers will remain with us and impose restrictions on our visual ability. The problems inevitably demand three-dimensional visual technology and viewing hardware. Until then multiple views and transformation will be and always will be in demand in a visualization system. To present multiple levels of complexity of fields and models, there is a need to develop visualization of multiple tensor fields simultaneously to study the structure and interaction of one field in the context of another or more. This calls for multivariate visualization methods in an integrated system. Visual representations become more complex when fields are unsteady. There are a number of hurdles in the quest for knowledge and information visualization. The missions are expected to add new dimensions in scientific visualizations and in modern theory development.

2

Field Descriptions and Kinematics

2.1 Lagrangian/Eulerian Description and Transformation

A reference frame is essential in the study of particles in motion or the inherent response of an object subjected to deformation and stresses. The position of a particle in space is specified by a set of independent variables in a chosen reference frame giving rise to coordinate representation of trajectories. The reference frame is a coordinate system and a set of synchronized clocks to define the spatial position at any time instant. The variables have dual nature depending on the viewpoint. Consider a scenario of fluids in motion. An observer is attached to an arbitrary field particle at a point in space and moves with it to a different position at a later time $t > t_0$. The spatial position of the label particle is defined as the point of intersection of three coordinate surfaces with respect to a curvilinear coordinate system ξ_i, also referred to as Lagrangian or material coordinate. The observer locates the position of particle she/he is riding with by referring to Cartesian frame x_i. Without loss of generality, the observer's position is defined by position vector $\vec{r}$ being a function of ξ_i and t known as Lagrangian variables. The position vector describes the path of a particle moving from one position to another in the continuum as time proceeds. It constitutes a continuous mapping of points in space from t-axis. In this Lagrangian viewpoint, attention is on a specific particle. We observe changes in field property associated with the particle by following its course of motion. At time $t = t_0$, a specific particle is labeled $\xi_i = \xi_0$ by a Lagrangian coordinate system. This is equivalent to specifying

Visualization of Fields and Applications in Engineering, First Edition. Stephen Tou.
© 2011 John Wiley & Sons, Ltd. Published 2011 by John Wiley & Sons, Ltd.

the initial condition for the time dependent labeled particle by its instantaneous location.

$$\bar{r} = \bar{r}(\xi_i, t) = \sum x_i \hat{e}_i \tag{2.1a}$$

where

$$x_i = x_i(\xi_i, t) \tag{2.1b}$$

Initial condition: $t = t_0$, $\xi_i = \xi_0$

$$\bar{r} = \bar{r}(\xi_0, t) \tag{2.1c}$$

$$x_i = x_i(\xi_0, t)$$

Or

$$x = x(\xi_0, t) = x(x_0, y_0, z_0, t)$$
$$y = y(\xi_0, t) = y(x_0, y_0, z_0, t)$$
$$z = z(\xi_0, t) = z(x_0, y_0, z_0, t)$$

$\hat{e}_i$ is the i-th Cartesian coordinate unit vector. The velocity of particle is determined by definition from material derivatives of position vector holding ξ_0 constant, that is,

$$V = \frac{d\bar{r}}{dt}$$

The above kinematics defines motion for a continuum. This is the particle method of material description in Lagrangian coordinates, which provides historical information about an individual particle moving along the path line as a function of time. The subject method of analysis is equivalent to the concept of a closed system.

In a different scenario, an observer is attached to a point x_i in space with reference to the Cartesian frame at some instant $t = t_0$. There is a flow of particles passing through any given point continuously in time. Field properties such as velocity changes as the point may take on different particles. The velocity with respect to an observer fixed to the point is a function of x_i and t known as Eulerian variables in Lagrangian coordinate, that is,

$$V = V(\bar{r}, t) = \sum v_i \hat{e}_i$$

or

$$v_x = v_x(x, y, z, t) = \frac{dx}{dt}$$

$$v_y = v_y(x, y, z, t) = \frac{dy}{dt}$$

$$v_z = v_z(x, y, z, t) = \frac{dz}{dt}$$

The above function defines field property V for a continuum in terms of Eulerian variables. Specifically, the path line equation traces movement of a particle as a function of time. For a given instant or steady state, it reduces to trajectory equation (Equation 2.29). In this Eulerian viewpoint, attention is focused at a fixed point or region in space as field particles pass through continuously in time rather than what happens to a particular particle or its history. This is the field method of spatial description in Cartesian coordinates, which depicts change of field property in a series of instantaneous photographs regardless of particles as time proceeds. This method of analysis is equivalent to the concept of an open system or a control volume. By applying the initial condition, the solution of path line equation leads back to Lagrangian description, that is,

$$\vec{r} = \int_{t_0}^{t} V dt = \vec{r}(\xi_0, t)$$

In principle, any one of these descriptions can always be derived from the other in a mapping process. For a given physical phenomenon, the two descriptions are equivalent through exchange of variables in one to one correspondence.

$$x_i = x_i(\xi_i, t); \quad \xi_i = \xi_i(x_i, t)$$

In field analysis, it is assumed that motions and continuum are continuous, smooth and single-valued except for a finite number of singular points, such that functions are differentiable and uniquely defined. The basic mathematical idea of mapping between coordinate systems is developed from the geometric concept of point-to-point transformation in space. To provide the necessary conditions in point transformation, it is assumed that mapping functions are one-to-one correspondence and that the Jacobian determinant of coordinate transformation has finite value other than zero such that the inverse of functions exists.

$$J = \frac{\partial(x_1, x_2, x_3)}{\partial(\xi_1, \xi_2, \xi_3)} = \begin{vmatrix} \dfrac{\partial x_1}{\partial \xi_1} & \dfrac{\partial x_1}{\partial \xi_2} & \dfrac{\partial x_1}{\partial \xi_3} \\ \dfrac{\partial x_2}{\partial \xi_1} & \dfrac{\partial x_2}{\partial \xi_2} & \dfrac{\partial x_2}{\partial \xi_3} \\ \dfrac{\partial x_3}{\partial \xi_1} & \dfrac{\partial x_3}{\partial \xi_2} & \dfrac{\partial x_3}{\partial \xi_3} \end{vmatrix} = \left[\frac{\partial(\xi_1, \xi_2, \xi_3)}{\partial(x_1, x_2, x_3)} \right]^{-1} \tag{2.2}$$

The mapping process is interpreted as geometric transformation of a point between Eulerian Cartesian and Lagrangian coordinates. A point is specified by Eulerian variables or Lagrangian variables in the respective coordinate. Equation 2.1a represents the parametric equation of a curve in space with t as a parameter by labeling a particle located at point ξ_0 at time instant $t = t_0$. The curve is a path line traced by the label particle in motion from point to point. By choosing a particle at a different location $t = t_0$, a different path line is traced out. The inverse defines the Lagrangian coordinate of the label particle ξ_i with reference to Eulerian frame. Without loss of generalization, any property of a field particle in space may be observed and determined in either coordinate system as a function of respective variables. Consider Eulerian/Lagrangian transformation of a property function F. The property function may be expressed and treated in the same way as position property is. The following transformation is obtained based on the theorem of continuity of composite functions and tensor invariant with respect to coordinate systems.

$$F(\xi_i, t) = F[\xi_i(x_i, t), t] = F(x_i, t) \tag{2.3a}$$

Conversely,

$$F(x_i, t) = F[x_i(\xi_i, t), t] = F(\xi_i, t) \tag{2.3b}$$

Physically this says that the value of the property seen by an observer riding with the particle at time t has the same unique value at the position it occupies at that time. The relation between material derivatives of a field property in Lagrangian and in Eulerian descriptions is obtained from chain rules:

$$\frac{dF(\xi_i, t)}{dt} = \frac{\partial F(x_i, t)}{\partial t} + (V \cdot \nabla)F(x_i, t)$$

The left and right side of the equation represent Lagrangian and Eulerian viewpoints respectively. Lagrangian description gives the rate of change of property observed by moving with the particle while keeping $\xi_i = \xi_0$ constant. It is called material, total or sustentative derivative. Eulerian description consists of two parts. The first term expresses the rate of change of property at a fixed local point. The second term describes the convective change associated with the bulk stream passing through the same local point. When F represents field property in an infinitesimal volume, then the left side is interpreted as the rate of change of field property in a closed control volume and the right side is the rate of change in an open control volume plus the net property flux across it. The above expression represents property balance at a point. There are pros and cons in each method of description. The choice between them depends on the task at hand.

2.2 Curvilinear Coordinates

The concept of curvilinear coordinate systems and their transformation is useful when dealing with the specification of boundary conditions in domains having irregular boundaries. This is because to express boundary conditions in a simple way, it is necessary to have coordinate surfaces (or curves) that fit the physical boundaries of the problems. For instance, an axially symmetric domain in two-dimensions can be conveniently phrased in polar coordinates to take advantage of geometric symmetry such that the system is independent of polar angle with a reduction in spatial variables. On the other hand, graphs in polar coordinates are suitable for describing phenomena sensitive to change in directions in terms of polar angle otherwise it becomes difficult in Cartesian coordinates. Curvilinear transformations between Cartesian coordinate and other common orthogonal coordinate systems such as cylindrical and spherical are all familiar examples. Such transformation opens the door from Euclidean space into non-Euclidian space.

Because components of a tensor depend on coordinate systems, the choice of a suitable coordinate is of crucial consideration. Certain coordinate systems are more appealing to the users because they are convenient and effective when it comes to performing the task. New coordinate systems and transformations are desirable to extend our visual capability. Field patterns in different coordinate systems provide information supplementary to each other. There is a need to transform the original system to another coordinate system where field analysis/visualization could be carried out and information is transformed back to the original plane at will. The type of transformation depends on the objectives of the task.

A coordinate system in three-dimensions is specified in terms of coordinate surfaces. It is best to consider the location of a point in any reference coordinate system as the intersection of three mutually orthogonal surfaces represented by constant values of the coordinate variables. In Euclidian space, the Cartesian coordinate of the first degree is the simplest of all orthogonal coordinate systems that define a point in space by the intersection of three mutually orthogonal planes associated with the respective coordinate, namely $x_1 = C_1$, $x_2 = C_2$ and $x_3 = C_3$. Each coordinate plays exactly equal roles in a symmetric manner. Mathematically the unit of measurements could be different for each coordinate axis. In order to represent the shape of geometric figures correctly, it is necessary to have the same unit for Cartesian coordinate axes so that a square or slope will resemble a true square or true slope exactly in Euclidean space. In non-Euclidean space, a curvilinear coordinate system in Figure 2.1 is defined by three coordinate curves produced by the intersecting surfaces represented by Equation 2.4a when each of them assumes a parametric constant $\xi_i = C_i$. As the parametric constants vary, a system of three intersecting surfaces is produced. A coordinate curve ξ_i is determined by two intersecting surfaces being held constant. ξ_i represents a new set of variables, which may be any physical quantities other than lengths. $\partial\xi_i$ represents differential arc length along each coordinate curve. The coordinate directions are defined by the unit vectors $\hat{\xi}_i$ drawn tangentially through the point of intersection along each coordinate curve and points in the direction of increasing differential arc length. The coordinate directions may vary from point to point as the coordinate curves are allowed to stretch or bend. This unique feature differs from the Cartesian coordinate system with straight coordinate curves and fixed unit vectors $\hat{e}_i$. As a result, transformation involving curvilinear coordinate systems is likely to be non-linear. A curvilinear coordinate system is generally regarded as a material form description based on certain transformations of the Cartesian coordinate. The change of coordinates from Cartesian to curvilinear systems requires the specification of a point in Cartesian frame by one-to-one correspondence with its image point in curvilinear coordinates. This point transformation process is established from Cartesian coordinates to curvilinear coordinates by a set of mapping functions defined with ξ_i and x_i assuming the role of dependent and

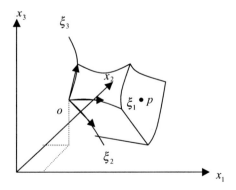

Figure 2.1 A general curvilinear coordinate.

independent variables respectively. The number of mapping functions is the same as the number of independent variables.

$$\begin{aligned}
\xi_1 &= \xi_1(x_1, x_2, x_3)\\
\xi_2 &= \xi_2(x_1, x_2, x_3)\\
\xi_3 &= \xi_3(x_1, x_2, x_3)
\end{aligned} \qquad (2.4a)$$

Each mapping function describes a ξ_i-coordinate surface implicitly when ξ_i assumes the role of contour surface parameters. The point of intersection of the system of ξ_i-coordinate surfaces is determined by specifying the parametric constants $\xi_i = C_i$. Hence, a point in space is associated with each triplet ξ_i and the mapping functions establish the correspondence of triplet in Cartesian coordinate. In this scenario, if we think of x_i as the coordinates of a point P and ξ_i another point Q, then point P is said to be mapped into image point Q. The domain D in Cartesian coordinate corresponds to the image set of points B or the range of mapping in curvilinear coordinate. These three sets of coordinate surfaces will cover the domain and form a global curvilinear network. Conversely,

$$\begin{aligned}
x_1 &= x_1(\xi_1, \xi_2, \xi_3)\\
x_2 &= x_2(\xi_1, \xi_2, \xi_3)\\
x_3 &= x_3(\xi_1, \xi_2, \xi_3)
\end{aligned} \qquad (2.4b)$$

Inverse mapping if it exists, then converts curvilinear coordinates back to Cartesian coordinates. Thus, curvilinear coordinates are an alternative and equivalent coordinate system for defining a point in space. When coordinate curves form orthogonal bases, they are called orthogonal curvilinear coordinates representing an important class of curvilinear coordinate systems such that:

$$\hat{\xi}_i \cdot \hat{\xi}_j = \sum_{k=1}^{3} \frac{\partial x_k}{h_i \partial \xi_i} \frac{\partial x_k}{h_j \partial \xi_j} \hat{e}_k \hat{e}_k = \sum_{k=1}^{3} \frac{\partial x_k}{h_i \partial \xi_i} \frac{\partial x_k}{h_j \partial \xi_j} = \delta_{ij} = \begin{cases} 1 & i = j \\ 0 & i \neq j \end{cases}$$

Adoption of orthogonal curvilinear coordinate systems requires certain information essential for the construction of mathematical models and systems:

(a) The Lame coefficients as a measure of relative differential coordinate lengths in curvilinear coordinates and Cartesian coordinates.
(b) Transformation of unit vectors and derivatives between the two coordinate systems.
(c) The differential operators and spatial rate of change of unit vectors in curvilinear coordinates.

Firstly, chain rule is invoked to yield the mapping relations for the differential coordinate lengths between two coordinate systems, that is,

$$dx_k = \sum_{j=1}^{3} \frac{\partial x_k}{\partial \xi_j} d\xi_j$$

$$= \sum_{j=1}^{3} J_{kj} d\xi_j$$

where

$$J_{kj} = \left[\frac{\partial x_k}{\partial \xi_j}\right] = \begin{bmatrix} \dfrac{\partial x_1}{\partial \xi_1} & \dfrac{\partial x_1}{\partial \xi_2} & \dfrac{\partial x_1}{\partial \xi_3} \\ \dfrac{\partial x_2}{\partial \xi_1} & \dfrac{\partial x_2}{\partial \xi_2} & \dfrac{\partial x_2}{\partial \xi_3} \\ \dfrac{\partial x_3}{\partial \xi_1} & \dfrac{\partial x_3}{\partial \xi_2} & \dfrac{\partial x_3}{\partial \xi_3} \end{bmatrix}$$

J_{kj} is the Jacobian matrix of coordinate transformation. It describes partial derivatives of Cartesian coordinates with respect to another coordinate resulting in a tensor of the next higher order. A line element dr is a scalar invariant in the two coordinate systems expressed in terms of metric tensor. It implies vector length is preserved in transformation.

$$dr = \sqrt{d\vec{r} \cdot d\vec{r}} = \sqrt{(dr)^2}$$

where

$$(dr)^2 = \sum_{k=1}^{3} dx_k dx_k$$

$$= \sum_{k=1}^{3} \frac{\partial x_k}{\partial \xi_i} \frac{\partial x_k}{\partial \xi_j} d\xi_i d\xi_j$$

$$= g_{ij} d\xi_i d\xi_j$$

$$\vec{r} = \sum_{k=1}^{3} x_k \hat{e}_k = x_1 \hat{i} + x_2 \hat{j} + x_3 \hat{k} = x_1(\xi_1, \xi_2, \xi_3)\hat{i} + x_2(\xi_1, \xi_2, \xi_3)\hat{j}$$

$$+ x_3(\xi_1, \xi_2, \xi_3)\hat{k}$$

$\vec{r}$ is the position vector defining an arbitrary point in space. The metric tensor is defined as:

$$g_{ij} = \sum_{k=1}^{3} \frac{\partial x_k}{\partial \xi_i} \frac{\partial x_k}{\partial \xi_j} = g_{ji} = h_i h_j \delta_{ij} \tag{2.5}$$

The metric tensor relates distance to the infinitesimal coordinate arc lengths. For orthogonal coordinate systems, the off-diagonal components vanish in metric tensor. The diagonal components are related to Lame coefficients, which represent the scaling ratio of distance between coordinate differential lengths in Euclidean space, that is,

$$g_{ij} = \begin{cases} 0 & i \neq j \\ h_i^2 > 0 & i = j \end{cases}$$

$$\text{Hence, } h_i = \sqrt{g_{ii}} = \sqrt{\sum_{k=1}^{3} \left(\frac{\partial x_k}{\partial \xi_i} \right)^2} = \left(\sqrt{\sum_{k=1}^{3} \left(\frac{\partial \xi_i}{\partial x_k} \right)^2} \right)^{-1} \tag{2.6}$$

$$= \left| \frac{\partial \vec{r}}{\partial \xi_i} \right|$$

Because curvilinear coordinates may have any physical quantities other than lengths in Cartesian coordinates, it is necessary to ensure dimensional homogeneity in coordinate transformation. Lame coefficients are used to convert linear infinitesimal coordinate lengths between two coordinate systems with the same dimensional form. As the line element coincides successively with each Cartesian coordinate direction, it yields differential coordinate length components in the respective coordinate axis. This is because the partial derivatives of the line element with respect to any coordinate are to be taken by keeping the other two coordinate variables constant in that coordinate set.

$$dr = \sqrt{g_{ij} d\xi_i d\xi_j} = \sqrt{h_i^2 \delta_{ij} d\xi_i d\xi_j} = h_i d\xi_i = dx_i \quad \text{No summation} \tag{2.7}$$

Alternatively, we recall the definition of tangent vector and note that the tangent vector to ξ_i-coordinate curve is aligned with each Cartesian coordinate direction by keeping the other two coordinate variables constant. Such that,

$$\frac{\partial \vec{r}}{\partial \xi_1} = \left| \frac{\partial \vec{r}}{\partial \xi_1} \right| \hat{i} = h_1 \hat{i}$$

$$\frac{\partial \vec{r}}{\partial \xi_2} = \left| \frac{\partial \vec{r}}{\partial \xi_2} \right| \hat{j} = h_2 \hat{j}$$

$$\frac{\partial \vec{r}}{\partial \xi_3} = \left| \frac{\partial \vec{r}}{\partial \xi_3} \right| \hat{k} = h_3 \hat{k}$$

Hence, the same results are obtained.

$$d\vec{r} = \frac{\partial \vec{r}}{\partial \xi_1}d\xi_1 + \frac{\partial \vec{r}}{\partial \xi_2}d\xi_2 + \frac{\partial \vec{r}}{\partial \xi_3}d\xi_3$$

$$= h_1 d\xi_1 \hat{i} + h_2 d\xi_2 \hat{j} + h_3 d\xi_3 \hat{k} = dx_1 \hat{i} + dx_2 \hat{j} + dx_3 \hat{k}$$

$$\rightarrow h_i d\xi_i = dx_i$$

To extend the transformation for an elemental area between two coordinate systems, we consider the parallelogram formed by the cross product of differential coordinate vectors on (x_1, x_2) plane,

$$dx_1\hat{i} \times dx_2\hat{j} = \left(\frac{\partial x_1}{\partial \xi_1}d\xi_1\hat{i} + \frac{\partial x_1}{\partial \xi_2}d\xi_2\hat{j}\right) \times \left(\frac{\partial x_2}{\partial \xi_1}d\xi_1\hat{i} + \frac{\partial x_2}{\partial \xi_2}d\xi_2\hat{j}\right)$$

$$\rightarrow dx_1 dx_2 \hat{k} = \begin{vmatrix} \hat{i} & \hat{j} & \hat{k} \\ \frac{\partial x_1}{\partial \xi_1}d\xi_1 & \frac{\partial x_1}{\partial \xi_2}d\xi_2 & 0 \\ \frac{\partial x_2}{\partial \xi_1}d\xi_1 & \frac{\partial x_2}{\partial \xi_2}d\xi_2 & 0 \end{vmatrix}$$

$$\text{Or } dx_1 dx_2 = \begin{vmatrix} \frac{\partial x_1}{\partial \xi_1} & \frac{\partial x_1}{\partial \xi_2} \\ \frac{\partial x_2}{\partial \xi_1} & \frac{\partial x_2}{\partial \xi_2} \end{vmatrix} d\xi_1 d\xi_2$$

$$= \frac{\partial(x_1, x_2)}{\partial(\xi_1, \xi_2)}d\xi_1 d\xi_2 = J d\xi_1 d\xi_2 = h_1 h_2 d\xi_1 d\xi_2$$

In general, $dx_i dx_j = \frac{\partial(x_i, x_j)}{\partial(\xi_i, \xi_j)}d\xi_i d\xi_j = h_i h_j d\xi_i d\xi_j ; i \neq j$

Transformation of an elemental volume between two coordinate systems follows:

$$dV = \frac{\partial x_i}{\partial \xi_n}d\xi_n \cdot \left(\frac{\partial x_j}{\partial \xi_m}d\xi_m \times \frac{\partial x_k}{\partial \xi_l}d\xi_l\right)$$

$$= \varepsilon_{ijk}\frac{\partial x_i}{\partial \xi_1}d\xi_1 \frac{\partial x_j}{\partial \xi_2}d\xi_2 \frac{\partial x_k}{\partial \xi_3}d\xi_3$$

Or

$$dV = dx_1 dx_2 dx_3 = \begin{vmatrix} \dfrac{\partial x_1}{\partial \xi_1} & \dfrac{\partial x_1}{\partial \xi_2} & \dfrac{\partial x_1}{\partial \xi_3} \\[2mm] \dfrac{\partial x_2}{\partial \xi_1} & \dfrac{\partial x_2}{\partial \xi_2} & \dfrac{\partial x_2}{\partial \xi_3} \\[2mm] \dfrac{\partial x_3}{\partial \xi_1} & \dfrac{\partial x_3}{\partial \xi_2} & \dfrac{\partial x_3}{\partial \xi_3} \end{vmatrix} d\xi_1 d\xi_2 d\xi_3 = \frac{\partial(x_1, x_2, x_3)}{\partial(\xi_1, \xi_2, \xi_3)} d\xi_1 d\xi_2 d\xi_3$$

$$= J \, d\xi_1 d\xi_2 d\xi_3$$

$$= h_1 h_2 h_3 d\xi_1 d\xi_2 d\xi_3$$

Jacobian determinant of coordinate transformation is related to the determinant of metric tensor, that is,

$$J = \frac{\partial(x_1, x_2, x_3)}{\partial(\xi_1, \xi_2, \xi_3)} = \sqrt{\det |g_{ij}|}$$

Jacobian determinant of coordinate transformation expresses mathematically an approximation of a differentiable function. One important aspect of it is a scalar quantity that gives information about field topology and behavior near a given point. As Jacobian determinant varies from point to point, curvilinear transformation could be applied to distort the relative sizes of a large figure by mapping it from Cartesian coordinates of the same units of measurements. It also provides the venue for the change of variables in multiple integrals.

Curvilinear transformation has an important interpretation in the representation of deformation or fluids in motion. In a geometric viewpoint, we think of a material substance as spread out at an initial instant over a region and then undergoing deformation in another region. The initial configuration of substance is represented by $d\xi_1 d\xi_2 d\xi_3$ with reference to a curvilinear coordinate in Figure 2.2. The image region of substance covers a region different from the initial region representing the current configuration $dx_1 dx_2 dx_3$ in Cartesian coordinates. The relation between initial and current states of material regions is given by Jacobian determinant as a measure of volumetric ratio or density ratio in terms of a dimensionless quantity.

$$J = h_1 h_2 h_3 = \frac{dx_1 dx_2 dx_3}{d\xi_1 d\xi_3 d\xi_3} = \frac{dV}{d\xi_1 d\xi_2 d\xi_3}$$

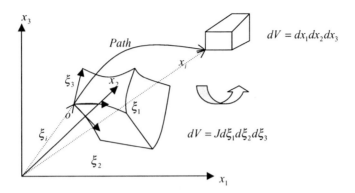

Figure 2.2 Transformation of an infinitesimal volume.

Otherwise, Jacobian is interpreted as the representation of a differential volume in non-Euclidean space to ensure dimensional homogeneity in coordinate transformation.

$$dV = J\,d\xi_1 d\xi_2 d\xi_3$$

The roles of Jacobian determinant are:

$J > 0$, mapping preserves orientation. No change in direction of rotation.
$J < 0$, mapping reverses orientation. There is a change in the sense of direction from right hand system (counterclockwise) to left hand (clockwise) and conversely.
$J > 1$, mapping results in volumetric expansion.
$J < 1$, mapping results in volume shrinking.
$J = 1$, mapping is isochronic with no change in volume or density.
$J = -1$, reflection mapping with no change in volume or shape but reverse orientation.
$J = C$, mapping is linear.
$J = \infty$ or $J = 0$, inverse fails to exist.

From chain rule of partial differentiation,

$$\frac{\partial}{\partial x_i} = \sum_{j=1} \frac{\partial}{\partial \xi_j}\frac{\partial \xi_j}{\partial x_i}$$

$$= \frac{\partial}{\partial \xi_1}\frac{\partial \xi_1}{\partial x_i} + \frac{\partial}{\partial \xi_2}\frac{\partial \xi_2}{\partial x_i} + \frac{\partial}{\partial \xi_3}\frac{\partial \xi_3}{\partial x_i}$$

It is understood that the partial derivatives with respect to any coordinate are to be taken by keeping other two coordinate variables constant in that coordinate set when ∇-operator is to be defined in a chosen coordinate system. This implies line element coincides successively with each coordinate direction, such that

$$\frac{\partial}{\partial x_1} = \frac{1}{h_1}\frac{\partial}{\partial \xi_1}; \quad \frac{\partial}{\partial x_2} = \frac{1}{h_2}\frac{\partial}{\partial \xi_2}; \quad \frac{\partial}{\partial x_3} = \frac{1}{h_3}\frac{\partial}{\partial \xi_3}$$

By choosing Cartesian coordinates or curvilinear coordinates, we obtain the expression of ∇-operator in each of them from its definition, that is,

$$\nabla^2 = \frac{1}{h_1 h_2 h_3}\sum_i \frac{\partial}{\partial \xi_i}\left(\frac{h_1 h_2 h_3}{h_i^2}\frac{\partial}{\partial \xi_i}\right)$$

$$\nabla \cdot V = \frac{1}{h_1 h_2 h_3}\sum_i \frac{\partial}{\partial \xi_i}\left(\frac{h_1 h_2 h_3}{h_i} v_i\right)$$

$$\nabla \times V = \frac{1}{h_1 h_2 h_3}\begin{vmatrix} h_1\hat{\xi}_1 & h_2\hat{\xi}_2 & h_3\hat{\xi}_3 \\ \partial/\partial\xi_1 & \partial/\partial\xi_2 & \partial/\partial\xi_3 \\ h_1 v_1 & h_2 v_2 & h_3 v_3 \end{vmatrix}$$

Cartesian coordinate:

$$\nabla \equiv \sum_{i=1} \frac{\hat{e}_i \partial}{h_i \partial x_i} = \hat{i}\frac{\partial}{\partial x_1} + \hat{j}\frac{\partial}{\partial x_2} + \hat{k}\frac{\partial}{\partial x_3}; \quad h_i = 1$$

Curvilinear coordinate:

$$\nabla \equiv \sum_{i=1} \frac{\hat{\xi}_i \partial}{h_i \partial \xi_i} = \hat{\xi}_1\frac{\partial}{h_1 \partial \xi_1} + \hat{\xi}_2\frac{\partial}{h_2 \partial \xi_2} + \hat{\xi}_3\frac{\partial}{h_3 \partial \xi_3}$$

The relations between unit vectors in orthogonal curvilinear coordinate and Cartesian coordinate systems are:

$$\hat{\xi}_i = \sum_{k=1} \frac{\partial x_k}{h_i \partial \xi_i}\hat{e}_k$$

$$\hat{e}_k = \sum_{i=1} \frac{h_i \partial \xi_i}{\partial x_k}\hat{\xi}_i$$

The spatial rates of change of unit vectors in curvilinear coordinates are obtained accordingly.

$$\frac{\partial \hat{\xi}_i}{\partial \xi_j} = \frac{\hat{\xi}_j}{h_i}\frac{\partial h_j}{\partial \xi_i} - \delta_{ij}\sum_{k=1} \frac{\hat{\xi}_k \partial h_i}{h_k \partial \xi_k}$$

2.2.1 Polar Coordinate

Polar coordinate is a type of curvilinear coordinate for two-dimensional domains used to define the position of a point as the point of intersection of a polar coordinate circle and a radial line. Polar coordinate is a convenient way to study motions sensitive to change in direction or depending on distance from a fixed point or subject to a central force. In polar coordinate, a point is defined by the radius vector, which is the polar representation of a position vector. The radius vector is specified by radial distance $r = \xi_1$ of the polar coordinate circle from a fixed point (origin) and by an angular measurement (polar angle) $\theta = \xi_2$ with respect to positive θ-axis anticlockwise. Positive values of r measure directed radial distance from origin and is equal to the magnitude of radius vector. It could have negative values. By placing polar coordinate origin and θ-axis to coincide with Cartesian coordinate origin and x-axis, the change of variables from Cartesian to polar coordinate is obtained from the following mapping functions based on geometric relation:

$$r = r(x, y) = \sqrt{x^2 + y^2}; \ \theta = \theta(x, y) = \tan^{-1}\left(\frac{y}{x}\right)$$

Conversely,

$$x = x(r, \theta) = r\cos\theta; \ y = y(r, \theta) = r\sin\theta$$

$$\begin{bmatrix} x \\ y \end{bmatrix} = [T(\theta)]\begin{bmatrix} r \\ 0 \end{bmatrix}$$

with $[T(\theta)] = \begin{bmatrix} \cos\theta & -\sin\theta \\ \sin\theta & \cos\theta \end{bmatrix}$

$$\vec{r} = r\hat{r} = \begin{bmatrix} r \\ 0 \end{bmatrix} = [T(\theta)]^{-1}\begin{bmatrix} x \\ y \end{bmatrix}; \ |\vec{r}| = |r\hat{r}| = r$$

Velocity and acceleration are transformed accordingly. The transformation is viewed as a rotation of coordinates about z-axis through polar angle and is written in matrix representation.

$$V = \begin{bmatrix} v_r \\ v_\theta \end{bmatrix} = \begin{bmatrix} \cos\theta & \sin\theta \\ -\sin\theta & \cos\theta \end{bmatrix}\begin{bmatrix} v_x \\ v_y \end{bmatrix} = \begin{bmatrix} \dot{r} \\ r\dot{\theta} \end{bmatrix} \tag{2.8a}$$

$$a = \begin{bmatrix} a_r \\ a_\theta \end{bmatrix} = \begin{bmatrix} \cos\theta & \sin\theta \\ -\sin\theta & \cos\theta \end{bmatrix}\begin{bmatrix} a_x \\ a_y \end{bmatrix} = \begin{bmatrix} \ddot{r} - r\dot{\theta}^2 \\ r\ddot{\theta} + 2\dot{r}\dot{\theta} \end{bmatrix} \tag{2.8b}$$

where

$$v_x = \dot{x} = \dot{r}\cos\theta - r\dot{\theta}\sin\theta$$
$$v_y = \dot{y} = \dot{r}\sin\theta + r\dot{\theta}\cos\theta$$
$$a_x = \ddot{x} = \ddot{r}\cos\theta - 2\dot{r}\dot{\theta}\sin\theta - r\dot{\theta}^2\cos\theta - r\ddot{\theta}\sin\theta$$
$$a_y = \ddot{y} = \ddot{r}\sin\theta + 2\dot{r}\dot{\theta}\cos\theta - r\dot{\theta}^2\sin\theta + r\ddot{\theta}\cos\theta$$
$$\dot{r} \equiv \frac{dr}{dt}, \ddot{r} \equiv \frac{d^2r}{dt^2}, \text{ etc.}$$

Polar coordinate is one of important set of curvilinear coordinates on x-y plane on which circles and radial lines cross at right angle everywhere. The radial component unit vector $\hat{r}$ and transverse component unit vector $\hat{\theta}$ are defined by mapping of unit vectors from Cartesian coordinates.

$$\hat{r} = \frac{1}{h_1}\left(\frac{\partial x}{\partial r}\hat{i} + \frac{\partial y}{\partial r}\hat{j}\right) = \cos\theta\hat{i} + \sin\theta\hat{j}$$

$$\hat{\theta} = \frac{1}{h_2}\left(\frac{\partial x}{\partial \theta}\hat{i} + \frac{\partial y}{\partial \theta}\hat{j}\right) = -\sin\theta\hat{i} + \cos\theta\hat{j}$$

Or $\begin{bmatrix} \hat{r} \\ \hat{\theta} \end{bmatrix} = [T(\theta)]^{-1} \begin{bmatrix} \hat{i} \\ \hat{j} \end{bmatrix} = \begin{bmatrix} \cos\theta & \sin\theta \\ -\sin\theta & \cos\theta \end{bmatrix} \begin{bmatrix} \hat{i} \\ \hat{j} \end{bmatrix}; \begin{bmatrix} \hat{i} \\ \hat{j} \end{bmatrix} = [T(\theta)] \begin{bmatrix} \hat{r} \\ \hat{\theta} \end{bmatrix}.$

The radial unit vector $\hat{r}$ in polar representation is given by Euler's formula and points in the direction of increasing r, that is,

$$\hat{r} = \exp(i\theta) = \cos\theta + i\sin\theta; \ i \equiv \sqrt{-1}$$
$$\hat{\theta} = \frac{\partial\hat{r}}{\partial\theta} = -\sin\theta + i\cos\theta = \hat{r}i$$

Transverse unit vector $\hat{\theta}$ is defined by the operation of imaginary number, which rotates unit vector $\hat{r}$ through a right angle counterclockwise and points in the direction of increasing arc length. Both unit vectors vary in direction according to polar angle. The spatial rate of change of polar coordinate unit vectors results in rotation through a right angle anticlockwise:

$$\frac{\partial\hat{r}}{\partial\theta} = -\sin\theta\hat{i} + \cos\theta\hat{j} = \hat{\theta}; \quad \frac{\partial\hat{r}}{\partial t} = \frac{\partial\theta}{\partial t}\hat{\theta}$$
$$\frac{\partial\hat{\theta}}{\partial\theta} = -\cos\theta\hat{i} - \sin\theta\hat{j} = -\hat{r}; \quad \frac{\partial\hat{\theta}}{\partial t} = -\frac{\partial\theta}{\partial t}\hat{r}$$

$$\frac{\partial \hat{r}}{\partial r} = 0; \quad \frac{\partial \hat{\theta}}{\partial r} = 0$$

$$\frac{\partial \hat{r}}{\partial t} \cdot \hat{r} = 0; \quad \frac{\partial \hat{\theta}}{\partial t} \cdot \hat{\theta} = 0$$

Lame coefficients and Jacobian determinant are determined accordingly. It is observed that Lame coefficients in addition to scaling, ensure dimensional homogeneity in coordinate transformation.

$$h_1 = \sqrt{\sum_{k=1}^{2} \left(\frac{\partial x_k}{\partial \xi_1}\right)^2} = \sqrt{\left(\frac{\partial x}{\partial r}\right)^2 + \left(\frac{\partial y}{\partial r}\right)^2} = \sqrt{\cos^2 \theta + \sin^2 \theta} = 1$$

$$h_2 = \sqrt{\sum_{k=1}^{2} \left(\frac{\partial x_k}{\partial \xi_2}\right)^2} = \sqrt{\left(\frac{\partial x}{\partial \theta}\right)^2 + \left(\frac{\partial y}{\partial \theta}\right)^2} = \sqrt{r^2 \sin^2 \theta + r^2 \cos^2 \theta} = r$$

$$J = \frac{\partial(x, y)}{\partial(r, \theta)} = \begin{vmatrix} \frac{\partial x}{\partial r} & \frac{\partial x}{\partial \theta} \\ \frac{\partial y}{\partial r} & \frac{\partial y}{\partial \theta} \end{vmatrix} = r \cos^2 \theta + r \sin^2 \theta = r = h_1 h_2$$

$$= \sqrt{\det |g_{ij}|} = \sqrt{(h_1 h_2)^2} = h_1 h_2$$

$$dxdy = Jd\theta dr = rd\theta dr$$

$$g_{11} = h_1^2 \delta_{11} = 1; \quad g_{22} = h_2^2 \delta_{22} = r^2; \quad g_{21} = g_{12} = h_1 h_2 \delta_{12} = 0$$

Similarly, transformation of derivatives follows.

$$\frac{\partial}{\partial x_i} = \sum_{j=1} \frac{\partial}{\partial \xi_j} \frac{\partial \xi_j}{\partial x_i}$$

$$\frac{\partial}{\partial x} = \frac{\partial}{\partial r} \frac{\partial r}{\partial x} + \frac{\partial}{\partial \theta} \frac{\partial \theta}{\partial x} = \frac{\partial}{\partial r} \frac{x}{\sqrt{x^2 + y^2}} + \frac{\partial}{\partial \theta} \frac{-y}{x^2 + y^2} = \cos \theta \frac{\partial}{\partial r} - \frac{\sin \theta}{r} \frac{\partial}{\partial \theta}$$

$$\frac{\partial}{\partial y} = \frac{\partial}{\partial r} \frac{\partial r}{\partial y} + \frac{\partial}{\partial \theta} \frac{\partial \theta}{\partial y} = \frac{\partial}{\partial r} \frac{y}{\sqrt{x^2 + y^2}} + \frac{\partial}{\partial \theta} \frac{x}{x^2 + y^2} = \sin \theta \frac{\partial}{\partial r} + \frac{\cos \theta}{r} \frac{\partial}{\partial \theta}$$

Or

$$\begin{bmatrix} \frac{\partial}{\partial x} \\ \frac{\partial}{\partial y} \end{bmatrix} = [T(\theta)] \begin{bmatrix} 1 & 0 \\ 0 & \frac{1}{r} \end{bmatrix} \begin{bmatrix} \frac{\partial}{\partial r} \\ \frac{\partial}{\partial \theta} \end{bmatrix} = [T(\theta)] \begin{bmatrix} \frac{1}{h_1} & 0 \\ 0 & \frac{1}{h_2} \end{bmatrix} \begin{bmatrix} \frac{\partial}{\partial r} \\ \frac{\partial}{\partial \theta} \end{bmatrix} = [T(\theta)] \begin{bmatrix} \frac{\partial}{\partial r} \\ \frac{\partial}{r \partial \theta} \end{bmatrix}$$

The ∇-operator is obtained in polar coordinate by definition and leads to Laplacian operator.

$$\nabla \equiv \sum_{i=1} \frac{\hat{\xi}_i \partial}{h_i \partial \xi_i} = \hat{r}\frac{\partial}{\partial r} + \hat{\theta}\frac{\partial}{r \partial \theta}$$

$$\nabla^2 = \nabla \cdot \nabla = \frac{\partial^2}{\partial r^2} + \frac{\partial}{r \partial r} + \frac{\partial^2}{r^2 \partial \theta^2} = \left(\frac{1}{r}\right)^2 \left[r\frac{\partial}{\partial r}\left(r\frac{\partial}{\partial r}\right) + \frac{\partial^2}{\partial \theta^2}\right]$$

To obtain information about particles in motion with reference to a polar co-ordinate, we need to know the position of a particle from which the particle's trajectory, velocity and acceleration are determined in terms of radial and transverse components.

$$V = \frac{d\vec{r}}{dt} = \frac{d(r\hat{r})}{dt} = \frac{dr}{dt}\hat{r} + r\frac{d\theta}{dt}\hat{\theta}$$
$$= v_r\hat{r} + v_\theta\hat{\theta} \tag{2.9a}$$

$$a = \frac{dV}{dt} = \left(\frac{d^2r}{dt^2} - r\left(\frac{d\theta}{dt}\right)^2\right)\hat{r} + \left(\frac{rd^2\theta}{dt^2} + \frac{2dr}{dt}\frac{d\theta}{dt}\right)\hat{\theta} = \left(\frac{d^2r}{dt^2} - r\left(\frac{d\theta}{dt}\right)^2\right)\hat{r}$$
$$+ \left(\frac{d}{rdt}\left(\frac{r^2d\theta}{dt}\right)\right)\hat{\theta} \tag{2.9b}$$
$$= a_r\hat{r} + a_\theta\hat{\theta}$$

Motions may involve rotation in addition to the translation component or linear momentum. The angular momentum of a particle with mass m at position $\vec{r}$ about origin is defined as:

$$Y = m\vec{r} \times V$$

For plane motions, the z-component angular momentum about z-axis is given by:

$$Y_z = mr^2\frac{d\theta}{dt} = mr v_\theta$$

It can be shown that vector length is preserved independent of coordinate systems.

$$|V| = \sqrt{v_r^2 + v_\theta^2} = \sqrt{v_x^2 + v_y^2}$$
$$|a| = \sqrt{a_r^2 + a_\theta^2} = \sqrt{a_x^2 + a_y^2}$$

Special types of motions:

(1) Motion in a straight line.

$$a_\theta = 0; \ \frac{d\theta}{dt} = 0; \rightarrow a_r = \frac{d\,|V|}{dt} = \frac{d^2 r}{dt^2}$$

$$v_r = \frac{dr}{dt}; \ v_\theta = 0; \ Y = mr \times V = mr^2 \frac{d\theta}{dt}\hat{k} = 0$$

$$a \times V = 0$$

(2) Circular motion.
This is an important type of plane motion on r-θ plane with polar origin placed at the circle center. The position vector is always perpendicular to velocity and assumes constant radius equal to radius of curvature. The direction of angular momentum Y is perpendicular to the plane of motion.

$$a_r = -r\left(\frac{d\theta}{dt}\right)^2 = -\frac{v_\theta^2}{r}; \ a_\theta = \frac{r\,d^2\theta}{dt^2}$$

$$|V| = v_\theta = r\frac{d\theta}{dt}; \ v_r = \frac{dr}{dt} = 0$$

$$V \cdot \vec{r} = \begin{bmatrix} 0 & r\dfrac{d\theta}{dt} \end{bmatrix} \cdot \begin{bmatrix} r & 0 \end{bmatrix} = 0$$

$$Y = mr \times V = mr^2 \frac{d\theta}{dt}\hat{k}$$

If circular motion is at constant speed, acceleration and velocity are perpendicular to each other everywhere. This orthogonal condition provides a means to trace a particle moving in a circle at constant speed. Furthermore, angular momentum and kinetic energy are conserved.

$$V \cdot a = V \cdot \frac{dV}{dt} = \frac{1}{2}\frac{d(V \cdot V)}{dt} = \frac{d\,V^2}{2dt} = 0$$

$$= \begin{bmatrix} 0 & v_\theta \end{bmatrix} \cdot \begin{bmatrix} a_r & a_\theta \end{bmatrix}$$

$$\rightarrow a_\theta = 0; \ \rightarrow \frac{d\theta}{dt} = \text{constant}$$

$$|V| = v_\theta = r\frac{d\theta}{dt} = \text{constant}; \ a_r = -\frac{v_\theta^2}{r} = \text{constant}$$

$$Y = mr \times V = mr^2 \frac{d\theta}{dt}\hat{k} = \text{constant}\,\hat{k}$$

(3) Motion in constant acceleration fields, $a = C$

$$V = \int_{t_0}^{t} a\,dt$$

$$\vec{r} = \int_{t_0}^{t} V\,dt$$

There is a different version of polar coordinates on which r and θ represent rectangular coordinates on r-θ plane. Circle of radius $r = C$ and radial line $\theta = C'$ are mapped as a pair of straight lines each parallel to respective coordinate axis.

2.2.2 Streamline (Flux Line) Coordinates

Curvilinear coordinates provide a reference frame to describe motions of a particle by its position in coordinate representation. In practice, it is necessary to change the reference frame from curvilinear coordinates or Cartesian coordinates to take advantage of certain property such as intrinsic representation or geometric symmetry. This is because such coordinate systems may become inconvenient to use. Alternatively, it is possible to describe motions using a particle's trajectory in intrinsic representation. Intrinsic property is one that depends on only the state of a moving particle in relation to the geometric shape of trajectories independent of coordinate systems. Information about motions of particles may be obtained by observing the behavior of a material point with reference to the instantaneous structure of trajectories or it remains implicit/hidden in other coordinate systems. The concept of intrinsic representation presumes trajectory is already known such that velocity of a particle always points in the trajectory tangent direction along which the particle advances. When trajectory is defined as a parametric equation of arc length, it is called intrinsic representation of trajectory. Such parametric trajectory in one degree of freedom is ideal to express forces acting on a particle in the tangential and normal direction. Otherwise, it may become difficult in coordinate systems with several independent variables. An arc length is defined as the distance between a fixed reference point and a point in motion along a trajectory. Although trajectories are generally not known beforehand, the convenient use of streamlines for two-dimensional flow field descriptions in a steady state leads to the introduction of streamline coordinates in hydrodynamics. A streamline coordinate serves as an intrinsic coordinate when one of its coordinate curves coincide with a streamline. The origin of a streamline coordinate is attached to the particle of interest and travels with it along a streamline while the other coordinate curve designated as normal trajectory crosses it orthogonally at the particle. The particle is free to rotate about the local vorticity line oriented perpendicular to the plane of motion. The motion trajectory is viewed as a succession of circular paths of ever-changing curvature and center (instantaneous center of rotation), which constitute non-uniform circular motions. Streamline coordinate is a class of orthogonal curvilinear coordinate systems constructed with a network of streamlines and normal trajectories. It renders flow fields one-dimensional locally at a point parallel to streamline tangent. By fitting physical boundaries, they are particularly suitable for simplified flow analysis/visualization near or along boundaries serving as streamlines of specific shapes. Streamline coordinate is a member of local coordinate systems but could be employed as global coordinates with smooth flows. The choice of a reference frame and the combination of local coordinates and global coordinates play a key part in model study and analysis. There is no fixed rule to be applied but it depends on the manner in which the field is generated or on the geometry of boundary. It is an art in science.

Consider plane flows on (x, y) plane spanned by unit vectors $\hat{i}$ and $\hat{j}$ in Figure 2.3. The plane of motion contains these two unit vectors and is perpendicular to unit vector $\hat{k}$. The trajectory, velocity and acceleration vectors all lie on this osculating plane. The location of a fluid particle could be chosen arbitrarily and is defined by a Cartesian coordinate pair (x_0, y_0) on the plane of motion. In this local coordinate, forces acting on the particle can be decomposed conveniently into components in streamline tangent and normal directions. Transformation of a point between Cartesian coordinates and streamline coordinates is defined mathematically by the following mapping

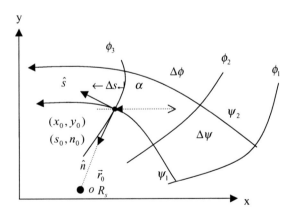

Figure 2.3 A streamline (flux line) coordinate.

functions as the point of intersection between two orthogonal coordinate curves on each coordinate system:

$$s = s(x, y) \tag{2.10a}$$

$$n = n(x, y) \tag{2.10b}$$

Conversely,

$$x = x(s, n) \tag{2.10c}$$

$$y = y(s, n) \tag{2.10d}$$

$\hat{s}$ and $\hat{n}$ are tangent unit vector and normal unit vector drawn through the particle at the point of intersection of coordinate curves tangent to streamline and normal trajectory respectively. The tangent vector defines the direction of motion while the normal vector gives the direction in which the curve is turning. $s = \xi_1$ assumes a spatial variable along s-coordinate curve coinciding with a streamline. $n = \xi_2$ represents the other spatial variable along normal n-coordinate curve. $\partial \xi_1 = \partial s$ and $\partial \xi_2 = \partial n$ represent differential arc lengths along each coordinate curve. The direction of trajectory $\hat{s}$ along which arc length increases is taken as positive direction. It corresponds to the direction along which particles advance. The direction of unit normal vector $\hat{n}$ is, by convention, obtained by a counterclockwise rotation of $\hat{s}$ through a right angle in right-handed coordinate systems and points in the direction of increasing arc length. Unit vectors of streamline coordinate are subject to change in directions from point to point as a function of position. A local streamline coordinate system is completed by defining the binormal unit vector $\hat{k}$.

$$\hat{k} = \hat{s} \times \hat{n} = \hat{i} \times \hat{j} = \begin{bmatrix} 0 & 0 & 1 \end{bmatrix} = \hat{\xi}_3 = \hat{z}$$
$$\xi_3 = z; \ \partial \xi_3 = \partial z; \ \rightarrow h_3 = 1$$

Binormal unit vector is perpendicular to (x, y) osculating plane and coincides with z-axis pointing in the same positive direction. It is the intersection line on rectifying plane and normal plane. The local streamline coordinate is referred to as a moving trihedral (Serret-Frenet triad).

The structure and shape of streamlines and normal trajectories are described geometrically in terms of torsion, tangents and radii of curvature. For plane fields, the former vanishes while the latter is given by the intrinsic equation of a curve independent of coordinate systems. It specifies the rate at which direction changes with respect to distance at each location along streamline or normal trajectory, that is by definition,

$$R_n = \frac{1}{\dfrac{d\alpha}{dn}}; \quad R_s = \frac{1}{\dfrac{d\alpha}{ds}} = \frac{1}{\left|\dfrac{d\hat{s}}{ds}\right|} = \frac{1}{\left|\dfrac{d^2\vec{r}}{ds^2}\right|}$$

Or in parametric representation.

$$\frac{1}{R_s} = \frac{d\alpha/dt}{ds/dt} = \frac{(dx/dt)(d^2y/dt^2) - (dy/dt)(d^2x/dt^2)}{[(dx/dt)^2 + (dy/dt)^2]^{3/2}} = \frac{|V \times a|}{|V|^3} \tag{2.11}$$

where

$$V = \frac{d\vec{r}}{dt} = \left[\frac{dx}{dt} \quad \frac{dy}{dt}\right]$$

$$a = \frac{dV}{dt} = \left[\frac{d^2x}{dt^2} \quad \frac{d^2y}{dt^2}\right]$$

α denotes streamline tangent angle defined with respect to positive x-axis counter-clockwise. R_s and R_n are the radii of curvature of streamline and normal trajectory. R_s (or R_n) lies on the osculating plane and defines the distance between a point of tangent and the corresponding trajectory curvature center in the direction of principal normal vector. For given curvature and boundary conditions of a trajectory, the solution of following system of intrinsic equations derived from definitions of curvature and tangent yields the position vector of trajectory in parametric form:

$$R_s \frac{d^2x}{ds^2} + \frac{dy}{ds} = 0; \quad R_s \frac{d^2y}{ds^2} - \frac{dx}{ds} = 0$$

$$\rightarrow \vec{r}(s) = \left[x(s) \quad y(s)\right] = \vec{r}[s(t)] = \vec{r}(t)$$

The equations above constitute a mapping of s-axis onto parametric trajectory. Like implicit representation, parametric curves are versatile and flexible in describing complex trajectories. It requires two equations to define a plane curve instead of one in implicit or explicit form. In this approach, we think of both coordinates x and y as a function of a third independent variable or parameter like arc length. The same trajectory may have different parametric representations through change of parameter such as time t. Similarly, the

t-axis is said to define a mapping from time scale onto the corresponding trajectory. The two axes are connected by the definition of arc length.

$$s = s(t) = \int_{t_0}^{t} \sqrt{V \cdot V} dt \qquad (2.12a)$$

where

$$V = \frac{d\vec{r}}{dt} = \frac{d\vec{r}}{ds}\frac{ds}{dt} = \hat{s}\frac{ds}{dt} = \hat{s}|V| = \hat{s}R_s\frac{d\alpha}{dt}$$

$$\rightarrow v_s = \frac{ds}{dt} = |V|; \quad v_n = \frac{dn}{dt} = 0$$

Parametric representation not only gives a set of points that describe a trajectory but also the location at a point on the curve specified by the values of parameter and the direction in which the trajectory advances in the same order as on an axis. The choice of parameter is to bring about simplification and depends on the problems.

However, streamline structure is not known a priori, but follows flow fields. From geometry and direction cosines definition, there exist relations between partial derivatives of variables in streamline coordinates and Cartesian coordinates of the same linear dimensions. In this respect, streamline coordinate is an example of a more general concept of local coordinates originating from the theory of differential geometry. It recalls the definition of tangent unit vector in terms of its direction cosines.

$$\hat{s} = \frac{d\vec{r}}{ds} = \left[\frac{dx}{ds} \quad \frac{dy}{ds} \right] = \left[\cos\alpha \quad \sin\alpha \right] = \left[\cos(x_1, s) \quad \cos(x_2, s) \right]$$

where

$$\frac{\partial x_i}{\partial \xi_j} = \cos(x_i, \xi_j)$$

The above represents the direction cosines between coordinate directions of Cartesian and curvilinear coordinate systems. $\hat{s}$ and $\vec{r}$ represent streamline tangent unit vector and position vector of a point on a streamline. α is the angle between unit vector $\hat{s}$ and positive x-axis counterclockwise. $d\vec{r}$ is an elemental displacement tangent to streamline trajectory. The direction cosines of the tangent pointing in the direction of increasing arc length are:

$$\frac{dx}{ds} = \frac{dx}{\sqrt{(dx)^2 + (dy)^2}} = \cos\alpha \qquad (2.12b)$$

$$\frac{dy}{ds} = \frac{dy}{\sqrt{(dx)^2 + (dy)^2}} = \sin\alpha \qquad (2.12c)$$

The direction cosines of a normal unit vector are obtained by rotating the positive direction of tangent through a right angle counterclockwise.

$$\frac{dx}{dn} = -\sin\alpha; \quad \frac{dy}{dn} = \cos\alpha$$

$$\hat{n} = \frac{d\vec{r}}{dn} = \left[\begin{array}{cc} \dfrac{dx}{dn} & \dfrac{dy}{dn} \end{array}\right] = \hat{k} \times \hat{s} = \left[\begin{array}{cc} -\dfrac{dy}{ds} & \dfrac{dx}{ds} \end{array}\right] = \left[\begin{array}{cc} -\sin\alpha & \cos\alpha \end{array}\right]$$

The normal unit vector could also be obtained from Frenet's formulas, that is,

$$\hat{n} = R_s \frac{d\hat{s}}{ds} = \frac{d\hat{s}}{d\alpha} = \left[\begin{array}{cc} -\sin\alpha & \cos\alpha \end{array}\right]$$

Hence,

$$\hat{s} \cdot \hat{n} = 0$$
$$\frac{\partial z}{\partial s} = \cos\left(\frac{\pi}{2}\right) = 0; \quad \frac{\partial z}{\partial n} = \cos\left(\frac{\pi}{2}\right) = 0$$
$$\frac{\partial x}{\partial z} = \frac{\partial y}{\partial z} = \cos\left(\frac{\pi}{2}\right) = 0; \quad \frac{\partial z}{\partial z} = \cos(0) = 1$$

The above system of equations leads exactly to the so-called Cauchy-Riemann relations, which ensure mapping is conformal between two coordinates:

$$\frac{\partial x}{\partial s} = \frac{\partial y}{\partial n}; \quad \frac{\partial x}{\partial n} = -\frac{\partial y}{\partial s}$$

$$\frac{\partial s}{\partial x} = \frac{\partial(s, y)}{\partial(x, y)} = \frac{\dfrac{\partial(s, y)}{\partial(s, n)}}{\dfrac{\partial(x, y)}{\partial(s, n)}} = \frac{\dfrac{\partial y}{\partial n}}{J}; \quad \frac{\partial n}{\partial y} = \frac{\dfrac{\partial x}{\partial s}}{J}; \quad \frac{\partial s}{\partial y} = -\frac{\dfrac{\partial x}{\partial n}}{J}; \quad \frac{\partial n}{\partial x} = -\frac{\dfrac{\partial y}{\partial s}}{J}$$

$$(2.13a)$$

where the Jacobian determinant of coordinate transformation is defined as:

$$J = \frac{\partial(x, y)}{\partial(s, n)} = \left|\begin{array}{cc} \dfrac{\partial x}{\partial s} & \dfrac{\partial x}{\partial n} \\[2mm] \dfrac{\partial y}{\partial s} & \dfrac{\partial y}{\partial n} \end{array}\right|$$

$$= \frac{\partial x}{\partial s}\frac{\partial y}{\partial n} - \frac{\partial x}{\partial n}\frac{\partial y}{\partial s} = \cos^2\alpha + \sin^2\alpha = 1$$

Hence, $\quad \dfrac{\partial s}{\partial x} = \dfrac{\partial n}{\partial y}; \quad \dfrac{\partial s}{\partial y} = -\dfrac{\partial n}{\partial x}$ $\qquad (2.13b)$

$$h_1 = \sqrt{\sum_{k=1} \left(\frac{\partial x_k}{\partial \xi_1}\right)^2} = \sqrt{\left(\frac{\partial x}{\partial s}\right)^2 + \left(\frac{\partial y}{\partial s}\right)^2} = \sqrt{\cos^2 \alpha + \sin^2 \alpha} = 1 = \sqrt{g_{11}}$$

$$h_2 = \sqrt{\sum_{k=1} \left(\frac{\partial x_k}{\partial \xi_2}\right)^2} = \sqrt{\left(\frac{\partial x}{\partial n}\right)^2 + \left(\frac{\partial y}{\partial n}\right)^2} = \sqrt{\sin^2 \alpha + \cos^2 \alpha} = 1 = \sqrt{g_{22}}$$

$$J = h_1 h_2 = 1$$

With both Lame coefficients equal to Jacobian determinant of unity, the transformation is conformal and isochronic. The geometric interpretation of streamline unit tangent is connected to the physical interpretation of velocity field as the rate of change of particle displacement.

$$\hat{s} = \frac{d\bar{r}}{dt}\frac{dt}{ds} = \frac{V}{|V|} = \left[\frac{v_x}{|V|} \quad \frac{v_y}{|V|}\right] \tag{2.14a}$$

$$\tilde{\theta} = \tan^{-1}\left(\frac{v_y}{v_x}\right) = \tan^{-1}\left(\frac{dy}{dx}\right) = \alpha \tag{2.14b}$$

V denotes velocity of a particle at a given point. The velocity angle $\tilde{\theta}$ being the inclination angle with respect to positive x-axis counterclockwise is identical to the streamline tangent angle everywhere. The velocity direction points in the tangential direction $\hat{s}$ of increasing arc length corresponding to the instantaneous direction of motion. In polar representation, streamline unit tangent vector defines a point on the unit circle at an angle equal to streamline tangent angle through the Euler formula, that is,

$$\hat{s} = \cos \alpha + i \sin \alpha = \exp(i\alpha); \quad i \equiv \sqrt{-1}$$

$$\hat{n} = R_s \frac{\partial \hat{s}}{\partial s} = \frac{\partial \hat{s}}{\partial \alpha} = -\sin \alpha + i \cos \alpha = \hat{s}i$$

$$|\hat{s}| = |\exp(i\alpha)| = 1 \text{ for all } \alpha$$

$$V = \begin{bmatrix} v_x & v_y \end{bmatrix} = \sqrt{v_x^2 + v_y^2}\exp(i\alpha) = |V|\exp(i\alpha)$$

A particle will be subject to a force equal to the product of acceleration vector and its mass. The acceleration vector is conveniently decomposed into streamline tangential and normal components since each component has a specific role to play. The tangential and normal components give rise to a change of speed and change of motion direction respectively. Speed increases or decreases when tangential acceleration is in the same or opposite direction of velocity. Speed will assume constant when acceleration and velocity are mutually orthogonal. The normal component is always directed normal to the path towards the center of curvature and is referred to as centripetal acceleration. It has a magnitude equal to the product of the square of speed and curvature and requires a force to keep the particle on the curve in accordance with Newton's law.

$$a = \frac{dV}{dt} = \frac{d}{dt}\left(\frac{ds}{dt}\hat{s}\right) = \frac{V^2}{R_s}\hat{n} + \frac{d^2s}{dt^2}\hat{s}$$

or

$$a = \frac{dV}{dt} = \frac{d^2\vec{r}}{ds^2}\left(\frac{ds}{dt}\right)^2 + \frac{d\vec{r}}{ds}\frac{d^2s}{dt^2}$$
$$= \frac{V^2}{R_s}\hat{n} + \frac{d\,|V|}{dt}\hat{s}$$

Transformation of streamline coordinates and Cartesian coordinates in three-dimensions is carried out by expanding two-dimension systems. Consider transformation of unit vectors.

$$\hat{s} = \frac{1}{h_1}\left(\frac{\partial x}{\partial s}\hat{i} + \frac{\partial y}{\partial s}\hat{j} + \frac{\partial z}{\partial s}\hat{k}\right)$$
$$= \cos\alpha\,\hat{i} + \sin\alpha\,\hat{j} \tag{2.15a}$$

$$\hat{n} = \frac{1}{h_2}\left(\frac{\partial x}{\partial n}\hat{i} + \frac{\partial y}{\partial n}\hat{j} + \frac{\partial z}{\partial n}\hat{k}\right)$$
$$= -\sin\alpha\,\hat{i} + \cos\alpha\,\hat{j} \tag{2.15b}$$

$$\hat{k} = \frac{1}{h_3}\left(\frac{\partial x}{\partial z}\hat{i} + \frac{\partial y}{\partial z}\hat{j} + \frac{\partial z}{\partial z}\hat{k}\right) = \hat{k} \tag{2.15c}$$

Or

$$\begin{bmatrix} \hat{i} \\ \hat{j} \\ \hat{k} \end{bmatrix} = [T(\alpha)] \begin{bmatrix} \hat{s} \\ \hat{n} \\ \hat{k} \end{bmatrix}$$

where $[T(\alpha)] = \begin{bmatrix} \cos\alpha & -\sin\alpha & 0 \\ \sin\alpha & \cos\alpha & 0 \\ 0 & 0 & 1 \end{bmatrix}$; $|T(\alpha)| = 1$

Similarly, the derivatives and velocity vector are transformed accordingly,

$$
\begin{bmatrix} \dfrac{\partial}{\partial x} \\ \dfrac{\partial}{\partial y} \\ \dfrac{\partial}{\partial z} \end{bmatrix} = \begin{bmatrix} \cos\alpha & -\sin\alpha & 0 \\ \sin\alpha & \cos\alpha & 0 \\ 0 & 0 & 1 \end{bmatrix} \begin{bmatrix} \dfrac{\partial}{\partial s} \\ \dfrac{\partial}{\partial n} \\ \dfrac{\partial}{\partial z} \end{bmatrix}
$$

$$
\begin{bmatrix} v_x \\ v_y \\ v_z \end{bmatrix} = \begin{bmatrix} \cos\alpha & -\sin\alpha & 0 \\ \sin\alpha & \cos\alpha & 0 \\ 0 & 0 & 1 \end{bmatrix} \begin{bmatrix} v_s \\ v_n \\ v_z \end{bmatrix} \tag{2.16}
$$

The transformation is an isometry and can be viewed as a simple rotation of a coordinate system about z-axis through an angle α.

The ∇-operator in streamline coordinate is written as:

$$
\nabla \equiv \sum_{i=1}^{} \frac{\hat{\xi}_i \partial}{h_i \partial \xi_i} = \hat{s}\frac{\partial}{h_1 \partial s} + \hat{n}\frac{\partial}{h_2 \partial n} + \hat{k}\frac{\partial}{h_3 \partial z} = \hat{s}\frac{\partial}{\partial s} + \hat{n}\frac{\partial}{\partial n} + \hat{k}\frac{\partial}{\partial z}
$$

Recall Frenet's formulas for plane curves with zero torsion.

$$
\frac{\partial \hat{s}}{\partial s} = \frac{1}{R_s}\hat{n}; \quad \frac{\partial \hat{n}}{\partial n} = -\frac{1}{R_n}\hat{s} \tag{2.17a}
$$

$$
\frac{\partial \hat{s}}{\partial n} = \frac{1}{R_n}\hat{n}; \quad \frac{\partial \hat{n}}{\partial s} = -\frac{1}{R_s}\hat{s} + \tau\hat{k} = -\frac{1}{R_s}\hat{s} \tag{2.17b}
$$

$$
\frac{\partial \hat{k}}{\partial s} = -\tau\hat{n} = \frac{\partial \hat{k}}{\partial n} = \frac{\partial \hat{k}}{\partial z} = 0 \tag{2.17c}
$$

$$
\frac{\partial \hat{s}}{\partial z} = 0; \quad \frac{\partial \hat{n}}{\partial z} = 0 \tag{2.17d}
$$

The above information could be used to establish Laplacian operator (∇^2-operator) in streamline coordinates. For plane fields,

$$
\nabla^2 = \nabla \cdot \nabla = \left(\hat{s}\frac{\partial}{\partial s} + \hat{n}\frac{\partial}{\partial n} + \hat{k}\frac{\partial}{\partial z}\right) \cdot \left(\hat{s}\frac{\partial}{\partial s} + \hat{n}\frac{\partial}{\partial n} + \hat{k}\frac{\partial}{\partial z}\right)
$$

$$
= \frac{\partial^2}{\partial s^2} - \frac{1}{R_s}\frac{\partial}{\partial n} + \frac{1}{R_n}\frac{\partial}{\partial s} + \frac{\partial^2}{\partial n^2}
$$

Special types of motions:

(1) Motion in a straight line.

$$R_s = \infty; \ a_n = 0; \ v_n = 0; \ |V| = v_s$$
$$a \times V = 0$$

(2) Circular motion.

Circular motion is an important type of trajectory in the family of conic sections in addition to the property of axis-symmetry. The circular trajectory is described completely by circle radius and position of circle center. Polar coordinate is ideal for circular motion in general and has relations with streamline coordinate in particular. By placing polar origin to coincide with the circle center, it can be shown that the polar angle specifies the trajectory tangent within a constant and that the radius vector defined by the polar coordinate circle radius is equal to the radius of trajectory curvature.

$$v_s = \frac{ds}{dt} = \left|\frac{d\vec{r}}{dt}\right| = \left|R_s\frac{d\hat{r}}{dt}\right| = \left|R_s\frac{d\theta}{dt}\hat{\theta}\right| = R_s\frac{d\theta}{dt} = r\frac{d\theta}{dt} = v_\theta; \ \rightarrow ds = R_s d\theta$$

$$r = R_s$$

$$v_r = \frac{dr}{dt} = 0 = v_n$$

$$-a_r = r\left(\frac{d\theta}{dt}\right)^2 = \frac{v_\theta^2}{r} = \frac{V^2}{R_s} = a_n = \frac{\left|\dfrac{d\vec{r}}{dt} \times \dfrac{d^2\vec{r}}{dt^2}\right|}{\left|\dfrac{d\vec{r}}{dt}\right|} = \frac{|V \times a|}{|V|}$$

$$a_\theta = \frac{d}{r\,dt}\left(r^2\frac{d\theta}{dt}\right) = \frac{d}{dt}\frac{ds}{dt} = \frac{d\,|V|}{dt} = a_s = \frac{\dfrac{d\vec{r}}{dt} \cdot \dfrac{d^2\vec{r}}{dt^2}}{\left|\dfrac{d\vec{r}}{dt}\right|} = \frac{V.a}{|V|}$$

$$\hat{s} = \frac{d\vec{r}}{ds} = R_s\frac{d\hat{r}}{d\theta}\frac{d\theta}{ds} = R_s\frac{\hat{\theta}}{R_s} = \hat{\theta}; \ \frac{\hat{n}}{R_s} = \frac{d\hat{s}}{ds} = \frac{d\hat{\theta}}{d\theta}\frac{d\theta}{ds} = -\hat{r}\frac{1}{R_s}; \ \rightarrow \hat{n} = -\hat{r}$$

$$\frac{d\hat{s}}{dt} = \frac{|V|}{R_s}\hat{n} = \frac{d\alpha}{dt}\hat{n} = \frac{d\theta}{dt}\hat{n}$$

$$\frac{d\hat{n}}{dt} = \frac{d\hat{n}}{d\alpha}\frac{d\alpha}{dt} = -\hat{s}\frac{d\alpha}{dt} = -\frac{d\theta}{dt}\hat{s}$$

$$Y = mr \times V = mr^2\frac{d\theta}{dt}\hat{k}$$

$$\theta = \alpha - \pi/2; \quad \frac{d\alpha}{dt} = \frac{d\theta}{dt}$$

$$\begin{bmatrix} \hat{s} \\ \hat{n} \\ \hat{k} \end{bmatrix} = [T(\alpha)]^{-1} \begin{bmatrix} \hat{i} \\ \hat{j} \\ \hat{k} \end{bmatrix} = \begin{bmatrix} -\sin\theta & \cos\theta & 0 \\ -\cos\theta & -\sin\theta & 0 \\ 0 & 0 & 1 \end{bmatrix} \begin{bmatrix} \hat{i} \\ \hat{j} \\ \hat{k} \end{bmatrix}$$

Hence, the transformation of unit vectors between polar coordinate and streamline coordinate in plane circular motion is given by:

$$\begin{bmatrix} \hat{s} \\ \hat{n} \end{bmatrix} = [T(\alpha)]^{-1} [T(\theta)] \begin{bmatrix} \hat{r} \\ \hat{\theta} \end{bmatrix}$$

$$= \begin{bmatrix} \cos(\alpha - \theta) & \sin(\alpha - \theta) \\ -\sin(\alpha - \theta) & \cos(\alpha - \theta) \end{bmatrix} \begin{bmatrix} \hat{r} \\ \hat{\theta} \end{bmatrix} = \begin{bmatrix} 0 & 1 \\ -1 & 0 \end{bmatrix} \begin{bmatrix} \hat{r} \\ \hat{\theta} \end{bmatrix}$$

$$= \begin{bmatrix} \hat{\theta} \\ -\hat{r} \end{bmatrix}$$

$\frac{|V|}{R_s}$ is interpreted as rotational rate of particle equal to the rate of change of streamline tangent angle. $\hat{s}$ and $\hat{n}$ are unit vector of trajectory in tangent and normal direction respectively. R_s is trajectory radius of curvature. If circular motion is at constant speed, then acceleration and velocity are perpendicular to each other and conversely. Both centripetal acceleration and angular momentum are constant.

$$a_s = 0; \quad a \cdot V = 0$$

$$\frac{d\theta}{dt} = \text{constant}; \quad a_n = \frac{v_\theta^2}{R_s} = \text{constant}$$

$$v_s = r\frac{d\theta}{dt} = \text{constant}$$

$$Y = mr \times V = mr^2\frac{d\theta}{dt}\hat{k} = \text{constant}\,\hat{k}$$

Transformation of plane fields or two-dimensional fields is of special interest in a class of conformal coordinates with equal Lame coefficients. Such transformation is connected with harmonic functions and the theory of functions of complex variables. To explore conformal representation of streamline coordinates (or orthogonal curvilinear coordinates) in relation to functions of complex variables, the following complex ζ-plane is introduced with ξ-axis parallel to x-axis of same unit of measurements.

$$\zeta - \text{plane} : \zeta = \xi + i\eta \tag{2.18}$$

$$z - \text{plane} : z = x + iy \tag{2.19}$$

Mapping functions: $z = z(\zeta); \zeta = \zeta(z)$

or

$$\xi = \xi(x, y)$$
$$\eta = \eta(x, y)$$

The important role played by analytic functions is to transform Cauchy-Riemann systems between z-plane and ζ-plane. Analytic functions offer a simple method of generating orthogonal curvilinear coordinate systems in two-dimensions by conformal mapping between complex planes. Any analytic mapping function defines a conformal map serving as a new set of orthogonal curvilinear coordinate systems. We are interested in the conformal representation of fields in curvilinear coordinates and the requirement of orthogonality on complex planes, which can be interpreted as meaning that any plane coordinate pair must satisfy conformal transformation. The Cauchy-Riemann relations ensure that all curves passing through a point will be scaled by magnification factor and rotated through the same twist angle independent of directions. Each pair of (x, y) rectangular coordinate axes is mapped to an image pair of (ξ, η) coordinate curves crossing at right angle in ζ-plane and conversely. This means at an arbitrary point z_0, there is a pair of orthogonal curvilinear directions along coordinate curves $\xi_0 = C_1$ and $\eta_0 = C_2$ defined implicitly by the mapping functions on z-plane. They may be taken as a set of coordinate axes at point ζ_0 on ζ-plane. At image point ζ_0, the tangent to $\xi_0 = C_1$ curvilinear coordinate curve is determined by the spatial rate of change of an infinitesimal displacement with respect to arc length $\partial\xi$ while keeping η constant, that is,

$$dx\,|_{\xi_0=C_1} = \frac{\partial x}{\partial \xi}d\xi + \frac{\partial x}{\partial \eta}d\eta = \frac{\partial x}{\partial \xi}d\xi$$

$$dy\,|_{\xi_0=C_1} = \frac{\partial y}{\partial \xi}d\xi + \frac{\partial y}{\partial \eta}d\eta = \frac{\partial y}{\partial \xi}d\xi$$

Tangent slope: $\dfrac{dy}{dx}\,|_{\xi_0=C_1} = \left(\dfrac{\partial y}{\partial \xi}\right) \Big/ \left(\dfrac{\partial x}{\partial \xi}\right) = \dfrac{\cos(y, \xi)}{\cos(x, \xi)} = \dfrac{\sin(x, \xi)}{\cos(x, \xi)} = \tan(x, \xi)$

The direction cosines of the tangent have been introduced into the above. Similarly, by keeping ξ constant, we obtain the tangent to $\eta_0 = C_2$ curvilinear coordinate curve.

$$dx\,|_{\eta_0=C_2} = \frac{\partial x}{\partial \xi}d\xi + \frac{\partial x}{\partial \eta}d\eta = \frac{\partial x}{\partial \eta}d\eta$$

$$dy\,|_{\eta_0=C_2} = \frac{\partial y}{\partial \xi}d\xi + \frac{\partial y}{\partial \eta}d\eta = \frac{\partial y}{\partial \eta}d\eta$$

$$\text{Tangent slope: } \frac{dy}{dx}\Big|_{\eta_0=C_2} = \left(\frac{\partial y}{\partial \eta}\right) \Big/ \left(\frac{\partial x}{\partial \eta}\right) = \frac{\cos(y, \eta)}{\cos(x, \eta)} = \frac{\cos(x, \xi)}{-\sin(x, \xi)}$$

$$= -1/\tan(x, \xi)$$

Hence,

$$\left(\frac{dy}{dx}\Big|_{\eta_0=C_2}\right)\left(\frac{dy}{dx}\Big|_{\xi_0=C_1}\right) = -1; \quad \xi_0 = C_1 \bot \eta_0 = C_2 \qquad (2.20)$$

This completes the proof that $\xi = C_1$ and $\eta = C_2$ form an orthogonal curvilinear coordinate on ζ-plane. By keeping η constant, we have,

$$\arg |d\xi_0 + i d\eta_0| - \arg |dx_0 + i dy_0| = \arg |\zeta'(z_0)|$$
$$\arg |d\xi_0| - \arg |dx_0| = \arg |\zeta'(z_0)|$$

where

$$|\zeta'(z)| = \sqrt{\left(\frac{\partial \xi}{\partial x}\right)^2 + \left(\frac{\partial \eta}{\partial x}\right)^2}$$

By keeping ξ constant,

$$\arg |i d\eta_0| - \arg |i dy_0| = \arg |\zeta'(z_0)|$$
$$(\arg(i) + \arg |d\eta_0|) - (\arg(i) + \arg |dy_0|) = \arg |\zeta'(z_0)|$$
$$\arg |d\eta_0| - \arg |dy_0| = \arg |\zeta'(z_0)|$$

Hence,

$$\arg |d\eta_0| - \arg |d\xi_0| - \arg |dy_0| \quad \arg |dx_0|$$

All curves issuing from a given point are rotated through the same twist angle. The angle between any two lines is the same when mapped to image plane.

$$d\xi = |\zeta'(z)| dx; \quad d\eta = |\zeta'(z)| dy$$
$$d\xi d\eta = |\zeta'(z)|^2 dxdy = \frac{dxdy}{J}$$

where

$$\sqrt{J} = \sqrt{\frac{\partial(x, y)}{\partial(\xi, \eta)}} = \sqrt{\frac{\partial x}{\partial \xi}\frac{\partial y}{\partial \eta} - \frac{\partial y}{\partial \xi}\frac{\partial x}{\partial \eta}} = |z'(\zeta)| = \sqrt{(\det |g_{ij}|)^{1/2}} = \sqrt{h_1 h_2}$$

$$\frac{\partial x}{\partial \xi} = \frac{\partial y}{\partial \eta}; \quad \frac{\partial y}{\partial \xi} = -\frac{\partial x}{\partial \eta}; \quad |\zeta'(z)| = \frac{1}{|z'(\zeta)|}$$

The equations above show that inverse mapping exists except at critical/ singular points and that ζ-plane is a conformal coordinate with equal Lame coefficients. Infinitesimal squares are transformed into infinitesimal curvilinear squares scaled by Jacobian with angle and local shape preserved. Conformal representation naturally sets up the link between metric tensor, magnification factor and Jacobian determinant of coordinate transformation in two-dimensions through Cauchy-Riemann relations.

Consider conformal mapping of complex potential between Cartesian co-ordinates and streamline coordinates defined by a pair of conjugate harmonic functions.

$$F = \phi + i\psi$$
$$\phi(x, y) = \phi[x(s, n), y(s, n)] = \phi(s, n)$$
$$\psi(x, y) = \psi[x(s, n), y(s, n)] = \psi(s, n)$$

The velocity components in streamline coordinates are derived from directional derivatives of velocity potential, that is,

$$v_s = \frac{\partial \phi}{\partial s} = \frac{\partial \phi}{\partial x}\frac{\partial x}{\partial s} + \frac{\partial \phi}{\partial y}\frac{\partial y}{\partial s} = v_x \cos\alpha + v_y \sin\alpha$$

$$= \hat{s} \cdot \nabla\phi = \hat{s} \cdot (\nabla\psi \times \hat{k}) = \nabla\psi \cdot (\hat{k} \times \hat{s}) = \nabla\psi \cdot \hat{n} = \frac{\partial \psi}{\partial n}$$

$$v_n = \frac{\partial \phi}{\partial n} = \frac{\partial \phi}{\partial x}\frac{\partial x}{\partial n} + \frac{\partial \phi}{\partial y}\frac{\partial y}{\partial n} = -v_x \sin\alpha + v_y \cos\alpha$$

$$= \hat{n} \cdot \nabla\phi = \hat{n} \cdot (\nabla\psi \times \hat{k}) = -\nabla\psi \cdot (\hat{n} \times \hat{k}) = -\nabla\psi \cdot \hat{s} = -\frac{\partial \psi}{\partial s}$$

The above expression is the generalization of Cauchy-Riemann relations in streamline coordinates. In theory, conjugate harmonic functions satisfy Cauchy-Riemann relations in any orthogonal curvilinear coordinates and it ensures that transformation is conformal. Since no flux is allowed to cross a streamline, the normal component vanishes locally at points along streamlines just as solid boundaries do. The flow fields are regarded as one-dimensional locally in the sense that velocity direction is parallel to a streamline unit tangent vector at a particle with vanishing normal component. The flux passing through a stream surface per unit depth vanishes.

$$\frac{Q}{\Delta z} = \int_s V \cdot \hat{n} ds = \int_s -\frac{\partial \psi}{\partial s} ds = 0$$
$$\rightarrow \psi \mid_s = C$$

It implies the normal and tangential derivatives of stream function will yield velocity components identical to those derived from velocity potential.

$$v_n = -\frac{\partial \psi}{\partial s}; \; v_s = \frac{\partial \psi}{\partial n}$$

2.2.3 Potential-Stream Function Coordinates

Curvilinear coordinates could be extended to the adoption of potential-stream function coordinates in a class of natural (or intrinsic) coordinates. Since potential function and stream function form an orthogonal system, they are ideal candidates to serve as a two dimensional orthogonal curvilinear coordinate system. A potential-stream function coordinate is a version of streamline coordinate on a rectangular coordinate system with potential ϕ and stream function ψ assuming the role of horizontal and vertical axes respectively. Both axes have the same unit of length pointing in the respective increasing directions. Streamlines and equipotentials are mapped as a system of rectangular coordinate nets in potential-stream function coordinates representing uniform flow fields parallel to horizontal axis. It is regarded as one type of orthogonal curvilinear coordinates with fixed coordinate unit vectors like Cartesian coordinates. A point defined by a pair of ϕ and ψ on (x, y) plane corresponds to the image pair of x and y represented on (ϕ, ψ) plane and vice versa. For a given equipotential and stream function on Cartesian coordinate, inverse mapping can be applied directly to reverse the role of variables from Cartesian to potential-stream function coordinate.

$$\phi = \phi(x, y) \tag{2.21a}$$

$$\psi = \psi(x, y) \tag{2.21b}$$

Conversely,

$$x = x(\phi, \psi) \tag{2.22a}$$

$$y = y(\phi, \psi) \tag{2.22b}$$

Displaying information from reverse perspectives could be thought of mathematically as inverse mapping. Inverse mapping arises frequently and naturally when images obtained in one plane are ready to be transformed back to the original plane. Such mapping reverses the roles of image and original by exchanging variables from independent to dependent and vice versa. Specifically, when a point is mapped from original coordinate to its image in another coordinate, the inverse will map the image back to its original. Inverse mapping exists when Jacobian of transformation is uniquely defined other than zero. To obtain inverse mapping functions, we invoke mapping of elemental lengths between potential-stream function coordinates and Cartesian coordinates through chain rule,

$$\begin{bmatrix} dx \\ dy \end{bmatrix} = [J] \begin{bmatrix} d\phi \\ d\psi \end{bmatrix}$$

where $[J] = \begin{bmatrix} \dfrac{\partial(x, y)}{\partial(\phi, \psi)} \end{bmatrix} = \begin{bmatrix} \dfrac{\partial x}{\partial \phi} & \dfrac{\partial x}{\partial \psi} \\ \dfrac{\partial y}{\partial \phi} & \dfrac{\partial y}{\partial \psi} \end{bmatrix}.$

The inverse is given by:

$$\begin{bmatrix} d\phi \\ d\psi \end{bmatrix} = \frac{1}{J} \begin{bmatrix} \dfrac{\partial y}{\partial \psi} & -\dfrac{\partial x}{\partial \psi} \\ -\dfrac{\partial y}{\partial \phi} & \dfrac{\partial x}{\partial \phi} \end{bmatrix} \begin{bmatrix} dx \\ dy \end{bmatrix}$$

where $J = \dfrac{\partial(x, y)}{\partial(\phi, \psi)} = \begin{vmatrix} \dfrac{\partial x}{\partial \phi} & \dfrac{\partial x}{\partial \psi} \\ \dfrac{\partial y}{\partial \phi} & \dfrac{\partial y}{\partial \psi} \end{vmatrix} = h_1 h_2 = \dfrac{1}{V^2}.$

$$h_1 = \left(\sqrt{\left(\frac{\partial \phi}{\partial x}\right)^2 + \left(\frac{\partial \phi}{\partial y}\right)^2} \right)^{-1} = \frac{1}{|V|}; \; h_2 = \left(\sqrt{\left(\frac{\partial \psi}{\partial x}\right)^2 + \left(\frac{\partial \psi}{\partial y}\right)^2} \right)^{-1} = \frac{1}{|V|}$$

Since, $\begin{bmatrix} d\phi \\ d\psi \end{bmatrix} = \begin{bmatrix} \dfrac{\partial \phi}{\partial x} & \dfrac{\partial \phi}{\partial y} \\ \dfrac{\partial \psi}{\partial x} & \dfrac{\partial \psi}{\partial y} \end{bmatrix} \begin{bmatrix} dx \\ dy \end{bmatrix}.$

Comparisons of two systems of equations show that the corresponding matrix components must be identical. We have,

$$\begin{bmatrix} \dfrac{\partial \phi}{\partial x} & \dfrac{\partial \phi}{\partial y} \\ \dfrac{\partial \psi}{\partial x} & \dfrac{\partial \psi}{\partial y} \end{bmatrix} = \frac{1}{J} \begin{bmatrix} \dfrac{\partial y}{\partial \psi} & -\dfrac{\partial x}{\partial \psi} \\ -\dfrac{\partial y}{\partial \phi} & \dfrac{\partial x}{\partial \phi} \end{bmatrix}$$

By equating the corresponding components, we obtain the following integral expressions for the inverse mapping functions from (ϕ, ψ) plane to (x, y)

plane without actually solving the forward mapping functions explicitly, that is,

$$y = \int J \frac{\partial \phi}{\partial x} \partial \psi = \int \frac{\partial x}{\partial \phi} \partial \psi; \; y = \int -J \frac{\partial \psi}{\partial x} \partial \phi = \int -\frac{\partial x}{\partial \psi} \partial \phi$$

$$x = \int -J \frac{\partial \phi}{\partial y} \partial \psi = \int -\frac{\partial y}{\partial \phi} \partial \psi; \; x = \int J \frac{\partial \psi}{\partial y} \partial \phi = \int \frac{\partial y}{\partial \psi} \partial \phi$$

However, integrations are not necessary trial. For the same reason, it is generally not possible to find inverse mapping functions explicitly in closed analytical forms by solving directly the forward mapping functions unless they are simple functions. Alternatively, one could solve a pair of Laplace equations in terms of x and y subject to prescribed boundary conditions, that is,

$$\frac{\partial^2 x}{\partial \phi^2} + \frac{\partial^2 x}{\partial \psi^2} = 0; \; \frac{\partial^2 y}{\partial \phi^2} + \frac{\partial^2 y}{\partial \psi^2} = 0$$

Boundary conditions:

$$x = x(\phi, \psi); \; y = y(\phi, \psi)$$
$$\frac{\partial x_i}{\partial \phi} = g_i(\phi, \psi); \; \frac{\partial x_i}{\partial \psi} = f_i(\phi, \psi)$$

In addition to the above analytical methods, contour graphs of $x = C_1'$ and $y = C_2'$ could be constructed with x and y variables assuming the role of parameters on ϕ, ψ coordinate. This graphical procedure for plotting contours of Equation 2.22 on potential-stream function coordinates takes a series of different x with y being held constant. By repeating the process of reading off values of ϕ and ψ and inserting them at corresponding points on (ϕ, ψ) coordinates, we obtain a family of curves $y = C_2'$ on potential-stream function coordinates. Similarly, the corresponding curves for $x = C_1'$ are obtained by reversing x with y. As a result, a set of rectangular coordinate lines is mapped from z-plane into continuous orthogonal curves on potential-stream function coordinates. This implies a Direchlet boundary condition that fits an equipotential or a streamline is mapped as a vertical or horizontal straight coordinate line accordingly. Conversely, rectangular coordinate lines $\phi = C_1$ and $\psi = C_2$ are transformed to an orthogonal curvilinear net in Cartesian coordinate. Furthermore, the following Cauchy-Riemann relation also holds in potential-stream function coordinate and leads back to Laplace equation:

$$\frac{\partial x}{\partial \phi} = \frac{\partial y}{\partial \psi}; \; \frac{\partial y}{\partial \phi} = -\frac{\partial x}{\partial \psi} \qquad (2.23)$$

When problems are formulated with scalar potential and vector potential as independent variables, they are best equipped to deal with situations involving inverse design or boundaries corresponding to a streamline or equipotential. An irregular or curved boundary of Direchlet type will be mapped to a straight horizontal or vertical line to bring about simplification of boundary conditions. The above results suggest potential and stream function are to be combined in an analytic function to define mapping between z-plane and complex F-plane in the context of complex function theory. This is the same idea of a complex variable formed by combining two variables. By assigning potential function to horizontal axis and stream function to an imaginary vertical axis, a complex potential is constructed as follows:

$$F = f(z) = \phi(x, y) + i\psi(x, y)$$

The above complex potential expands the idea of potential-stream function coordinates in the context of harmonic functions and takes the advantage of complex function theory. Any conjugate pair of harmonic functions could be used to define a system of conformal coordinates. Such that contours of stream function and equipotential will form a curvilinear square network by conformal mapping except at singular/critical points. Furthermore, the complex potential is connected to the conjugate/hodograph plane through differentiation and hence the vector fields. Complex F-plane models uniform vector fields and facilitate the task of counting numbers of flux tubes and work tubes in the assessment of field domain characteristics. In particular, singular points, if any are mapped to points at infinity on F-plane making it possible to evaluate shape factor graphically. Suppose the magnitude of a complex potential is constant at a given point in the domain, that is,

$$|f(z)| = C$$

$$|f(z)|^2 = \phi^2 + \psi^2 = C^2$$

$$\phi\frac{\partial\phi}{\partial x} + \psi\frac{\partial\psi}{\partial x} = 0$$

$$\phi\frac{\partial\phi}{\partial y} + \psi\frac{\partial\psi}{\partial y} = 0$$

$$(\psi^2 - \phi^2)\frac{\partial\phi}{\partial x}\frac{\partial\psi}{\partial x} + \phi\psi\left[\left(\frac{\partial\phi}{\partial x}\right)^2 - \left(\frac{\partial\psi}{\partial x}\right)^2\right] = 0$$

$$\rightarrow \frac{\partial\phi}{\partial x} = \frac{\partial\psi}{\partial x} = 0$$

Similarly,

$$\frac{\partial \phi}{\partial y} = \frac{\partial \psi}{\partial y} = 0$$

Hence,

$$\begin{aligned}
\phi &= C_1; \ \psi = C_2 \\
|f(z)| &= \sqrt{\phi^2 + \psi^2} = C \\
|f(z)|_{\text{boundary}} &= C_{\text{max}} \\
|f(z)| &< C_{\text{max}}
\end{aligned} \qquad (2.24)$$

Every pair of streamline and equipotential that satisfies Equation 2.24 is mapped to the circle of radius C about origin on (ϕ, ψ) plane. The maximum modulus principle ensures that the function $|f(z)|$ is constant and attains its maximum on the boundary. The value of a harmonic function at a point is equal to the mean value on any circle with center at that point. For circles about origin, the mean value for each harmonic function is zero and hence $\phi = 0$ and $\psi = 0$ are mapped to origin on (ϕ, ψ) plane. The field at a point is specified by a pair of streamline and equipotential instead of position. Since reverse transformation is also conformal, we have,

$$\frac{dz}{dF} = \frac{\partial x}{\partial \phi} + i \frac{\partial y}{\partial \phi}$$

where

$$\frac{\partial x}{\partial \phi} = J \frac{\partial \psi}{\partial y} = J v_x; \ \frac{\partial y}{\partial \phi} = -J \frac{\partial \psi}{\partial x} = J v_y$$

The above interpretation is a vector representation of fictitious flow on (ϕ, ψ) plane. It is then scaled by Jacobian of determinant to yield the actual velocity corresponding to the image pair of x and y coordinates on the physical plane.

$$\frac{dz}{dF} = \frac{1}{\overline{V}} = \frac{1}{v_x - i v_y} = \frac{1}{|V|} \left(\frac{v_x}{|V|} + i \frac{v_y}{|V|} \right)$$

$\frac{v_x}{|V|}$ and $\frac{v_y}{|V|}$ are the direction cosines of the velocity vector on z-plane equal to the fictitious flow unit vector. The modulus of fictitious flow vector is the reciprocal of speed. Inverse mapping naturally leads to hodograph method and applications (Equation 5.20a) by mapping the flow fields from complex plane. Useful information could be obtained from graphs in potential-stream function coordinates. Consider the tangent along an arbitrary trajectory on (ϕ, ψ) plane in Figure 5.9.

$$\tan \beta = \frac{d\psi}{d\phi} = \frac{\dfrac{\partial \psi}{\partial x}dx + \dfrac{\partial \psi}{\partial y}dy}{\dfrac{\partial \phi}{\partial x}dx + \dfrac{\partial \phi}{\partial y}dy} = \frac{-\dfrac{v_y}{v_x} + \dfrac{dy}{dx}}{1 + \dfrac{v_y}{v_x}\dfrac{dy}{dx}}$$

$$= \tan(\chi - \alpha)$$

$$\rightarrow \beta = \chi - \alpha$$

where $\dfrac{v_y}{v_x} = \tan \alpha = \tan \tilde{\theta}$.

$$\tan \chi \equiv \frac{dy}{dx} \tag{2.25}$$

β is the angle between tangents at a point along the trajectory with respect to positive ϕ-axis on potential-stream function coordinate anticlockwise. χ is the image angle between tangents at image point with respect to positive x-axis on the physical plane. α is the tangent angle between streamline tangent and positive x-axis. α angle (or $\tilde{\theta}$) defines the direction of vector field at the given point. It is convenient to work with $y = C_2$ curves, such that the tangent is horizontal everywhere in the physical plane and above reduces to:

$$\chi = \tan^{-1}\left(\frac{dy}{dx}\right)\Big|_{y=C_2} = 0; \ \beta = -\alpha$$

Thus, direction of vector fields can be determined from β angle on potential-stream function coordinates. β angle could be obtained by measuring the tangent slope graphically. If $x = C_1$ trajectories are used, then

$$\chi = \tan^{-1}\left(\frac{dy}{dx}\right)\Big|_{x=C_1} = \frac{\pi}{2}; \ \beta = \frac{\pi}{2} - \alpha$$

The method indeed works for any curve by evaluating the corresponding tangent angle. This means we could always determine vector field direction at a point in the physical plane by its image on potential-stream function coordinates.

2.3 Field Kinematics and Visual Attributes

Visual representations include the use of graphs, icons, diagrams, charts and tables to display the structure and pattern of data with or without colors/ textures. The connection between visual representations and users is critical in the development of visual attributes and an effective visual communication/comprehension system. One of the basic elements of visual attributes is the mathematical description of trajectory in space. In this development, the ideas of geometry and the concepts in physics led us naturally to parametric representation or implicit representation of curves.

2.3.1 Field Line Trajectory

Field kinematics is concerned with the study of a field particle in motion in terms of trajectory, which describes changes in the field from point to point. Field line trajectories are geometric curves in space useful for particle motion visualization. They track the positions of particles traveling continuously in one degree of freedom. For every point along a trajectory, the trajectory tangent defines the direction in which the field lines runs to the next point parallel to the vector fields. Each point belongs to one and only one field line trajectory. The distributions of field line trajectories provide a global view of field structure and transport phenomena. Figure 2.4 shows a plane field line trajectory traced by position vector $\vec{r}$, which is defined by particle path function in Eulerian variables:

$$\vec{r} = \vec{r}(x_i, t) \tag{2.26}$$

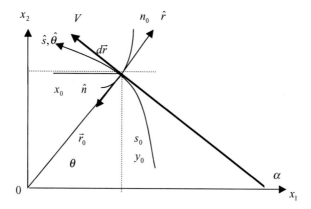

Figure 2.4   A plane field line trajectory.

To represent two-dimensional vector fields at an arbitrary time instant by curves, field line trajectory equation is invoked to translate the geometrical idea into mathematical forms. It is regarded as the differential representation of a curve written as:

$$V \times d\vec{r} = 0$$
$$\rightarrow \begin{bmatrix} 0 & 0 & v_x dx_2 - v_y dx_1 \end{bmatrix} = \begin{bmatrix} 0 & 0 & 0 \end{bmatrix} \tag{2.27}$$

$$\frac{dx_2}{dx_1} = \frac{v_y}{v_x} = f(x_1, x_2) \tag{2.28}$$

$d\vec{r}$ is an elemental displacement tangent to field line trajectory through a given point. V represents vector fields. The cross product of two vectors is zero only if they are parallel except at critical/singular points. It is merely a geometrical statement of two parallel lines. The trajectory tangents correspond to the concept of directional fields represented by a system of tangent vectors along an arbitrary trajectory. The graphical solution of trajectory equation is a visual attribute that traces the shape of trajectories as they are fitted through the fields of tangent vectors. The existence of field line trajectories is independent of sources or nature of the fields. They are a member of the field lines family. Field line trajectory is ready to be extended to three-dimensional space, that is,

$$\frac{dx_1}{v_x} = \frac{dx_2}{v_y} = \frac{dx_3}{v_z} \tag{2.29}$$

2.3.2 Field Line Integral Curves

Implicit representation of curves/surfaces has its root in implicit functions. An implicit function encodes positions in space into scalar values equal to the level (parametric) constant useful for complex mathematical description. The use of implicit functions is advantageous, as they always exist while explicit expressions may be difficult or fail. It expedites the determination of points whether they are on the curve or one side of it. Similarly, each trajectory is conveniently visualized and identified according to the value of parameter constant especially for conic curves. Solution to the above three-dimensional field line trajectory equation is the geometrical intersections of two-parameter family of level surfaces in space expressed by a pair of implicit functions. In this respect, trajectories produced by the intersections of surfaces represent the same idea of vector field representation by curves referred to as field line integral curves.

$$f(\psi, \lambda) = 0$$

where

$$\lambda(x_1, x_2, x_3) = C_1 \tag{2.30}$$
$$\psi(x_1, x_2, x_3) = C_2 \tag{2.31}$$

Implicit representation retains interpretation and engagement of stream function in terms of two stream surfaces; each defined by a scalar function. C_1 and C_2 are parameter constants denoting contour levels for each family of stream surfaces. As the constants vary, a system of integral curves in space is generated wherever two families of contour surfaces intersect each other. Each of them could be identified by the corresponding values of parametric constants. A unique field line integral curve is determined if two families of contour surfaces intersect only once. Figure 2.5 depicts the vector fields defined by the tangents along field line integral curves. It shows that a general vector field is derivable from a pair of vector functions ($\nabla\psi$ and $\nabla\lambda$) perpendicular to the respective contour surfaces and hence their cross product at the line of intersection of two contour surfaces is tangent to the integral curve so defined. The existence of field line integral curves requires that vector fields are solenoidal. Field line integral curves are a member of the field lines family.

$$V = (\nabla\psi \times \nabla\lambda)$$
$$= \left[\left(-\frac{\partial\lambda}{\partial x_2}\frac{\partial\psi}{\partial x_3} + \frac{\partial\lambda}{\partial x_3}\frac{\partial\psi}{\partial x_2} \right) \left(-\frac{\partial\lambda}{\partial x_3}\frac{\partial\psi}{\partial x_1} + \frac{\partial\lambda}{\partial x_1}\frac{\partial\psi}{\partial x_3} \right) \left(-\frac{\partial\lambda}{\partial x_1}\frac{\partial\psi}{\partial x_2} + \frac{\partial\lambda}{\partial x_2}\frac{\partial\psi}{\partial x_1} \right) \right]$$
$$\nabla \cdot V = \nabla\lambda \cdot (\nabla \times \nabla\psi) - \nabla\psi \cdot (\nabla \times \nabla\lambda) = 0 \tag{2.32}$$

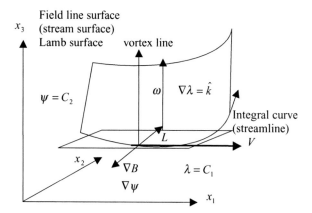

Figure 2.5   Field line surface and integral curve.

The flux passing through a surface with unit normal $\hat{b}$ bounded by a closed path c is evaluated by flux integral:

$$Q = \int_A V \cdot \hat{b} \, dA$$

$$= -\int_A \nabla \times (\lambda \nabla \psi) \cdot \hat{b} \, dA \tag{2.33}$$

$$= -\oint_c \lambda \nabla \psi \cdot \hat{l} \, dl = -\oint_c \lambda \, d\psi = \Delta \lambda \Delta \psi$$

However, visualization of vector fields in three-dimensional space is a difficult task. The mathematical complexity associated with the employment of two implicit functions has hampered their use in practice. There appears little mathematical advantage with the employment of dual function, which consists of a pair of scalar functions in place of three independent components. Moreover field lines in three-dimensions are not unique if two surfaces intersect more than once unless they are mutually orthogonal everywhere. Nevertheless, representation of vector fields by dual scalar functions provides a general approach to describe vector fields in multi-dimensions. In practice, there is a need to reduce the dual function formulation to the familiar Lagrangian or Stokes stream function by identification of one of the functions arbitrarily in terms of the relevant analysis plane. In this respect, we consider plane flow fields by setting one of independent variables say,

$$x_3 = \lambda(x_1, x_2) = C_1$$

The λ-stream surface degenerates onto a coordinate (x_1, x_2) plane of motions with a surface normal unit vector $\nabla \lambda = \hat{k}$ parallel to z-axis and cuts it at C_1. The thickness of the plane of motion is negligible compared with the other two dimensions in a class of plane stress problems. Each parallel coordinate (x_1, x_2) plane is a plane of symmetry. The remaining ψ-stream surface degenerates into cylindrical surface like an upright landscape. Such a surface is a type of ruled surfaces with one degree of freedom generated by a straight line normal to the plane of motion as it moves parallel to itself along a curve. If the straight generating line approaches infinite, then the three dimensional effects at cylinder ends are considered negligible like plane strain problems. The intersection of cylindrical surface and a coordinate plane yields a plane integral curve representing a streamline corresponding to a given C_2. By varying C_2, a system of streamlines is generated on the coordinate plane of motion and is bounded by cylindrical surfaces. The fields on the plane of motion are a class of plane fields similar to plane stress problems.

$$\psi(x_1, x_2, x_3 = C_1) = C_2 \qquad (2.34)$$

Above defines the stream function at a point on a plane parallel to the plane of motions and yields the same value at all points in space along each line normal to the plane. It is regarded as a mapping of a three-dimensional surface (stream surface) onto two-dimensional plane curves (streamlines) or vice versa. Plane curves have the property of zero torsion and that any line lying in it is perpendicular to the normal of the plane. As C_1 varies, the stream surface in Figure 2.5 resembles the union of parallel copies of plane integral curves on the coordinate plane moving infinitely parallel to z-axis. Hence, integral curves could be described with one single implicit function having two independent spatial variables and one dummy variable. The dummy variable defines the plane of motion as the plane of reference. A plane field structure is represented by cylindrical surfaces in space and is the same in any plane parallel to the plane of motion. Curves on the plane of motion are to be interpreted as longitudinal cross sections bounded by infinite cylindrical surfaces. The three-dimensional effects at the ends of an infinitely long cylinder can be neglected similar to plane strain problems. In the case of two-dimensional fields, field properties are considered a function of two independent spatial variables. In both two-dimensional fields and plane fields, there are no field components in the direction perpendicular to the planes. The two fields are then equivalent mathematically and represented geometrically by the plane of motion when the dummy variable of the plane fields is specified or remains an arbitrary constant.

Restriction to plane fields is a practical necessity partly because of the limitations of complex function theory in two-dimensional domains. With this constraint, field line integral curves are suitable to describe vector fields when they possess a certain symmetry. When symmetry exists, it is indeed possible to select, for example λ-function to suit the symmetry and to reduce the dual functions to one. By setting $\nabla\lambda = \hat{k}$ to assume the coordinate plane of symmetry, field line surfaces are then mutually orthogonal at all points of intersection. The intersections give rise to a system of field line integral curves lying on the coordinate plane of symmetry moving infinitely along z-axis. Equation 2.32 reduces to:

$$V = \nabla\psi \times \hat{k} = \nabla \times (\psi\hat{k})$$

$$= \left[\frac{\partial\psi}{\partial x_2} \quad -\frac{\partial\psi}{\partial x_1} \right] \qquad (2.35a)$$

where

$$\psi(x_1, x_2) = C_2 \tag{2.35b}$$

By differentiating the above integral curves, it yields,

$$\frac{dx_2}{dx_1} = -\frac{\partial\psi}{\partial x_1}\bigg/\frac{\partial\psi}{\partial x_2} = \frac{v_y}{v_x}$$

Hence, the geometric interpretation of an implicit function is tangent vectors along the integral curves. This is because an implicit function is a general solution to the two-dimensional trajectory equation. Since implicit function is an exact differential, it is also called complete integral of differential equation. Each constant value corresponds to an integral curve. Field line integral curves then specify vector fields uniquely except at singular/critical points. Unlike field line trajectories, fluxes bounded between field line integral curves are invariant. The volumetric rate of flow per unit depth $\Delta\lambda = \Delta z$ per unit time reduces to:

$$\frac{Q}{\Delta z} = \Delta\psi \tag{2.36}$$

Field line integral curves and field line trajectories are equivalent in solenoidal fields.

2.3.3 Field Lines, Material Lines and Path Lines

To connect field lines with physical reality, the concept of material lines is perhaps one step closer to materialize field lines in physical forms and to make them come "alive" in the same way as Faraday's line of forces depicted by the arrangement of electric dipoles. A point that coincides with the same particle at all times is called a material point or particle. Each particle in motion is in the same position relative to the rest of the body. The property possessed by a particle is independent of motion of the observer. The concept of a material point applies similarly to lines/curves, surfaces or volume of regions. Under the notion of a macroscopic continuum, a material line consists of a collection of material particles moving in space as a continuous function of time. Material lines can form open or closed loops.

The material description immediately links to the mathematical idea of a point in motion in Lagrangian description. It is described completely by the particle path function, that is,

$$\vec{r} = \vec{r}(x_i, t) = \vec{r}[x_i(\xi_i, t), t] = \vec{r}(\xi_i, t)$$
$$\text{At } t = t_0, \ \xi_i = \xi_0$$

$\vec{r}(\xi_i, t)$ is interpreted in Lagrangian description as the parametric equation of a curve in space as parameter t varies. Its geometric representation is a path line presented in Figure 2.6. The material curve goes through an initial point ξ_0 where the material point is located at an arbitrary time t_0. The Lagrangian coordinate ξ_i is referred to as the material coordinate. It is necessary to specify such an initial point when defining a vector field in Lagrangian description. This enables us to identify a specific material point at time t_0 and to follow its course of motion. The material coordinate moving with the material point will do just as well and as such, it is also called convected coordinate. Without loss of generality, consider an arbitrary velocity field V derived from trajectory.

$$V = \frac{d\vec{r}}{dt} = \lim_{\Delta t \to 0} \frac{\Delta \vec{r}}{\Delta t}$$

$$\frac{dx_2}{v_y} = \frac{dx_1}{v_x} = dt \qquad (2.37)$$

For a given initial position of a particle $\xi_i = \xi_0$ at a time instant $t = t_0$ and a velocity vector field, the solution of path line equation yields path line trajectory as a function of time. The path trajectory traced in space coincides with the same moving material particle as time varies. For steady fields, path line equation becomes identical to field line trajectory equation. Thus, path line is a material trajectory of a streamline in steady state fields. The basic feature common to both instantaneous path lines and steady field lines is the fact that vector fields are tangent to both sets of lines everywhere. A connection between field line trajectories, integral curves and material lines or steady path lines is established in spite of the differences in origins and concepts. These unique families of lines are continuous and non-intersecting in space except at critical/singular points. Path lines, material lines and field

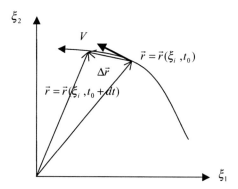

Figure 2.6   A two-dimensional path line.

line trajectories are regarded as local icons since there is generally no relation between trajectories of their own.

2.3.4 Streamlines (Flux Lines)

Because the above systems of lines are based on kinematics and geometric considerations, they lack information about the physics of vector fields. It is necessary to circumvent these drawbacks with something, which has its root on physical grounds. Recall equation of continuity for incompressible fluids flowing in solenoidal fields where divergence of velocity fields vanishes, that is,

$$\nabla \cdot V = 0$$

The above is satisfied by the integral curve, which represents the definition of Lagrangian stream function (or flux function) with reference to Cartesian coordinate. Along a streamline,

$$\frac{\partial \psi}{\partial s} = \frac{\partial \psi}{\partial x}\frac{\partial x}{\partial s} + \frac{\partial \psi}{\partial y}\frac{\partial y}{\partial s}$$
$$= V(-\sin\alpha\cos\alpha + \cos\alpha\sin\alpha) = 0 \qquad (2.38)$$
$$\rightarrow \psi\,|_{\text{streamline}} = C_2(n)$$

Each parameter constant defines a unique streamline. It is observed that streamlines are identical to field line integral curves and that both exist in solenoidal fields. Physical vector fields and geometric vector fields are connected in spite of the differences in their origins and concepts.

In polar coordinate (r, θ) with $h_1 = 1$ and $h_2 = r$, stream function is defined as:

$$V = \left[\frac{\partial \psi}{r\,\partial \theta} \quad -\frac{\partial \psi}{\partial r} \right] \qquad (2.39)$$

In axis-symmetric fields, one of the dual functions is taken to serve as plane of symmetry. In cylindrical coordinate system (r, θ, z), this function takes on the rotational axis of symmetry, that is, z-axis with $(\nabla\lambda = \hat{\theta})$. The rotational axis of symmetry must lie on $\theta = C$ plane and coincide with a streamline. The vector fields are independent of azimuthal angle θ and could be described by Stokes stream function. Each constant value of Stokes function is a stream surface enclosing a streamtube inside which the volume flux is invariant. Tangents on streamtube surfaces define the directions of velocity fields. Stokes function has a different unit from those in plane fields and is defined as:

$$V = \left[\frac{\partial \Psi}{r \partial z} \quad -\frac{\partial \Psi}{r \partial r} \right] \tag{2.40}$$

In a spherical coordinate system (r, θ, φ), the rotational axis of symmetry (z-axis) lies on $\theta = C$ plane and coincides with a streamline. The vector fields are independent of azimuthal angle θ and could be derived from Stokes stream function to result in:

$$V = \left[\frac{\partial \Psi}{r^2 \sin \varphi \partial \varphi} \quad -\frac{\partial \Psi}{r \sin \varphi \partial r} \right]$$

Thus, stream function and streamlines represent an entirely different concept with physical significance. Their existence is based on physical laws. Each streamline may be imagined as if emanating from a source and ending on a sink on the boundary or external to it. Flux function exists only in solenoidal fields. Nevertheless, it is not subject to the requirements of an irrotational field. Along a streamline, the material rate of change of stream function vanishes in steady state, since velocity normal to a streamline/surface is zero.

$$\frac{D\psi}{dt} = \frac{\partial \psi}{\partial t} + (V \cdot \nabla)\psi = \frac{DC_2}{dt}$$
$$= 0 + (V \cdot \hat{n})|\nabla \psi|$$
$$\rightarrow V \cdot \hat{n} = 0$$

Or

$$\frac{D\psi}{dt} = \frac{\partial \psi}{\partial t} + (V \cdot \hat{s})|\nabla \psi| = 0$$
$$\rightarrow \psi |_s = C$$

$\hat{n}$ and $\hat{s}$ are unit vectors in streamline normal and tangent directions. The physical interpretation of $(V \cdot \nabla)$ is the rate of change of field property moving with flows. The zero value indicates no change of stream function in the direction of motion. The function is conserved for the motion of fluids. It is a constant by definition along a streamline, which always consists of the same fluid particles.

In the study of local behavior of trajectory in the neighborhood of a point, it leads to the geometric ideas of tangent, normal and radius of curvature.

They are useful visual attributes derivable from trajectory. Tangent plane is a linear approximation to a surface at a point (x_0, y_0) given by:

$$\frac{\partial \psi}{\partial x}\Big|_{\substack{x=x_0 \\ y=y_0}} (x - x_0) + \frac{\partial \psi}{\partial y}\Big|_{\substack{x=x_0 \\ y=y_0}} (y - y_0) = 0$$

The positive direction of tangent points in the direction of vector fields. The equation of normal at a point (x_0, y_0) along a streamline is:

$$\frac{\partial \psi}{\partial y}\Big|_{\substack{x=x_0 \\ y=y_0}} (x - x_0) - \frac{\partial \psi}{\partial x}\Big|_{\substack{x=x_0 \\ y=y_0}} (y - y_0) = 0$$

The positive direction of normal is defined by the positive direction of tangent by rotating it counterclockwise at the point through a right angle. The radius of streamline curvature is given by the rate of change of angle of inclination α with respect to the change in arc length s:

In implicit form: $\psi = \psi(x, y)$

$$R_s = \frac{ds}{d\alpha} = \frac{\left(\left(\frac{\partial \psi}{\partial x}\right)^2 + \left(\frac{\partial \psi}{\partial y}\right)^2\right)^{3/2}}{\begin{vmatrix} \partial^2\psi/\partial x^2 & \partial^2\psi/\partial x \partial y & \partial\psi/\partial x \\ \partial^2\psi/\partial y\partial x & \partial^2\psi/\partial y^2 & \partial\psi/\partial y \\ \partial\psi/\partial x & \partial\psi/\partial y & 0 \end{vmatrix}_{\substack{x=x_0 \\ y=y_0}}} \tag{2.41}$$

In explicit form: $y = f(x)$

$$R_s = \frac{\left(1 + \left(\frac{dy}{dx}\right)^2\right)^{3/2}}{\left|\frac{d^2y}{dx^2}\right|}\Bigg|_{\substack{x=x_0 \\ y=y_0}}$$

In parametric form: $x = x(t)$; $y = y(t)$

$$R_s = \frac{[(dx/dt)^2 + (dy/dt)^2]^{3/2}}{(dx/dt)(d^2y/dt^2) - (dy/dt)(d^2x/dt^2)}$$

In polar coordinate: $r = r(\theta)$

$$R_s = \frac{\left[r^2 + \left(\frac{dr}{d\theta}\right)^2\right]^{3/2}}{r^2 + 2(dr/d\theta)^2 - rd^2r/d\theta^2}$$

Or from motion information,

$$R_s = \frac{v_s^2}{a_n}; \; R_s = \frac{|V|^3}{|V \times a|}$$

The coordinates of the center of curvature (x_c, y_c) are obtained accordingly.

$$x_c = x_0 + \cfrac{\dfrac{\partial \psi}{\partial x}\left(\left(\dfrac{\partial \psi}{\partial x}\right)^2 + \left(\dfrac{\partial \psi}{\partial y}\right)^2\right)}{\begin{vmatrix} \partial^2 \psi/\partial x^2 & \partial^2 \psi/\partial x \partial y & \partial \psi/\partial x \\ \partial^2 \psi/\partial y \partial x & \partial^2 \psi/\partial y^2 & \partial \psi/\partial y \\ \partial \psi/\partial x & \partial \psi/\partial y & 0 \end{vmatrix}_{\substack{x=x_0 \\ y=y_0}}} \tag{2.42a}$$

$$y_c = y_0 + \cfrac{\dfrac{\partial \psi}{\partial y}\left(\left(\dfrac{\partial \psi}{\partial x}\right)^2 + \left(\dfrac{\partial \psi}{\partial y}\right)^2\right)}{\begin{vmatrix} \partial^2 \psi/\partial x^2 & \partial^2 \psi/\partial x \partial y & \partial \psi/\partial x \\ \partial^2 \psi/\partial y \partial x & \partial^2 \psi/\partial y^2 & \partial \psi/\partial y \\ \partial \psi/\partial x & \partial \psi/\partial y & 0 \end{vmatrix}_{\substack{x=x_0 \\ y=y_0}}} \tag{2.42b}$$

To zoom into the local behavior of three-dimensional graphs $z = f(x, y)$, we could visualize the orientation of tangent plane and its normal vector $\hat{n}$ at a point of interest.

$$z = f(x_0, y_0) + \frac{\partial f}{\partial x}\Big|_{x_0, y_0}(x - x_0) + \frac{\partial f}{\partial y}\Big|_{x_0, y_0}(y - y_0) \tag{2.43a}$$

$$\hat{n} = \nabla f(x, y, z) = \begin{bmatrix} \dfrac{\partial f}{\partial x} & \dfrac{\partial f}{\partial y} & -1 \end{bmatrix}_{x_0, y_0} \tag{2.43b}$$

where $f(x, y, z) = f(x, y) - z = 0$.

Torsion of a trajectory at a point is an index describing the deviation of curve from a plane in the neighborhood of the point. It is zero for plane trajectories. In addition to critical/singular points, there are special points of interest, which also help to visualize trajectories. Inflection points are points where curvature changes its sign and both sides of tangent intersect streamline. Vertices are the points of trajectory where curvature has a maximum/minimum. A cuspidal point indicates the orientation of trajectories changes according to the position of tangent, which passes through the cuspidal point and shared by a trajectory on each side of the tangent. A point of contact is where two trajectories touch without crossing each other. There is only one tangent passing through each of these points. Such points may or may not satisfy critical point requirements depending on whether the vector field vanishes or not at the point.

To evaluate field transport processes, flux integrals were introduced. Flux integral may be interpreted when flux lines extended into the volume from outside are cut off at the surface, then the field inside volume can only be maintained at its original state by replacing an equivalent source inside the volume. Figure 2.7 shows the field fluxes per unit depth entering from the left side and bottom and contributing to the total flux bounded between two flux lines. In this connection, stream function could be used to facilitate the evaluation of field property transport passing through an arbitrary surface bounded between two streamlines by obviating flux integrals. In flow fields, the volumetric rate of flow per unit depth per unit time is given by:

$$\frac{Q}{\Delta z} = -\int_1^2 v_x dx_2 + \int_1^2 v_y dx_1$$

$$= \int_1^2 \nabla\psi \cdot (\hat{k} \times \hat{b})dl$$

$$= -\int_1^2 \nabla\psi \cdot \hat{l}dl$$

$$= \psi_1 - \psi_2 \tag{2.44}$$

$\hat{b}$ and $\hat{l}$ are surface unit normal and unit tangent vector respectively. Δz and dl are depth and differential arc length.

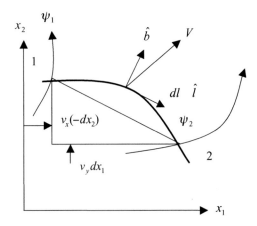

Figure 2.7 Evaluation of flux transport across a boundary bounded by two flux lines.

There is then a relation between flux line trajectories. In flow fields, the difference between two flux line values is equal to the volume flow rate per unit depth bounded by respective flux lines. Flux bounded between two flux lines is invariant with respect to coordinate transformation because no flow is allowed to cross a flux line and they can be considered as solid boundaries. The results are independent of the orientation of the surface unit normal vector. A change in flux line value can only occur across flux lines. Flux line values have no physical significance in themselves and are regarded as defined only within an arbitrary constant. The reference value of the flux function is arbitrary but is conveniently assigned to zero at line of symmetry or along a physical boundary. They serve to identify a particular trajectory in the fields. Flux line trajectories occupy a key position in field visualization and analysis.

The use of flux trajectories is fruitful and rewarding. The shapes and patterns of flux trajectories are unaltered when there is a change in transport rate in linear systems. That is to say if velocity V is a solution, so is cV and $c\psi$ (c is a constant). This means velocity vector will be changed by a factor of c everywhere in the flow fields. The change in flow rate will only introduce a change in velocity magnitudes but directions remain unaltered. Flow rate changes by the same factor. This is a useful observation as streamline patterns obtained from one flow rate can be applied to others. Flux functions can be added algebraically to describe resultant field patterns in accordance with the superposition principle for linear systems. Flux function possesses useful properties when it is represented graphically by a family of flux lines drawn on a sketch. These are best drawn so that the increment in flux function between pairs of adjacent lines is kept constant. This makes the flux invariant between each pair of lines, so that flux magnitude is easily visualized, being inversely proportional to lateral spacing. When flux lines "squeeze," high field flux density is anticipated towards that region. Flux lines do not cross except at critical/singular points. Each flux line has a starting point and a finishing point. A flux line originates from a source, which is either on the boundary or in the body of a region. A flux line terminates at a sink. Thus, flux lines must either form loops or extend to infinity. When flux lines close to form a "core," it generally indicates a poor transfer zone as field property is "trapped" within the core and transfer of field flux is not allowed to cross flux lines. To improve the transfer process, it is necessary to reduce the core size or, if possible, eliminate them. Flux functions also have several useful attributes as follows:

- Flux function is a constant along a flux line.
- Their spatial derivatives give rise to flux components in vector fields.
- They describe in algebraic form the geometry of field patterns.
- Together with equipotentials, flux lines can be used to construct field maps of curvilinear squares to facilitate field visualization and analysis.

3

Field Model, Representation and Visualization

3.1 Field Models and Concepts

Coulomb's law of electric force describes interaction between discrete point charges as a phenomenon of action at a distance. Recall Coulomb's law,

$$E = \lim_{Q_2 \to 0} \frac{F}{Q_2} = \frac{Q_1}{4\pi \varepsilon r^2} \hat{r}$$

ε is a dielectric constant. $\hat{r}$ is electric field unit vector. The limiting value of force per charge determines the direction and strength of line of forces by point charge, referred to as electric field intensity E, that is, electric field vector. It also gives a measure of electric field energy density at a point. To study the interaction between charged objects of finite sizes, it is necessary to evaluate the interaction of each pair of charges separately in order to determine the total contributions as a vector sum in the system. In practice, a model of electric fields is introduced in a manner similar to gravitational fields to circumvent the inherent shortcoming of Coulomb's law. The electric field model considers the creation of an affected space with charges and current in an electric field and it exerts a force on any other charge placed in it. In essence, charges do not simply act on each other through the intervening medium but that this intervening medium plays a critical part in the action.

Visualization of Fields and Applications in Engineering, First Edition. Stephen Tou.
© 2011 John Wiley & Sons, Ltd. Published 2011 by John Wiley & Sons, Ltd.

This intervening medium is a first order tensor field described by a family of field lines perpendicular to the charge object surface in analogy to gravitational fields. The tangent of field lines with respect to charge object defines the out-going or in-coming field direction by its positive or negative sign. It corresponds to the direction of the force it would exert on a positive test charge. In isotropic homogenous dielectrics with no space charge, electric field lines are interpreted as flux lines. This system of line of forces in Faraday's eyes is a visual representation of electric fields. They serve the same purpose as field lines in gravitation. This suggests that in lieu of Coulomb's law, field structure could be derived from flux line distributions in a two-dimensional field map by determining the corresponding electric flux density. In this graphical method, a plane surface is placed perpendicular to the flux line tangent at the local point of interest and the amount of dielectric flux Q crossing the plane surface per unit area defines the average flux density. The flux direction is parallel to the field line trajectory tangent. Since,

$$Q = D \cdot \hat{b} \Delta A \text{ and } \hat{b} \parallel E$$

$$\text{Flux density: } |D| = \lim_{\Delta A \to 0} \frac{Q}{\Delta A} = \lim_{dl \to 0} \left| \frac{\varepsilon(\psi_2 - \psi_1)}{dl} \right| \qquad (3.1)$$

where $\Delta A = dl \Delta z$, $D = \varepsilon E \, \hat{b}$, dl and Δz are the surface unit normal, length and depth of plane surface respectively. ψ represents flux lines bounding the plane surface length.

In order to describe and analyze field structure in multi-dimensions quantitatively, it is necessary to connect the physics of fields with mathematics and functions. In this respect, tensor theory and analysis provide the tools in the development of field models based on the invaluable concepts of potentials and fields. The idea of a potential is introduced to deal with phenomena involving fundamental forces in nature by a scalar potential function of position, which measures field energy levels at each point. This immediately rings the bell of energy being carried by fields in addition to the action of line of forces. Suppose $V = \phi(x_1, x_2, x_3)$ represents electric potential in three-dimensional static fields. The electric force exerted at a point charge Q_2 is derived from gradients of the potential referred to as electric field intensity,

$$F = Q_2 E$$

where

$$E = -\nabla V \qquad (3.2)$$

There are then electric field vectors set up by the distribution of electric potential pointing in the direction of decreasing potential. The electric field vectors correspond to the directions along which maximum spatial rates of change of potential occur. The gradient operation takes each point in the domain of scalar function onto the corresponding gradient vector in a gradient mapping process. Thus, potential offers an alternate way to derive vector fields of multi-dimensions from scalar fields analytically.

In mathematical language, a field is regarded as an infinite set containing all geometric points in a domain. It is defined to be a region of macroscopic continuum in which each point, except singular points, is characterized by some physical quantities being a continuous function of space and time. The amount of information carried in fields escalates rapidly as the order of tensors or domain dimension increases. Thus to describe fields literately would be difficult if not impossible. In a field, whether real or complex, multiplication and addition satisfy the distributive, associative and commutative laws. The nature of fields is classified under the notion of tensor orders depending on the physical theory in which they occur. The fundamental properties of fields are single-valued, bounded and continuous. Continuous distribution of field properties enables us to consider and analyze infinitesimal elements and implies functions are differentiable. To explore field property and structure, it is often desirable to manipulate/transform the nature and characteristics of a given field from one to another. This will provide us with new information and a better understanding of the original fields in different perspectives. In tensor analysis, the vector differential operator provides the basic tools for field operation and construction. It gives information about the behavior of fields in the neighbors of a point by a linear approximation of a differentiable function in a form of contiguous action. Field operations include but are not limited to, gradient, divergence, curl and dyadic. These operations are useful to capture the field sources and strengths and to construct specific fields of interest. The success of field theory and potential theory in ruling the macroscopic world has verified field concepts and models and laid the foundation for tensor field representation/transformation and visualization.

3.2 Scalar Fields and Representation

Scalar fields having one component without directional information are described by scalar functions and they share the property of tensor invariant with respect to coordinate transformation. $f(x, y, z)$ represents a scalar function of position in three independent variables and encodes each point in space with a unique number. A set of numbers constitutes a scalar field of physical significance like temperature, density and so on in a given domain. This is the coordinate representation of a scalar field, which may be visualized as a region of space filled with numbers, each associated with a particular point. Contour surfaces are appropriate surface icons for visualization of scalar fields in three-dimensions in terms of an implicit function

$f(x, y, z) = C$. Implicit representation allows us to depict the graphs using non-intersecting surfaces/curves except at critical/singular points. The parameter C assumes the role of dependent variable representing a scalar field of interest. It denotes the contour surface in a level set of points having the same numeric value. As the parameter varies, a family of contour surfaces is generated in space. However, to present the graphs of such functions requires four dimensions. With reduction in variables, it generally becomes easier to visualize the behavior/pattern of a function in simplified form. This is equivalent to reducing field dimensions by holding certain variables constant systematically. There are generally two standard ways to present graphs of functions in three-dimensions. The cross section method makes use of one of the coordinate planes such as z-plane by holding the corresponding spatial variable constant $(z = c)$. The method degenerates three-dimensional surfaces into plane curves when the coordinate plane cuts through the surface. The intersections of surfaces and coordinate plane define the contour curves on that coordinate plane. This then represents the graphs of function at cross section $z = c$. By varying parametric constant C, we obtain a system of non-intersecting contour curves on the cross section; each corresponds to a specific parametric constant. As an application example, the contour curves represent streamlines when the function and cross section assume the roles of stream function and a plane of motion in plane flow models. It is common practice to plot contours at equal intervals, so that we can obtain information about the spatial rate of change of contour levels from contour density. Numerical values of points in between contours could be interpolated with a good degree of accuracy. The method could be used to visualize surfaces by slicing a series of coordinate planes parallel to each other at various $z = c$. They are assembled to build three-dimensional landscape graph-like multilayers of carpets. The method could be applied by replacing z-plane with other two coordinate planes in sequence, that is, $f(x = c, y, z) = C$ and/or $f(x, y = c, z) = C$ at various cross sections of c. A pseudo three-dimensional field could then be constructed by staggering up a combination of planes along two orthogonal directions. This dimensional stacking method could be extended to display three mutually orthogonal coordinate planes at a particular point of interest.

The other method is to present contour surfaces in explicit form, that is, $z = f(x, y, C)$ to render three-dimensional surfaces in an elevation plot on a three-dimensional rectangular coordinate system (x, y, z). Two horizontal rectangular coordinates are used to represent x and y spatial variables and the vertical coordinate represents the dependent variable z. For a given parametric constant C, the resulting graph is a three-dimensional surface rendered by a two-dimensional geometric object setting in space in terms of functions of two independent variables. Following the idea of a cross section method, the shape of the surface can be pictured by slicing through parallel coordinate planes to produce intersection curves. In this landscape view, densely packed contours imply steepness of the surface with rapid change of spatial rate of contour. The widely spaced contours indicate a flat terrain. A long valley has parallel contour lines indicating rising/lowering elevations on both sides of the valley. It is, however, difficult to determine the dependent variable z accurately when three-dimensional graphs are drawn on a two-dimensional view plane. z-contour lines are drawn in to improve the accuracy. There are hidden spots as surfaces closer to the viewer obscure those behind. Hidden-line renderings may be introduced to enhance the presentation. Simple geometric transformation such as translation and/or rotation could be applied to make hidden spots visible by viewing it from different angles. To zoom onto the local behavior of the graphs for better accuracy, we call upon the tangent plane and its normal vector at a point of interest. We could then determine the true slope of tangent plane from the plane edge view. Alternatively, the structure of a surface may be discretized into a set of numeric data points arranged in a matrix table with x and y variable assuming the rows and columns of the table. By connecting points corresponding to rows and columns of the x-y matrix data table, a system of smooth meshes with connection components (wire-frame) is produced to project a three-dimensional effect of a surface in space. This method of numerical presentation could be further enhanced to incorporate an area-coloring corresponding to field strength and surface shading to encode the ranges of z-contour values. The resulting graphs are also called density plots.

A great number of three-dimensional fields in practice could be reduced to two dimensions in plane-parallel when the effects of one of the dimensions are negligible. The above techniques could be readily applied to such two-dimensional plane-parallel approximations on just one cross section with functions in two independent variables. Plane contour curves are appropriate line icons for visualization of two-dimensional fields. The implicit representation of plane curves is described by $f(x, y) = C$. To visualize its graphs in three-dimensions, we let the parametric constant assume the role of dependent variable z in explicit form $z = f(x, y)$. The graph of explicit function is then a set of points giving rise to a surface in space. Such graphs in an elevation plot (or rubber sheet) generally give a good idea about the function characteristics. In this representation, certain features may exhibit distinctive characteristics such as mountain peaks, saddle points or long valleys. It is often useful to construct a profile from a graph to show what a cross section of fields looks like between two points. By holding one of the independent variables constant, that is, $z = f(x = c, y)$ or $z = f(x, y = c)$, the cross sections describe the behavior of fields with x or y fixed corresponding to the rows or columns in the x-y data table. Elevation plots, plane contour curves and profiles may be displayed side by side or in multiple views to complement each other. In this respect, we can visualize the surface structure by transformation/correlation between displays as we follow each plane contour curve

C raised or lowered corresponding to the coordinate plane $z = C$ in elevation plots augmented by graphs of profiles.

3.3 Vector Fields and Representation

Vectors are the simplest class of tensors with one column of components; each represents information along mutually orthogonal directions. Steady vector fields are concerned with a region in space that varies continuously depending on spatial variables. A vector field in coordinate representation is defined by a vector point function in terms of its components with reference to a chosen coordinate system. Each component in turn is specified by a scalar point function of coordinate variables. Vector fields can be visualized geometrically as mappings of vector space from one coordinate to another. Alternatively, vector fields could be represented in parameter form by defining each vector component as a function of a single parameter. Parametric representation may have different forms such as time (t-axis) or arc length (s-axis) serving as a parameter. It constitutes a continuous mapping of points in space from an axis.

$$V = \begin{bmatrix} v_x(x, y, z) & v_y(x, y, z) & v_z(x, y, z) \end{bmatrix}$$

or

$$V = \begin{bmatrix} v_x(t) & v_y(t) & v_z(t) \end{bmatrix}$$
$$x = x(t); \quad y = y(t); \quad z = z(t)$$

Vector representations are closely related to the geometrical idea of a directed line segment drawn locally at a point. Geometric vectors (or spatial vectors) of these kinds based on kinematics merely describe changes in fields from point to point in some coordinate systems. They generally bear no physical significance. The magnitude of vector fields is a scalar quantity usually presented separately as field intensity contour surfaces. For two-dimensional fields, vector components define the orientation (or slope) of vector fields at a point by a discrete line segment marked by an arrowhead. The length or thickness of line segment is drawn proportional to its magnitude. This is the most straightforward vector field mapping by point icons making up a family of arrowhead line segments. A sensible number of line segments are to be utilized in this vector plot in order for it to convey useful information. Other point icons such as glyphs are also developed. Arrow icons are particularly useful to depict the state of motions by displaying both velocity and acceleration vector at a point. The arrowheads indicate directions of velocity and acceleration vector. The law of parallelogram could be applied to decompose the acceleration vector into streamline tangential and normal components. The tangential component determines whether flow is accelerating or decelerating when it is in the same or opposite direction of velocity, which always points in streamline tangent direction along which particle advances. If this component vanishes or the acceleration and velocity are mutually orthogonal, then motion is at constant speed. The normal component represents centripetal acceleration and always points towards the center of curvature indicating the direction along which motion veers. If this component vanishes, it may be a point of inflection or a straight path.

The other method is the graphical representation of vectors by curves (line icons), also known as vector lines or field line trajectories. For two-dimensional fields, field line trajectories degenerate into plane curves. A system of such plane curves provides a direct observation on the external characteristics and distributions of vector fields with continuity. Arrows drawn along field line trajectories indicate a global view of vector field directions. A local tangent vector along a trajectory defines the vector field direction at a point. For solenoidal fields, they are particularly suitable to visualize plane field structure when field lines are displayed at equal intervals, such that a fixed amount of flux is bounded throughout between two adjacent flux lines. Field intensity (or flux density) increases when spacing between two adjacent flux lines reduces and vice versa.

A simple extension of flux lines to flux tubes is fruitful as it enhances graphical representation of three-dimensional vector fields. This is a surface icon of arbitrary shape formed by a closed path to contain the same system of flux lines throughout. This tube of field line trajectories bounding a finite surface element is to be inserted at some points in field domains at right angles to field lines. The number of field lines (or flux) passing through the surface element is proportional to the product of field intensity and the surface area and remains constant since no field lines are allowed to enter or leave through the sides of tube. This amount of flux is a measure of strength of a stream tube. The size of surface element varies inversely proportional to field intensity. By displaying suitable tube bundles at selected regions of space, it aids to visualize three-dimensional smooth vector fields through one-dimensional tubes of varying cross sections.

3.4 Vector Icons and Classifications

An icon is defined as a geometric object/image that conveys information efficiently by encoding data through geometric characteristics and visible attributes. Icons or skeletons played the role of graphical elements of patterns just like regular polygons/polyhedra played the role of structural elements of shapes. Because icons express a relation between data information and its shape and appearance, they are effective means of communicating with human visual systems in a form of visual language. Icons are generally classified based on domain configurations, information levels or topological skeletons near critical points.

3.4.1 Classification Based on Domain Configurations

3.4.1.1 Vector Point Icons

Vector point icons such as arrows are local icons suitable for graphical representation of vector fields at selected locations. The disadvantage of point icons is that they generally create problems of visual cluster or loss of data continuity. To avoid such problems, usually a subset of data is selected for displays. Under some circumstances, point icons may become difficult to interpret when vector line segments cross each other because long segments are required to match with increasing magnitudes. Problems of visual cluster are also encountered when the range of variations of vector magnitude increases to compete with adequate graphical resolutions. There is generally no relation between point icons.

3.4.1.2 Vector Line Icons

Improved perception of continuous tensor fields is obtained using vector line icons instead of point icons. A tangent at a point along a line icon gives the field direction at that point. Vector line icons do not cross each other except at critical/singular points. This family of line icons covering the field domain continuously includes for instance, flux line trajectories and field line trajectories.

3.4.1.3 Surface Icons

Surface icons consist of a large number of surface particles assembled to make up a material line segment or a tube giving rise to the appearance of a continuous surface, that is, stream-surfaces, stream tubes or vortex tubes. When narrow surfaces are defined by two adjacent streamlines, they are called stream-ribbons.

3.4.1.4 Volume Icons

Various volume rendering techniques are used to visualize volumetric data sets with displays of information or images in three-dimensional icon texture. Examples, vortex cores.

3.4.2 Classification Based on Information Levels

3.4.2.1 Elementary Icons

Elementary icons are generally used to represent data at selected points strictly across the extent of spatial domains.

3.4.2.2 Local Icons

Local icons represent local distributions of a data set including gradients information. They give field directions at a point but lack magnitude information and continuity. There is generally no relation between icons themselves.

3.4.2.3 Global Icons

Global icons like streamlines describe field structure and patterns of a given data set quantitatively with direction and magnitude information continuously. They are best represented at equal contour intervals. Furthermore, they may bear physical significance and there is a relation between icons.

3.4.3 Classification Based on Topological Skeleton

Field patterns generated near critical points have unique features that reflect field behavior such as saddle points. They provide information about a local region rather than just a point. The mathematical method coupled with visualization laid the framework for critical point classification and analysis based on field topological skeletons.

3.5 Scalar Potential

The basic idea of potentials in all field applications is to generate a specific tensor using a scalar function to deal with systems involving a great number of unknowns or components. A scalar function such as stress function in elasticity is capable of describing tensor components in multi-dimensions and can be dealt with efficiently. They are versatile tools for deriving fields of great complexity. The concept of field potential may be interpreted in a manner to facilitate and broaden our perception and understanding of the phenomena in terms of some physical quantities such as field energy and work done. Because the convenience of using scalar functions is so great we use them whenever possible. This includes the use of tensor invariants, tensor contraction by dot/inner products and potential functions for tensor descriptions/visualizations. In potential theory, vector identity warrants the existence of a scalar potential in conservative fields. The theory is connected closely to the study of harmonic functions. It has provided a platform for study in a wide range of disciplines whenever the phenomena admit the definition of scalar potential.

The geometric interpretation of scalar potentials is a family of equipotential surfaces in space. Each surface has the same potential numerically and is described by a potential function of position implicitly:

$$\phi(x, y, z) = C_1 \tag{3.3}$$

C_1 is a potential level parameter also referred to as equipotential. As the parameter varies, a family of non-intersecting equipotential surfaces is displayed in space. The physical interpretation of scalar potentials is a measure of energy levels, which characterize field strength and transport capability with respect to a reference datum. Velocity potential and electric potential are examples of such function. The directional derivative of equipotentials is useful to reveal the spatial rate of potential change at a point especially when the direction is parallel to unit normal of equipotential surface $\hat{b}$. It determines the maximum rate of change of equipotentials equal to potential gradient. Thus, the line of force is everywhere perpendicular to equipotentials. Gradient mapping of a potential will do exactly that. In two-dimensional fields, equipotential surfaces degenerate into plane level curves. $\hat{b}$ is then the unit normal to the curve lying on the same plane. In a manner similar to flux trajectories, equipotential contours are best drawn so that the increment in potential function between pairs of adjacent contours is kept constant. This makes the change of potential invariant between each pair of contours, so that equipotential density is easily visualized, being inversely proportional to lateral spacing. Field strength is manifested and is reflected by the crowding of equipotentials.

To evaluate the change in potential between any two points, a line integral of vector fields is to be performed along an arbitrary path.

$$\Delta\phi = \int_1^2 V \cdot d\vec{r} = \int_1^2 \frac{\partial\phi}{\partial x}dx_1 + \frac{\partial\phi}{\partial y}dx_2$$

$$= \int_1^2 d\phi = \phi_2 - \phi_1 \tag{3.4}$$

$d\vec{r}$ is an elemental displacement tangent to the path. In potential theory, line integrals are generally interpreted as work giving rise to a change in potential between two points. Line integral represents a complete differential independent of path and is equal to the difference in equipotential values between two end points. Thus potential does not depend on its absolute value at a point but only that part of the potential, which gives rise to a potential difference between different points. For a closed path, the line integral of velocity fields in hydrodynamics yields circulation, which can be evaluated also from the knowledge of vorticity and Stokes theorem.

$$\Gamma = \oint_C V \cdot d\vec{r} = \int_A \omega \cdot \hat{b}dA \tag{3.5}$$

This means the fields must have no circulation (zero vector sources) for the existence of a scalar potential. The definition of any scalar potential can be changed by adding to it a constant without altering the fields at all.

When potential function exists, it indicates the process is optimal. Under this condition, it becomes possible for a field particle to move along an arbitrary path and back to the initial point without a change in potential. Transport processes are affected by domain configurations and boundary conditions. When these conditions are prescribed, equipotentials are useful to assess the effects of anisotropy of a medium and give an insight view of resistance distributions. Therefore, equipotentials serve as a reference for comparisons with actual distributions of field potentials. Anisotropic effects become more pronounced when two distribution patterns deviate more severely. Information of this kind provides physical interpretation of potential theory.

Another utility of potentials is to relate a scalar field to its scalar sources. By extracting scalar sources from fields, we are able to explore field internal structure through source distributions. The divergence of field gradients leads to the relation between potentials and sources in a contraction process, referred to as Poisson's equation.

$$\nabla \cdot (\nabla \phi) = \nabla^2 \phi = \nabla \cdot V = \rho \tag{3.6}$$

ρ represents scalar source density. The above reduces to Laplace equation in solenoidal fields. The solution to Poisson's equation is obtained by summing up sources through Green's theorem, that is,

$$\phi = \int_v \frac{\rho}{4\pi r} dv + \oint_A \frac{S \cdot \hat{b} dA}{4\pi r} \tag{3.7}$$

S is the source density per unit area enclosed by the surface. The first term represents contribution from a volume source. The connection between scalar field sources and its associated potentials means any one could be determined from the other. Alternatively, scalar potential can be found directly in terms of vector field components based on potential definition, that is,

$$\phi = \int \left(\frac{\partial \psi}{\partial x_2} \partial x_1 - \frac{\partial \psi}{\partial x_1} \partial x_2 \right) + C$$

Or

$$\phi = \int v_x \partial x_1 + f(x_2); \quad \phi = \int v_y \partial x_2 + g(x_1) \tag{3.8}$$

v_x and v_y represent components of the given vector fields with reference to Cartesian coordinates.

There are various types of scalar sources. Sources may exist within a region (volume source) or along a boundary (surface source) or external to it. Figure 3.1 shows field lines in an irrotational field, whether these lines originate due to a source (S) or are lost due to a sink (s) in the region or passing through

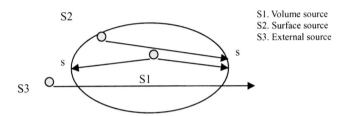

Figure 3.1   Fields generated by scalar sources.

region from one side to the other without the presence of any source. In the latter case, field is generated by sources external to it, a phenomenon referred to as diverging.

3.6 Vector Potential

Vector potential concept was introduced in the study of electromagnetics to circumvent the shortcoming of a magnetic scalar potential, which fails to work in the presence of currents. This is the problem immediately recognized when fields are rotational. The existence of vector sources motivates the search for vector potentials, which may serve to play similar roles as that of scalar potentials. The concept of scalar potential and its relation with sources could be applied to the development of vector potentials and associated vector sources. This analogy allows us to visualize vector fields using lines of constant vector potential but more importantly flux lines of the field generated from it. In this approach, three-dimensional vector fields are derived from a vector potential A:

$$V = \nabla \times A$$
$$= \left[\frac{\partial A_z}{\partial x_2} - \frac{\partial A_y}{\partial x_3} \quad \frac{\partial A_x}{\partial x_3} - \frac{\partial A_z}{\partial x_1} \quad \frac{\partial A_y}{\partial x_1} - \frac{\partial A_x}{\partial x_2} \right]$$

Hence,

$$\nabla \cdot V = \nabla \cdot (\nabla \times A) = 0$$

The vector identity warrants the existence of an arbitrary vector potential in solenoidal fields. In order that field is uniquely specified, the divergence of vector potential is required to vanish. This is generally referred to as Lorentz condition, that is,

$$\nabla \cdot A = 0$$

At this point, the introduction of vector potential seems to have no advantage since vector potential function is a vector itself replacing another vector with the same number of components. However, for two-dimensional problems, vector potential reduces to only one active component when one of the surfaces is a plane of symmetry say, (x_1, x_2) plane. Then,

$$v_z = \frac{\partial A_y}{\partial x_1} - \frac{\partial A_x}{\partial x_2} = 0 \tag{3.9}$$

$$\frac{\partial A_x}{\partial x_3} = \frac{\partial A_y}{\partial x_3} = 0$$

$$\nabla \cdot V = \frac{\partial^2 A_z}{\partial x_1 \partial x_2} - \frac{\partial^2 A_z}{\partial x_1 \partial x_2} = 0$$

In comparison with Equation 2.35a, we have

$$V = \left[\frac{\partial A_z}{\partial x_2} \quad -\frac{\partial A_z}{\partial x_1} \right] \tag{3.10}$$

where

$$A_z = \psi + C$$

Or

$$A = \left[A_x \ A_y \ \psi \right] \tag{3.11}$$

Thus, two-dimensional solenoidal vector fields could be derived from z-component vector potential, which plays the role of stream function.

The physical interpretation of a vector source is vorticity in hydrodynamics. To extract vector source from vector fields, we take the curl of it to result in a vector Poisson's equation. The utility of vector potentials is that it provides a convenient way to relate a vector field to its vector sources with physical interpretations.

$$\nabla \times V = \nabla \times (\nabla \times A) = \nabla(\nabla \cdot A) - \nabla^2 A = \omega$$

Or

$$\omega = -\nabla^2 A \tag{3.12}$$

For two-dimensional fields,

$$\omega_x = -\nabla^2 A_x = 0 \tag{3.13a}$$
$$\omega_y = -\nabla^2 A_y = 0 \tag{3.13b}$$
$$\omega_z = -\nabla^2 A_z = -\nabla^2 \psi \tag{3.13c}$$

A solution to Poisson's equation is obtained by summing up vector sources through Green's theorem,

$$A = \int_v \frac{\omega}{4\pi r}dv + \oint_A \frac{S \times \hat{b}dA}{4\pi r} \tag{3.14}$$

S is the surface source density. The first term represents contribution from a volume source. The connection between vector sources and its associated potentials means any one could be determined from the other. Alternatively, vector potential can be found directly in terms of vector field components based on flux function definition, that is,

$$\psi = \int \left(-\frac{\partial \phi}{\partial x_2} \partial x_1 + \frac{\partial \phi}{\partial x_1} \partial x_2 \right) + C$$

Or

$$\psi = \int v_x \partial x_2 + f(x_1); \quad \psi = -\int v_y \partial x_1 + g(x_2) \tag{3.15}$$

Vector sources may be present inside or outside a region. Figure 3.2 illustrates the field line characteristics of such vector fields. Flux lines generated from an internal vector source are continuous by forming loops and closing upon themselves giving rise to a non-vanishing circulation or they extend indefinitely. When sources are external to a closed region, all flux lines entering the region must leave it. Flux bounded between two flux lines is invariant. When fields are not solenoidal, vector potential does not exist and field line trajectories remain one of the basic techniques to visualize vector fields.

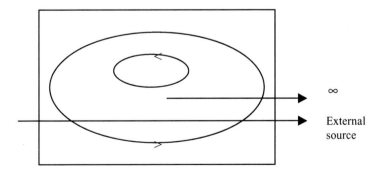

Figure 3.2 Fields generated by vector sources.

3.7 Vector Field Specification

3.7.1 Helmholtz's Theorem

Following the ideas of representing vector fields by potentials, one immediately faces with some important questions. What kind of potential is required to specify a vector field uniquely and completely? How does a field structure? Helmholtz's theorem has the answers. This fundamental theorem of vector fields could be interpreted through various perspectives. It is regarded as a representation of vector fields in terms of potentials on Euclidean geometry. The theorem states that an arbitrary vector field, which vanishes at infinity can be represented as the sum of an irrotational part q_1 and a solenoidal part q_2, each is generated by a scalar potential ϕ and a vector potential A respectively, that is,

$$q = q_1 + q_2 \qquad (3.16)$$

$$\text{Irrotational part: } q_1 = \nabla(\phi + c) = \nabla\phi \qquad (3.17)$$

$$\text{Solenoidal part: } q_2 = \nabla \times (A + \nabla C) = \nabla \times A \qquad (3.18)$$

c and C are arbitrary scalar constant and continuous function. The above expresses a mathematical fact that an irrotational field could be derived by taking the gradient of an arbitrary scalar function. A solenoidal field may be obtained by taking the curl of an arbitrary vector function. Thus, the basic characteristic of an irrotational field is that it accounts for all scalar sources in the fields and that a scalar potential exists. The basic characteristic of a solenoidal field is that it accounts for all vector sources in the fields and that a vector potential exists. Figure 3.3 depicts schematically a general vector field structure built on the divergence of a scalar potential and curl of a vector potential in accordance with Helmholtz's theorem.

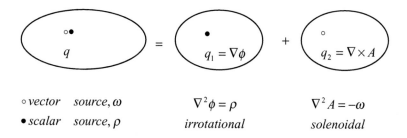

○ vector source, ω

● scalar source, ρ

$\nabla^2\phi = \rho$

irrotational

$\nabla^2 A = -\omega$

solenoidal

Figure 3.3 Vector field specification and structure.

The theorem also indicates, from a different perspective that vector fields compose of two types of sources with a scalar origin and a vector origin. Field sources describe field internal structure, which is characterized by the aspects of rotation (vector source), and dilatation (scalar source). When and only when both types of sources vanish, can vector fields be derived from either the irrotational part or solenoidal part and the fields are called harmonic. In this case, Poisson's equation reduces to the Laplace equation with each potential assuming the role of harmonic function. The field is then driven by sources outside the region; otherwise, the field is nil everywhere and serves no purpose. Under this condition, a unique field exists in the region subject to prescribed boundary conditions. In two-dimensional harmonic fields, there is a relation between two potentials known as Cauchy-Riemann equation. For right hand systems,

$$
q = \pm \begin{bmatrix} \dfrac{\partial \phi}{\partial x_1} \\[2ex] \dfrac{\partial \phi}{\partial x_2} \end{bmatrix} = \pm \begin{bmatrix} \dfrac{\partial \psi}{\partial x_2} \\[2ex] -\dfrac{\partial \psi}{\partial x_1} \end{bmatrix} \tag{3.19}
$$

Or in polar coordinates with $h_1 = 1$ and $h_2 = r$

$$
q = \pm \begin{bmatrix} \dfrac{\partial \phi}{\partial r} \\[2ex] \dfrac{\partial \phi}{r \partial \theta} \end{bmatrix} = \pm \begin{bmatrix} \dfrac{\partial \psi}{r \partial \theta} \\[2ex] -\dfrac{\partial \psi}{\partial r} \end{bmatrix} \tag{3.20}
$$

The $\pm$ sign signifies the field vector pointing in the direction of increasing scalar potential with positive or in decreasing direction with negative. When one of the potentials is known, the other can be determined accordingly. Furthermore,

$$
\nabla \phi \cdot \nabla \psi = \frac{\partial \phi}{\partial x_1}\frac{\partial \psi}{\partial x_1} + \frac{\partial \phi}{\partial x_2}\frac{\partial \psi}{\partial x_2} = \frac{\partial \phi}{\partial x_1}\frac{\partial \psi}{\partial x_1} - \frac{\partial \psi}{\partial x_1}\frac{\partial \phi}{\partial x_1} = 0 \tag{3.21}
$$

$$
|q| = |\nabla \phi| = |\nabla \psi| \tag{3.22}
$$

Thus, equipotentials and flux lines are orthogonal everywhere in harmonic fields. This is an important property shared by both systems of curves and makes them an ideal rectangular coordinate system referred to as potential-stream function coordinate. It has been suggested that these two potentials form a conjugate pair of harmonic functions and leads to the construction of complex potentials. Harmonic functions have unique properties summarized below:

- Any function that has continuous partial derivative of second order satisfying Laplace equation is a harmonic function.
- Harmonic functions having identical values upon a closed contour and without singularities within the contour are identical throughout the region bounded by the contour. This means for example, a temperature field within a region is determined uniquely by temperature on the boundaries. This corresponds to the specification of Direchlet conditions.
- Direchlet and Neumann boundary conditions are reciprocal for equipotentials and flux functions. This property is referred to as dual principle. The values in these two types of boundary conditions remain unchanged by conformal transform.
- The value of a harmonic function at a point is equal to the mean value on any circle with the center at that point. The maximum or minimum values assume on the boundary of a region in accordance with the maximum principle.
- A complex potential defined by a conjugate pair of harmonic functions is analytic. Given one of them, the other can be found from Cauchy-Riemann relation within a constant. The complex potential is also harmonic with respect to complex $F(\phi, \psi)$ plane. Alternatively, complex potential can be viewed as a conformal mapping between complex z-plane and complex F-plane.
- Stream function and equipotential could be interchanged in a complex potential to get solution of another problem. A harmonic function is mapped to another harmonic function in conformal mapping.

3.8 Tensor Contraction and Transport Process Visualization

The idea of using scalars for tensor descriptions/visualizations can be realized by tensor contraction. This is basically a version of tensor rank reduction technique in which two tensor fields are combined to yield a quantity of lower tensor rank through dot product or inner product. In a contraction process, useful information is extracted or new information is produced. For instance, divergence of a vector field yields a scalar source. A velocity vector field can be transformed into a scalar quantity in terms of kinetic energy in a single contraction. A vector and a tensor could be contracted by Cauchy's formula into a vector on a reference plane or at a point. It is possible to have tensor rank reduction by two orders. Strain energy is evaluated by double contraction of stress tensor and strain tensor through inner product. Contraction could also be performed by identifying two indices of a tensor trace so summing up on them gives a scalar quantity. We are concerned about the transformation of vector fields by contraction or by vector operation in order to obtain quantities of interest. We then apply the concepts of flux and stream function to visualize the transport trajectories and processes. This includes mechanical energy function, heatfunction, warping flux function and vorticity function.

3.8.1 Mechanical Energy Function and Heatfunction

Consider isochronic motion of velocity fields in a two-dimensional continuum as a typical example involving transport of momentum, mass and energy. Recall vector identities and governing equations of incompressible fluids in motion in a steady and conservative body force system,

$$\text{Continuity equation: } \nabla \cdot V = 0 \tag{3.23a}$$

$$\text{Thermal energy equation: } V \cdot \nabla T = \alpha \nabla^2 T = \frac{k}{\rho c_p} \nabla^2 T \tag{3.23b}$$

$$\text{Momentum equation: } \frac{\partial V}{\partial t} = -(V \cdot \nabla)V - \frac{\nabla p}{\rho} + \nu \nabla^2 V - \nabla U = 0 \tag{3.23c}$$

Or

$$L + \nabla B = \nu \nabla^2 V$$

$$\text{Angular momentum theorem: } \sum \Gamma = \frac{\partial Y}{\partial t} + \int_A \rho \vec{r} \times V(V \cdot \hat{n}) dA \tag{3.23d}$$

where

$$\text{Angular momentum per unit mass: } \frac{Y}{m} = \vec{r} \times V \tag{3.23e}$$

$$\nabla^2 V = \nabla \cdot \nabla V = \nabla(\nabla \cdot V) - \nabla \times (\nabla \times V) = -\nabla \times \omega$$

$$(V \cdot \nabla)V = \nabla \frac{V^2}{2} + \omega \times V = \sum_i v_i \frac{\partial v_k}{\partial x_i}; \ k = 1, 2, 3$$

$$= V \cdot \nabla V + V(\nabla \cdot V) = \nabla \cdot [VV]$$

$$\frac{1}{\rho} \nabla \cdot [\tau] = -\left(\frac{\lambda}{\rho} \nabla \cdot (\nabla \cdot V) \delta_{ij} + \nu \nabla \cdot [\nabla V] + \nu \nabla \cdot [\nabla V]^T \right)$$

$$= -\nu \nabla^2 V - \nu \nabla(\nabla \cdot V) = -\nu \nabla^2 V$$

$$= \nu \nabla \times \omega$$

$$\frac{1}{\rho} V \cdot (\nabla \cdot [\tau]) = \frac{1}{\rho} \nabla \cdot ([\tau] \cdot V) - \frac{1}{\rho} [\tau] : [\nabla V]$$

$$\omega = \nabla \times V; \; \mu = \rho \nu; \; \lambda = \kappa - \frac{2\mu}{3} \approx -\frac{2\mu}{3}$$

$$L = \omega \times V = 2\dot{\Omega} \times V = (\nabla \times V) \times V \qquad (3.24)$$

Γ and U represent torque and potential in conservative force fields. ω and T represent vorticity and temperature field respectively. μ is fluid dynamic viscosity and ρ is fluid density of homogenous fluids. p is pressure and V is fluid velocity. k and c_p denotes thermal conductivity and specific heat of field medium.

The vector operator in continuity equation transforms the vector fields into a scalar source, which vanishes for an incompressible fluid. It implies the net flux across a control volume is zero and stream function exists. By taking the dot product of velocity and momentum equation, the vector fields are contracted to yield the following mechanical energy equation:

$$V \cdot \nabla \frac{V^2}{2} + V \cdot \frac{\nabla p}{\rho} + \frac{1}{\rho} \nabla \cdot (V \cdot [\tau]) - \Pi + V \cdot \nabla U + V \cdot (\omega \times V) = 0 \quad (3.25)$$

where

$$\Pi = \frac{[\tau] : [\nabla V]}{\rho} \qquad (3.26)$$

$$= \frac{1}{\rho} [\tau] : ([\dot{\gamma}] + [\Omega])$$

$$= -[\frac{1}{\rho}(\lambda + 2\mu)\dot{I}_1^2 - 4\nu \dot{I}_2]$$

$$= -\frac{4}{3} \nu \dot{I}_1^2 + 4\nu \dot{I}_2 < 0$$

The above represents dissipation function. For incompressible fluids, it reduces to:

$$\Pi = -2\nu \dot{\gamma}_{ij} \dot{\gamma}_{ij}$$

or

$$\Pi = 4\nu \dot{I}_2 = -\frac{\nu}{2} \sum_{i=1} \sum_{j=1} \left[\frac{\partial v_i}{\partial x_j} + \frac{\partial v_j}{\partial x_i} \right]^2 \qquad (3.27)$$

The first term in mechanical energy equation represents kinetic energy by convection. The second term describes reversible rate of work done by pressure. The third term is interpreted as reversible rate of work done on fluids by the stress tensor due to deformation. It means energy will be released when a deformed body restores to its initial state; otherwise, energy is required to cause deformation depending on whether it is positive or negative. The fourth term Π represents energy dissipation function, which characterizes energy dissipation rate due to viscous effects. This sink term is negative and causes a drain of mechanical energy irreversibly from fluids into heat. The sum of the third and fourth term constitutes the total work done by the stress tensor. The fifth term represents work done by conservative force. The last term is the rate of work done by Lamb vector defined in Equation 3.24. This term vanishes, meaning that Lamb vector does no work since it is perpendicular to flow fields.

$$V \cdot (\omega \times V) = V \cdot L = \omega \cdot (V \times V) = 0 \qquad (3.28)$$

With $\nu = 0$,

$$V \cdot L = V \cdot \nabla B = 0$$

This means Bernoulli function does not change in the direction of flow. For irrotational flows, Lamb vector vanishes and hence,

$$\nabla B = \nu \nabla^2 V - L = 0$$

Bernoulli function does not change everywhere. The results are statements of Bernoulli's theorem.

For two-dimensional fields in Cartesian coordinate:

$$\frac{\partial q_x}{\partial x} + \frac{\partial q_y}{\partial y} - \Pi = 0 \qquad (3.29)$$

where

$$q_x = \left[B v_x + \frac{1}{\rho}(\sigma_x v_x + \tau_{xy} v_y) \right]$$

$$q_y = \left[B v_y + \frac{1}{\rho}(\sigma_y v_y + \tau_{xy} v_x) \right]$$

$$B \equiv \frac{V^2}{2} + \frac{p}{\rho} + U \qquad (3.30)$$

The above describes mechanical energy balance at a point in terms of net energy flux across a control volume and a sink inside of it. When viscous dissipation is negligible, mechanical energy function E is introduced to visualize energy transfer processes in a manner similar to stream function.

$$x\text{-direction: } \frac{\partial E}{\partial y} = q_x = Bv_x + \frac{\sigma_x v_x + \tau_{xy} v_y}{\rho} \qquad (3.31a)$$

$$y\text{-direction: } -\frac{\partial E}{\partial x} = q_y = Bv_y + \frac{\sigma_y v_y + \tau_{xy} v_x}{\rho} \qquad (3.31b)$$

Mechanical energy function is obtained by integration.

$$E = \int q_x \partial y - \int q_y \partial x + C$$

For inviscid fluids, the energy flux associated with viscous stress tensor vanishes and Bernoulli's constant is invariant in potential flows. Above yields,

$$E = \int B \frac{\partial \psi}{\partial y} \partial y + \int B \frac{\partial \psi}{\partial x} \partial x + C = B\psi + C$$

$$\frac{Q}{\Delta z} = \rho(E_2 - E_1) = \rho B(\psi_2 - \psi_1) \qquad (3.32)$$

Mechanical energy function is found to relate to stream function linearly within a constant and share the same contour patterns. It is given by the product of Bernoulli's constant and stream function. The tangent along mechanical energy contours gives energy transfer direction parallel to velocity vector everywhere. When adjacent streamlines squeeze together, mechanical energy flux lines do likewise. It indicates velocity is increasing and therefore kinetic energy increases as well while the sum of pressure energy and potential energy reduces such that the total mechanical energy remains invariant between two adjacent mechanical energy flux lines. There is transfer of energy from one form to another in a conservative system. The energy bounded between two mechanical energy flux lines per unit depth per unit time is given by the difference of mechanical energy contour values.

Kimura and Bejan (1983, Bejan 1995) introduced heatfunction to visualize heat transfer paths. It is an energy analog of stream function in hydrodynamics. Heatfunction is defined in terms of heat fluxes in two mutually orthogonal directions. From the thermal energy equation, we obtain heatfunction ψ_t by integrating a pair of partial differential equations.

$$\frac{\partial}{\partial x}[\rho c_p v_x(T - T_{ref}) - k\frac{\partial T}{\partial x}] + \frac{\partial}{\partial y}[\rho c_p v_y(T - T_{ref}) - k\frac{\partial T}{\partial y}] = 0 \quad (3.33)$$

$$x\text{-direction: } q_x^* = \left[\rho c_p v_x(T - T_{ref}) - k\frac{\partial T}{\partial x}\right] = \frac{\partial \psi_t}{\partial y}$$

$$\text{y-direction: } q_y^* = \left[\rho c_p v_y (T - T_{ref}) - k\frac{\partial T}{\partial y}\right] = -\frac{\partial \psi_t}{\partial x}$$

$$\psi_t = \int q_x^* \partial y - \int q_y^* \partial x + C \tag{3.34}$$

Distribution of heatlines is also presented at equal contour interval to deliver transport information in quantitative terms. The energy bounded between two heatlines per unit depth per unit time is given by the difference of heatline values.

$$\frac{Q}{\Delta z} = \psi_{t2} - \psi_{t1}$$

3.8.2 Strain Energy Trajectory and Strain Function

Strain energy transfer in elastic systems in the absence of body forces could be visualized by taking dot product of displacement vector fields U and equation of equilibrium in a contraction process, that is,

$$(\nabla \cdot [\tau]) \cdot U = 0 \tag{3.35}$$

or

$$\nabla \cdot q - 2\Pi_y = 0 \tag{3.36}$$

where

$$q = U \cdot [\tau]$$
$$q_x = [\sigma_x u_x + \tau_{xy} u_y + \tau_{zx} u_z]$$
$$q_y = [\tau_{xy} u_x + \sigma_y u_y + \tau_{yz} u_z]$$
$$q_z = [\tau_{zx} u_x + \tau_{yz} u_y + \sigma_z u_z]$$

Each term represents a component of energy flux vector associated with the stress tensor in respective orthogonal direction. The energy flux vector could be used in the trajectory equation to visualize energy transport paths. The above expresses energy balance between net energy flux across a control volume and strain energy storage inside of it. Strain energy is interpreted as work done at a point by internal forces in response to external loads as the structure deforms at infinitesimal speed. Strain energy storage is referred to as strain energy function defined as:

$$\Pi_y = \frac{1}{2}[\tau] : [\nabla U]$$
$$= \frac{1}{2}\left[\sigma_x \frac{\partial u_x}{\partial x} + \sigma_y \frac{\partial u_y}{\partial y} + \sigma_z \frac{\partial u_z}{\partial z} + \tau_{xy}\left(\frac{\partial u_x}{\partial y} + \frac{\partial u_y}{\partial x}\right)\right.$$
$$\left. + \tau_{zx}\left(\frac{\partial u_z}{\partial x} + \frac{\partial u_x}{\partial z}\right)\tau_{yz}\left(\frac{\partial u_y}{\partial z} + \frac{\partial u_z}{\partial y}\right)\right]$$

$$= \frac{1}{2}[\sigma_x \varepsilon_x + \sigma_y \varepsilon_y + \sigma_z \varepsilon_z + 2\tau_{xy}\gamma_{xy} + 2\tau_{xz}\gamma_{xz} + 2\tau_{yz}\gamma_{yz}]$$

$$= \frac{1}{2}[\tau]:[\gamma]$$

$$= \frac{\lambda}{2}\gamma_{ii}^2 + \mu\gamma_{ij}\gamma_{ij} \tag{3.37}$$

λ and μ are elastic constants. Strain energy in solids is analogous to the rate of viscous dissipation in fluids. However, the process is considered reversible. A bulk solid material can be distorted in many different ways: shearing, shrinking, bending, twisting and so on. Each mode of deformation contributes its own elastic energy to the materials. The total energy in the system is the sum of all contributions. Strain energy distributions and mechanical energy transfer paths in a system are useful information in the study of fracture and failure phenomena.

3.9 Multiple Fields

Multiple vector fields occur in mathematics and physical sciences. Multiple fields may or may not interact with one another. Interaction may take place between fields of different origin/nature and produce fields of another quantity. Examples include but are not limited to, Poynting vector derived from magnetic fields and electric fields. Magnetic force generated by magnetic fields and velocity fields. Lamb vector and helicity produced from vorticity and flow fields. By extracting salient information and key structural elements of fields from multiple data sets, we are able to gain an understanding of multiple fields. Consider a case of horizontal plane flows at steady state. The governing equations for incompressible ideal fluids are written in streamline coordinates with the aid of ∇-operator and Frenet's formula. The intrinsic representations introduce flow field structure in terms of radii of curvature of streamlines and equipotentials into the dynamics of the systems.

$$\text{Euler equation in } \hat{s}\text{-direction:} \quad \frac{\partial p}{\rho \partial s} + v_s \frac{\partial v_s}{\partial s} + \frac{\partial U}{\partial s} = v_n \omega_z = 0 \tag{3.38a}$$

$$\hat{n}\text{-direction:} \quad \frac{\partial p}{\rho \partial n} + v_s \frac{\partial v_s}{\partial n} + \frac{\partial U}{\partial n} = -v_s \omega_z \tag{3.38b}$$

Or

$$\frac{\partial p}{\rho \partial n} + \frac{\partial U}{\partial n} + \frac{v_s^2}{R_s} = 0 \tag{3.38c}$$

$$\hat{k}\text{ - direction :} \quad \frac{\partial p}{\rho \partial z} + \frac{\partial U}{\partial z} = 0 \tag{3.38d}$$

Continuity equation: $\nabla \cdot V = \left(\hat{s} \dfrac{\partial}{\partial s} + \hat{n} \dfrac{\partial}{\partial n} \right) \cdot (v_s \hat{s} + v_n \hat{n}) = 0$

$$\rightarrow \frac{\partial v_s}{\partial s} + \frac{v_s}{R_n} = 0$$

or

$$\nabla \cdot V = \frac{1}{h_1 h_2 h_3} \sum \frac{\partial}{\partial \xi_i} \left(\frac{h_1 h_2 h_3}{h_i} v_i \right)$$

$$= \frac{1}{h_1 h_2 h_3} \frac{\partial}{\partial s} \left(\frac{h_1 h_2 h_3}{h_1} v_s \right)$$

$$= \frac{v_s}{h_1 h_2 h_3} \frac{\partial}{\partial s} (h_2 h_3) + \frac{v_s}{h_1} \frac{\partial v_s}{\partial s}$$

$$= \frac{v_s}{R_n} + \frac{\partial v_s}{\partial s} = 0$$

where

$$\frac{1}{h_1 h_2 h_3} \frac{\partial}{\partial s} (h_2 h_3) = \frac{1}{R_n} = \frac{\partial \alpha}{\partial n}; \quad h_1 = 1; \quad v_n = 0; \quad v_z = 0$$

The plane geometric constraints play an important role in flow field structure and behavior. The direction of angular momentum of a particle about an arbitrary point is perpendicular to the plane of motion. Not only are velocity and position vector confined to the plane of motion but also circulation and vorticity activity with only one non-zero vorticity component. This vorticity component is conserved for the motion of inviscid fluids and vortex lines are infinite and straight perpendicular to the plane of motion. As stretching and tilting of vorticity lines fail to operate, complex flows become difficult if not impossible. Furthermore, with vanishing helicity everywhere, there is no mechanism to endow streamlines with torsion or twisting. Lamb vector field structure represented by Lamb surfaces are formed by an orthogonal network of streamlines and vortex lines and are bounded by cylindrical surfaces like plane strain with an infinite longitudinal axis parallel to vortex lines. The end effects are considered negligible. The plane of motion is a plane of symmetry of the Lamb cylindrical cross section perpendicular to the longitudinal axis and is approximated as plane stress in two dimensions. Lamb surfaces coincide with stream surfaces and Bernoulli surfaces moving together in steady plane flow fields. For horizontal plane flows, gravitational effects are operative only along vertical direction and pressure is required to follow hydrostatic distribution.

Vorticity kinematics depicts vorticity structure in terms of local velocity gradients at a point.

$$\omega = \nabla \times V$$

The non-zero vorticity component is given by Equation 6.15c,

$$\omega_z = -\frac{\partial v_s}{\partial n} + \frac{v_s}{R_s}$$

Or

$$\omega_z = -\nabla^2 \psi$$

$$= -\left(\frac{\partial^2}{\partial s^2} - \frac{1}{R_s} \frac{\partial}{\partial n} + \frac{1}{R_n} \frac{\partial}{\partial s} + \frac{\partial^2}{\partial n^2} \right) \psi \qquad (3.39)$$

$$= -\frac{\partial v_s}{\partial n} + \frac{v_s}{R_s}$$

Rate of shear: $\dot{\gamma}_{12} = \dot{\gamma}_{21} = \frac{1}{2} \left(\frac{\partial v_s}{\partial n} + \frac{v_s}{R_s} \right)$

where

$$\dot{\gamma}_{ij} = \frac{1}{2} \left[[\nabla V] + [\nabla V]^T \right]$$

$$[\nabla V] = \sum_i \sum_j \left(\frac{\hat{e}_i \partial(v_j \hat{e}_j)}{h_i \partial x_i} \right)$$

$$= \begin{bmatrix} \dfrac{\partial v_s}{\partial s} & \dfrac{v_s}{R_s} \\[2mm] \dfrac{\partial v_s}{\partial n} & \dfrac{v_s}{R_n} \end{bmatrix}$$

From kinematics and Frenet's formula, we find that two subtle components of forces emerge in fluid acceleration, that is, coriolis acceleration and centripetal acceleration. They act along streamline tangent and normal directions respectively.

$$V = v_s \hat{s} + v_n \hat{n}$$

$$a = a_s \hat{s} + a_n \hat{n} = \frac{dV}{dt} = \left(\frac{dv_s}{dt} - \frac{v_n v_s}{R_s} \right) \hat{s} + \left(\frac{dv_n}{dt} + \frac{v_s^2}{R_s} \right) \hat{n}$$

$$= \left(\frac{dv_s}{dt} - v_n \omega_z \right) \hat{s} + \left(\frac{dv_n}{dt} + \frac{v_s^2}{R_s} \right) \hat{n}$$

The term $v_n \omega_z = 2\dot{\Omega}_z v_n$ is referred to as coriolis acceleration acting in streamline tangential direction. This component of force is connected to vorticity and n-component velocity. The force magnitude is proportional to the product of two field strengths. Coriolis acceleration is generally not operative since velocity direction must align with trajectory tangent. The flow fields are rendered one-dimensional locally with vanishing n-component velocity. Motion in tangential direction then represents complete integral of Euler equation and leads to Bernoulli's theorem. The theorem states that Bernoulli function B interpreted as mechanical energy is constant for inviscid fluids at steady motion along a streamline whether flow is rotational or not. It is a statement of energy conservation principle applicable along a streamline or everywhere for irrotational flows. The geometric interpretation of Bernoulli function is an integral curve of constant B coinciding with a streamline.

$$\frac{\partial B}{\partial s} = \frac{\partial p}{\rho \partial s} + v_s \frac{\partial v_s}{\partial s} + \frac{\partial U}{\partial s} = 0$$

$$\rightarrow B = h_0 \equiv h + \frac{V^2}{2} + U = C(n)$$

In the normal direction of a curved streamline, there is another component v_s^2 / R_s recognized as centripetal acceleration always pointing towards the center of curvature of streamline. This component of acceleration depends on trajectory and flow field strength. It exerts a force on a particle in the normal direction to change its motion towards center of curvature. The force magnitude is proportional to the square of tangential velocity component and inversely to the radius of streamline curvature. Both coriolis and centripetal components are regarded as kinematic requirements deduced as necessary for curved trajectories. It is observed that tangential component of acceleration reflects time rate of change of speeds independent of trajectory's shape. The normal component of acceleration takes care of change of velocity directions and depends on the shape of trajectory. It requires other force(s) to maintain force balance in order to keep particles on curved tracks. Centripetal acceleration vanishes at an inflection point of trajectory due to infinite radius of curvature or in a straight line.

Motion in normal direction suggests flow behavior may become complex as velocity and vorticity interaction takes place giving rise to Lamb vector fields and helicity when rotational flow axis is not parallel to velocity vector. Lamb vector has important properties for tracking vorticity and flow fields and guides the interpretation of field phenomena. The cross product of vorticity and velocity yields a Lamb vector perpendicular to both of them in the direction determined by right hand rule. In a physical sense, the Lamb vector represents a force per unit mass deflecting the direction of motion. The axis of rotation (vortex line) and the direction of rotation (velocity) spin a tangent plane at each point known as Lamb plane. The tangent plane normal coincides with Lamb vector, which has only one non-zero Lamb vector component L_n in normal direction. Lamb vector magnitude depends on vorticity and flow field strength.

$$L_n = -\frac{\partial B}{\partial n} = v_s \omega_z = v_s \frac{v_s}{R_s} - v_s \frac{\partial v_s}{\partial n}$$

$$L = L_n \hat{n} = v_s \hat{n} \omega_z = -\frac{\partial B}{\partial n} \hat{n} = -\left(\frac{\partial B}{\partial s}\hat{s} + \frac{\partial B}{\partial n}\hat{n} + \frac{\partial B}{\partial z}\hat{k}\right) = -\nabla B \quad (3.40)$$

$$= -\left(\nabla \frac{V^2}{2} + \frac{\nabla p}{\rho} + \nabla U\right)$$

$$= -\nabla \frac{V^2}{2} + V \cdot (\nabla V)$$

Or in terms of stream function:

$$L = -\frac{\partial B}{\partial \psi}\frac{\partial \psi}{\partial n}\hat{n} = -\frac{\partial B}{\partial \psi}\nabla \psi$$

$$= -(\nabla \psi)\nabla^2 \psi = (\nabla \psi)\omega_z \quad (3.41)$$

Furthermore,

$$L \times \nabla \psi = (\nabla \psi \times \nabla \psi)\omega_z = 0$$

Lamb surfaces represent normal congruence of tangent planes and provide the geometric interpretation of Lamb vector. With Lamb vector parallel to streamline normal at each point, there is a Lamb surface coincide with the corresponding stream surface and move together. It is observed from Lamb vector structure that Lamb vector accounts for the combined effects of convective acceleration and centripetal acceleration in streamline normal direction in rotational flows. Lamb vector could exist due to either effect. Or the two effects do not cancel each other out. The nature of Lamb vector could be interpreted in the context of magnetic force produced by the interaction of magnetic fields and velocity fields of charged particles. Although Lamb vector does not affect motion in tangential direction, it exerts a force perpendicular to velocity and alters its path to veer to one side of it in the normal direction. Hence, Lamb vector does no work and does not change particle's energy or speed. This is similar to magnetic force, which deflects the directions of motion of charged particles but not magnitudes in a uniform constant magnetic field. Under this condition, magnetic force is the sole external force to maintain force balance with centripetal force in simple circular motions at constant speed. In fluid systems, motions are more complex especially in streamline normal direction because Lamb vector, convective acceleration, pressure gradient and body force may be present simultaneously and play a profound role in fluid flow behaviors as illustrated by forced vortex and free vortex motions.

Lamb vector has the dual nature of physical vector similar to Poynting vector, carrying force and energy gradient vector. To investigate the energy aspect of Lamb vector fields, the idea of integration along a streamline is useful not only to arrive at Bernoulli's theorem but to study entropy as well. It is found that the line integral of Lamb vector interpreted as work done by Lamb vector vanishes along a streamline because Lamb vector is perpendicular to velocity (or streamline tangent) everywhere. The vanishing work done by Lamb vector implies,

$$\int_s L \cdot d\vec{r} = \int_s L \cdot V dt = 0$$

or

$$\int_s L \cdot d\vec{r} = \int_s L \cdot \hat{s} ds = \int_s T \nabla S \cdot d\vec{r} - \int_s \nabla B \cdot d\vec{r} = \int_s T \nabla S \cdot \hat{s} ds = \int_s T \frac{\partial S}{\partial s} ds = 0$$

$$\rightarrow \frac{\partial S}{\partial s} = 0$$

$$S|_{\text{streamline}} = C(n); \ S = S(\psi)$$

Hence, entropy as well as mechanical energy (or total enthalpy h_0) are constant along a streamline just like stream function. Entropy will vary from one streamline to another in the direction normal to streamline at maximum spatial rate. It will vanish if the system is isentropic or Lamb vector is equal to mechanical energy gradient vector in opposite direction for inviscid fluids. Entropy gradient vector is produced due to viscous dissipation of energy from mechanical form irreversibly into heat in real fluids or due to energy transfer. When integral path is closed, circulation of Lamb vector will vanish in isothermal or isentropic systems or following a closed streamline path.

Along an arbitrary path: $\oint_c L \cdot d\vec{r} = \int_A \nabla \times (T \nabla S - \nabla B) \cdot \hat{b} dA = 0$ (3.42)

Along a closed streamline: $\oint_c L \cdot d\vec{r} = \oint_s L \cdot V dt = 0$

When Bernoulli's constant assumes the role of contour surface parameter, it provides geometric interpretation of Bernoulli's theorem in the context of a material surface structure. Bernoulli surfaces are material surfaces on which mechanical energy is constant. For plane flows, the equation of motion in vertical direction implies Bernoulli's surface is a ruled surface, that is, a cylindrical surface of infinite longitudinal axis. As the surface parameter varies, a system of Bernoulli surfaces is generated in space and partitions flow fields into smooth non-intersecting surfaces, each carrying constant mechanical energy equal to Bernoulli's constant. Since ∇B is the gradient vector normal to the surface of constant B, both velocity and vorticity vector are tangent to Bernoulli surface in the same manner as stream surface or Lamb surface.

Lamb vector is also interpreted as an energy gradient vector apart from a force vector. According to Crocco's theorem, there is a relation between mechanical energy gradient vector, entropy gradient vector and Lamb vector.

$$\text{Crocco's theorem} : L = T\nabla S - \nabla B \tag{3.43}$$

In order to isolate mechanical energy from other energy sources or forms, we consider non-viscous incompressible fluids in an isentropic system. Under this condition, exchange of energy is reversible in conservative systems. Furthermore, entropy will assume constant everywhere and entropy gradient vector vanishes accordingly. Lamb vector then determines the direction and magnitude of Bernoulli gradient vector along which maximum spatial rates of change of mechanical energy occurs. Lamb surface and Bernoulli's surface coincide with each other. The divergence of Lamb vector yields the sole energy source to change Bernoulli's constant across streamlines. In plane fields, Lamb vector lies on the plane of motion and Lamb surface coincides with intersecting systems of streamlines and vortex lines. Because velocities and vorticity vectors are at a tangent to streamlines and vortex lines respectively, Lamb surface is formed by a network of streamlines and vortex lines crossing each other at right angles continuously in space. For steady motions,

$$V \cdot \nabla B = V \cdot L = 0$$
$$\omega \cdot \nabla B = \omega \cdot L = 0$$
$$\rightarrow \frac{DB}{Dt} = \frac{\partial B}{\partial t} + V \cdot \nabla B = 0$$

Bernoulli's function is conserved in the motion just like the stream function. Not only is Bernoulli's constant invariant along streamlines, but is also constant along vortex lines. Figure 2.5 depicts one of the streamlines at the intersection of a plane of motions and a Lamb surface. Lamb surface intersects all planes parallel to the plane of motions orthogonally everywhere. Each parallel plane along binormal direction is a plane of symmetry. The intersections yield the same unique plane curve on each parallel plane, which coincides with a streamline since all streamlines lie on a Lamb surface. The union of parallel copies of streamlines along binormal direction defines the shape of a Lamb surface like an upright cylindrical landscape. The longitudinal cross section of cylindrical surface will bent and stretch in unsteady motions. Streamlines are geodesic curves on a Lamb surface because of zero helicity everywhere and of the fact that streamline geodesic curvature vanishes. Furthermore, torsion of streamline (plane curve) is not allowed. As the geodesic curvature of vortex line also vanishes, vortex lines are straight geodesic lines on a Lamb (cylindrical) surface parallel to the longitudinal axis. Both longitudinal axis and vortex lines extend to infinity such that the requirements of no stretching/tilting or twisting of vortex lines are satisfied and that smooth flows will prevail. Alternatively, a vortex line could generate the cylindrical surface as it moves along a streamline parallel to itself.

Bernoulli surfaces are functionally related to stream-surfaces through the distribution of vorticity.

$$\rightarrow B = -\int_{\psi_1}^{\psi_2} \omega_z \partial \psi = f(\psi) \tag{3.44}$$

The Bernoulli surface parameter will change when stream function parameter constant C_2 varies from one streamline to another. This calls for a change of mechanical energy across streamlines accordingly. Since both mechanical energy and stream function are invariant along streamlines, the space between two streamlines corresponds to a fixed ratio of ∂B and $\partial \psi$. Thus, vorticity must also have a fixed value between two streamlines, that is,

$$\omega_z = -\frac{\partial B}{\partial \psi} = C' \tag{3.45a}$$

Or

$$v_s \omega_z = -\frac{\partial B}{\partial n} \tag{3.45b}$$

It shows that Bernoulli's constant depends on the distribution of vorticity and remains invariant along a streamline. It also suggests particles retain vorticity in steady motions and that the strength of vortex tube is invariant. Indeed vortex lines move with flows of non-viscous fluids in accordance with Helmholtz vortex laws.

As velocity is perpendicular to Lamb vector, no flux passes through the material surface. The existence of such material surfaces is a necessary condition for steady motions. A Lamb surface specifies the position of a material surface for all time, the invariance of property requires that mechanical energy does not change when evaluated at fluid particle on a Lamb surface, which moves with flows and always consists of the same fluid (material) particles. It extends Bernoulli's theorem applicable not only along streamlines but also along vortex lines, that is, the Lamb surface of constant mechanical energy. Thus, the existence of Lamb vector implies multiple steady fields with the coincidence of Lamb surfaces, stream surfaces and Bernoulli surfaces moving together in isentropic rotational flows. Although Lamb vector, velocity and vorticity are mutually orthogonal, they are coupled due to the interaction of flow fields and vorticity through the cross product operation. One may think of a Lamb surface being swept out by streamlines and vortex lines. Correlation or relation may exist among these material surfaces in multiple fields. Correlation may prevail between the dual nature of Lamb vector fields carrying force/mechanical energy and flow fields carrying momentum/kinetic energy respectively. Kinematic relation does exist between vorticity and the effects of friction/dissipation through vector identity. The kinematic relation between vorticity and flow fields allows us to derive vorticity from the curl operation of flow fields or through the stream function-vorticity relation. To aid visualization of vorticity, a tangent plane on the napes of cylindrical surface will give rise to a line of tangent coinciding with a vortex line. The tangent vector of geodesic curve always points in the same direction along streamline. Streamlines are therefore unlikely to become chaotic since they cannot leave the cylindrical surface or reverse direction. Lamb vector fails to operate in irrotational flows. Or there is no interaction between flow fields and vorticity when the two vector fields are mutually orthogonal. In a local sense, Lamb vector vanishes at velocity or vorticity critical points. Lamb vector vanishes wherever the combined effects of centripetal acceleration and convective acceleration in streamline normal direction cancel each other out. Under any of the above conditions, Lamb surface vanishes and Bernoulli surface as well as its relation with stream function follow suit. Both Bernoulli's constant and entropy are invariant not only on stream surface but also everywhere for conservative systems.

To extend the analysis of Lamb vector, Lamb vector divergence is a useful parameter to visualize the internal characteristics of Lamb vector fields by capturing the energy sources and their distributions. It also reveals the peculiar effects of Lamb vector and its role in flow behavior in the realm of a multiple field system.

$$
\begin{aligned}
\nabla \cdot L &= \nabla \cdot (\omega \times V) \\
&= V \cdot (\nabla \times \omega) + (-\omega^2) \\
&= (-V \cdot (\nabla^2 V)) + (-\omega^2) \\
&= -\nabla^2 B
\end{aligned}
\tag{3.46}
$$

The above expression could also be derived by taking the divergence of Euler equation or Crocco's theorem in isentropic flows. It shows that Lamb vector divergence represents an energy source term in Poisson's equation of mechanical energy. The energy balance at a point depends on the work done by fluid strain rate (first term) subject to deformation and those by vorticity (second term) in rotation. The first term represents a reversible process and vanishes at velocity or vorticity critical points. The second term representing enstrophy density assumes the role of a sink. It describes the dissipation effects regardless of fluid viscosity. Enstrophy density is a measure of vorticity vector magnitude due to fluid rotation. It is defined as the square of vorticity per unit volume in analogy to kinetic energy density and it vanishes at vorticity critical points.

$$\frac{d\Phi_\omega}{dv} \equiv \frac{1}{2}\omega^2$$

Depending on the state of source balance in Lamb vector divergence, the sign of Lamb vector divergence switches between negative, zero or positive. It signifies local regions of non-homogeneities in momentum/mechanical energy distributions as fluid particles undergo a distinct change in dynamic characteristics of flow fields. When Lamb vector divergence is positive, it implies the source contributed by the first term is positive and represents the reversible rate of work done associated with the stress tensor as fluids restore from deformation and is greater than second term. The source indicates regions of high strain rate. When it is zero, strain rate and enstrophy are at equal strength and balance each other out. Or flow is irrotational. When the source assumes negative, it signifies an energy sink attributed to rotation and hence regions of strong vorticity. If the first term is also negative, then the region could be high strain rate as well subject to deformation.

Apart from Lamb vector and its divergence, the interaction of vorticity and velocity can take an entirely different form. In this case, the dot product of these two vector fields gives rise to a scalar quantity known as helicity. Helicity density per unit volume is defined as:

$$\frac{dH}{dv} \equiv \omega \cdot V$$
$$= [0 \quad 0 \quad \omega_z] \cdot [v_s \quad v_n \quad 0] = 0$$

It represents a topological measure of vorticity fields as the degree of knottiness of vortex lines. Helicity endows streamlines with torsion, twisting them into helical curves in the direction of fluid motion. In two-dimensional flow fields, helicity vanishes as vorticity is perpendicular to velocity everywhere. Helicity attains maximum strength in Beltrami flows. Lastly, the equation of motion in vertical direction implies that Bernoulli function varies only on the horizontal plane of motion for plane flows and that the pressure fields correspond to hydrostatic conditions.

$$\frac{\partial B}{\partial z} = -g\frac{\partial \varphi}{\partial z} = 0$$

where

$$\varphi = \frac{p}{\rho g} + \frac{U}{g}$$

4

Complex Analysis and Complex Potentials

4.1 Complex Variables/Functions and Applications

Real numbers have limitations, as they are only capable of defining points along real axis. The shortcoming of real numbers could be circumvented by the introduction of complex numbers. Complex numbers and analyzes arise naturally from solution of quadratic equations and from the study of plane Euclidean geometry. Complex numbers can be operated efficiently subject to the rules of complex numbers and is convenient for vector field representation. In fact, complex number operations and vector operations are equivalent. The physics of vector fields is then connected to the mathematics of complex numbers and can be developed using the theory of complex analysis. Complex variables possess two-dimensional characteristics, which allow us to describe a quantity with two independent variables into a single one. Complex variables define points in a plane by a one to one correspondence in an Argand diagram. These features make them an ideal device for displaying or mapping fields on various complex planes and become a useful mathematical technique for field problems in two dimensions. It is remarkable that complex numbers could contribute to boundary value problems and the solution of the Laplace equation. This is because when two complex planes are linked by analytic mapping functions, they remain solutions of Laplace equation. Complex variables extend the theory of potentials into a complex plane and allow regions with complicate boundaries to be visualized through conformal mapping.

Visualization of Fields and Applications in Engineering, First Edition. Stephen Tou.
© 2011 John Wiley & Sons, Ltd. Published 2011 by John Wiley & Sons, Ltd.

In connection with the idea of continuing real numbers into complex domains, the idea of continuing real functions into complex domains is a natural development. Because of the limitations of complex functions, complex analysis and theory are restricted to two-dimensions. Nonetheless, complex function theory relates a pair of conjugate harmonic functions in exactly the same form described by the Cauchy-Riemann relation. Complex potentials express an important geometric fact that the two families of potential contours are orthogonal and that they satisfy the Laplace equation. They lead us in a natural manner to visualize harmonic vector fields through complex functions. Any complex potential can be chosen to describe certain field phenomena in two-dimensions. The Cauchy-Riemann relation assures that complex potential is a class of analytic functions of complex variables and that mapping is conformal. The local behavior of an analytical function reveals information about global behavior. Complex potential is an elegant way of relating scalar potential and vector potential through the applications of complex function theory. The determination of an analytic complex potential will suffice to describe vector fields completely. Complex numbers focus on Euclidean geometry and representation whereas complex functions focus on the concept of mapping between complex planes. Together they offer a powerful tool for field problems in two dimensions. It is, however, impossible to draw graphs of a complex function. This is because to display two real functions of two real variables graphically would require four dimensions. We can visualize it by sketching its domain of definition in the z-plane and its image in the complex F-plane. The real or imaginary part or both may carry physical information. Generally, physical phenomena occur on the z-plane also known as the physical plane while its image in the F-plane describes uniform field patterns in two different views; one given by equipotential contours and the other given by flux line contours in a system of orthogonal rectangular coordinate lines. The two planes are connected by a complex function, that is, the mapping function.

Historically, complex theory and analysis are developed for a special class of functions under the notion of analyticity. This restriction is necessary to ensure derivatives of complex functions exist regardless of the direction of differentiation. Any analytic function of a complex variable could be considered a mapping function. To explore the nature of analytic functions, we consider a complex function, which defines a mapping from z-plane to F-plane, namely

$$F\text{-plane: } F = f(z) = \phi(x, y) + i\psi(x, y) \tag{4.1}$$

$$z\text{-plane: } z = x + iy \tag{4.2}$$

The complex variable z is represented in an Argand diagram in Figure 4.1. This is a complex z-plane defined by a two-dimensional rectangular

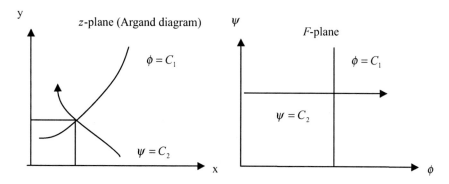

Figure 4.1 Representation of a complex function of complex variables.

coordinate system consisting of a horizontal (real) x-axis and a vertical (imaginary) y-axis. Connection between complex numbers and vectors is established by the definition of imaginary number i. It is regarded as an operator which, when applied to a vector, rotates it through 90 degrees counterclockwise. The real part, being one of the vector components in x-direction is presented along x-axis. The imaginary part, operated by the imaginary operator i is the other component in y-direction. x and y are coordinates of a point in z-plane. A vector field in this plane is depicted as a vector emanating from point z. The modulus of a complex number is:

$$|z| = \sqrt{z\bar{z}} = \sqrt{x^2 + y^2} \qquad (4.3)$$

The complex conjugate of z is defined as:

$$\bar{z} = x - iy \qquad (4.4)$$

Figure 4.1 shows a complex F-plane constructed with equipotential (scalar potential) and flux function (vector potential) as abscissa and ordinate respectively. Complex function always maps a complex variable into a real variable ϕ and ψ represented on F-plane. It maps a set of points from z-plane onto a set of corresponding points in F-plane. Flux lines and equipotentials are mapped to horizontal and vertical lines respectively in complex F-plane representing a uniform field and providing an interpretation of complex potentials. Field patterns in the two planes are related by mapping functions through the parametric constants of conjugate harmonic functions. If z is made to vary by moving a point tracing a flux line or an equipotential in z-plane, then the corresponding point will trace out an image line parallel to the horizontal or vertical axis in the F-plane respectively. A small change in z will produce a corresponding change in F. The ratio of these changes must be unique and unaltered regardless of the paths taken. A complex function as such combines two potentials into a single function and it allows fields to be described geometrically in four dimensions by two complex planes. The theory of complex functions makes complex potentials more versatile than working with each potential separately.

Complex potentials and mapping have practical applications. Flow modeling of source, sink and doublet are examples of complex potentials. They are the building blocks in field synthesizing processes where complex flows are constructed by superposition. Of particular significance is the fact that a field map derived from a complex potential for certain geometry may be applied directly to different physical fields sharing the same analogy. It also facilitates the assessment of field transport characteristics and its relation to domain geometry. Besides being a technique for solving equations and transforming problems into simple geometries by successive conformal mapping, it could be applied to wrap images while preserving angles and local shapes. Once an analytic complex potential is available, it is ready to facilitate field visualization and to derive a spectrum of information of interest by mapping through the action of differentiation, integration or composite functions. Thus, construction of complex potentials forms the central theme in harmonic field visualization, analysis and mapping.

4.2 Complex Analysis and Cauchy–Riemann Equation

An attractive feature of complex analysis is that it only requires a few basic operations such as differentiation and integration yet it is a powerful tool in vector field analysis. Thus, investigation of a complex function property involves the action of differentiation and integration from which vital information about fields could be obtained. Differentiation of complex potential is independent of the path taken and gives,

$$\frac{dF}{dz} = \frac{\partial \phi}{\partial x} + i\frac{\partial \psi}{\partial x} = \frac{\partial \psi}{\partial y} - i\frac{\partial \phi}{\partial y} \tag{4.5}$$

By equating the above real and imaginary parts, we obtain Cauchy–Riemann relation, which leads back to a pair of Laplace equations through differentiation.

$$\frac{\partial^2 \phi}{\partial x^2} = \frac{\partial^2 \psi}{\partial x \partial y}; \frac{\partial^2 \phi}{\partial y^2} = -\frac{\partial^2 \psi}{\partial y \partial x}$$

$$\frac{\partial^2 \psi}{\partial x^2} = -\frac{\partial^2 \phi}{\partial x \partial y}; \frac{\partial^2 \psi}{\partial y^2} = \frac{\partial^2 \phi}{\partial y \partial x} \tag{4.6}$$

$$\nabla^2 \phi = 0; \nabla^2 \psi = 0$$

There then exist a pair of conjugate harmonic functions that can be interchanged mathematically but they lead to the solution of different phenomena. For example, a free vortex becomes a source or vice versa. For prescribed boundary conditions, the roles of ϕ and ψ are not interchangeable. The real part of the complex function represents field strength (or intensity) in terms of a scalar potential ϕ. The imaginary part characterizes field directions given by vector potential ψ as field property transports along flux line trajectories. The Cauchy-Riemann equation expresses the relationship between ϕ and ψ and ensures that they can be combined to form a complex potential to describe a possible two-dimensional harmonic field irrespective of its form. Complex potential is an alternative specification of harmonic fields. An interpretation between complex functions and the physics of tensor fields is then established. The connections between scalar potential, vector potential and complex potential in plane problems are used to advantage in the application of complex theory and conformal mapping. It makes all resources of complex function theory available to the study of harmonic fields and opens the window of applications in field visualization and mapping. In general, any complex function, analytic or not, can be defined as per Equation 4.1. That is to say, any complex function of a complex variable can be represented by two-dimensional vector fields, but not every such function meets the Cauchy-Riemann requirements of analyticity. The analyticity of a complex function breaks down at singular/critical points at which the Cauchy-Riemann relation fails to hold.

4.3 Differentiation of Complex Function

Consider a velocity complex potential. Differentiation of velocity complex potential is a mapping of the complex F-plane to another complex plane referred to as conjugate hodograph $\bar{q}$-plane on which complex conjugate velocity is defined as a position vector from origin, that is,

$$\text{Conjugate hodograph plane:} \bar{q} = \frac{dF}{dz} = f'(z) = \frac{\partial \phi}{\partial x} + i \frac{\partial \psi}{\partial x} = q_x - i q_y$$

(4.7)

$$\text{Or in polar form:} \bar{q} = \left| f'(z) \right| \exp(i \arg \left| f'(z) \right|) = |q| \exp(-i\bar{\theta}) = |q| \exp(-i\alpha)$$

(4.8)

α is the flow angle between resultant velocity and positive x-axis equal to streamline tangent angle. Differentiation yields a conjugate velocity vector expressed in terms of twist angle and magnification factor and is interpreted as a geometric mapping process. The action of rotation and scaling of an infinitesimal object is entirely different and independent of each other. The combined action is a local phenomenon and is encoded in the complex derivative, which defines a mapping of an infinitesimal vector at a point between the z-plane and F-plane through multiplication operation of complex numbers, that is,

$$dF = f'(z)dz = |f'(z)| \exp(i \arg |f'(z)|)dz \qquad (4.9)$$

$$\text{Magnification factor: } |f'(z)| = \sqrt{\left(\frac{\partial \phi}{\partial x}\right)^2 + \left(-\frac{\partial \psi}{\partial x}\right)^2} = |q| = |\bar{q}| \quad (4.10)$$

$$\text{Twist angle: } \arg |f'(z)| = \tan^{-1}\left(\frac{\frac{\partial \psi}{\partial x}}{\frac{\partial \phi}{\partial x}}\right) = -\tan^{-1}\left(\frac{q_y}{q_x}\right) = -\tilde{\theta} = -\alpha \quad (4.11)$$

$$|f'(z)|^2 = \frac{\partial(\phi, \psi)}{\partial(x, y)} = \frac{1}{J}$$

The twist angle defines the direction of the conjugate velocity vector with respect to positive x-axis anticlockwise in a right-hand coordinate. It is a mirror reflection image of the actual velocity about x-axis. The distribution of twist angle gives a global view regarding field direction changes and the curvature of field trajectories. It is equivalent to isoclines depicting the mirror image of field directions reflected about x-axis. When the distribution contours squeeze towards a region, it indicates rapid change of field directions and trajectory curvature. Magnification factor represents modulus of complex derivative and defines the scaling ratio of infinitesimal line segments between the z-plane and F-plane. The physical significance of the magnification factor is the representation of vector field magnitude as a measure of field strength. Since complex derivative at a point is unique and independent of vector directions, all line segments passing through a given point are scaled by the same magnification factor and rotated through the same twist angle. To illustrate the mapping process, consider an arbitrary point p in the F-plane and its image p' in the z-plane in Figure 4.2. Suppose an infinitesimal object is defined by a pair of elemental vectors forming an angle β in the neighborhood of that point on the z-plane. The direction of each elemental vector is represented by a tangent, which is defined by the respective inclination angle $\arg |dz|$ with respect to positive x-axis. The lengths of both elemental vectors are scaled by magnification factors and are rotated through the twist angle of the same amount on the F-plane. The angle β remains unaltered in the mapping process.

$$|dF_1| = |f'(z)| |dz_1| ; |dF_2| = |f'(z)| |dz_2|$$
$$\arg |dF_2| = \arg |f'(z)| + \arg |dz_2| \qquad (4.12)$$
$$\arg |dF_1| = \arg |f'(z)| + \arg |dz_1|$$
$$\rightarrow \arg |dF_2| - \arg |dF_1| = \arg |dz_2| - \arg |dz_1| = \beta \qquad (4.13)$$

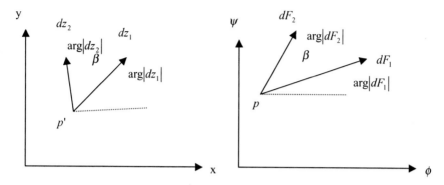

Figure 4.2 Mapping of an elemental object between complex F-plane and z-plane.

Hence, the corresponding figures in these two planes are scaled equally in all directions and are similar with the preservation of angle. This is the unique property of an analytic function referred to as conformal transformation. Because magnification factor and twist angle are a function of position, the amount of scaling and rotation that the elemental lengths undergo between two planes varies from point to point. Large figures tend to be distorted. The above mapping can be interpreted by affine transformation in terms of an isotropic dilation (scaling) matrix and a rotation matrix about the z-axis through the twist angle. In this respect, the multiplication operation of complex numbers and affine mapping may be established through matrix representation/transformation of infinitesimal vectors. By adopting the following scaling matrix and rotation matrix, it shows that affine mapping gives rise to similarity transformation and is conformal. One may observe the roles of scaling matrix and rotation matrix vs. magnification factor and twist angle in the context of geometric transformation. There will be no change in shape just size when the magnification factor differs from unity.

$$dF = [B][T(\theta)]dz$$

where

$$[B] = \begin{bmatrix} |f'(z)| & 0 \\ 0 & |f'(z)| \end{bmatrix}; \quad [T(\theta)] = \begin{bmatrix} \cos\theta & -\sin\theta \\ \sin\theta & \cos\theta \end{bmatrix}$$
$$\theta = \arg|f'(z)|$$

To interpret the action of differentiation of complex potential and its physical significance further, it can be shown that the acceleration vector could also be derived from complex potential and its differentiations through the multiplication operation of complex numbers.

$$\bar{a} = \frac{d\bar{q}}{dt} = \frac{d}{dt}f'(z) = f''(z)\frac{dz}{dt} = f''(z)\overline{f'(z)}$$

$$a = \overline{f''(z)}f'(z)$$

(4.14a)

$$|a| = \left|\overline{f''(z)}\right|\left|f'(z)\right| = |\bar{a}|$$

Or in Cartesian representation,

$$a = a_x\hat{i} + a_y\hat{j} = \frac{dq}{dt} = \frac{dx}{dt}\frac{\partial q}{\partial x} + \frac{dy}{dt}\frac{\partial q}{\partial y}$$

$$= \left(\frac{\partial q_x}{\partial x}\frac{dx}{dt} + \frac{\partial q_x}{\partial y}\frac{dy}{dt}\right)\hat{i} + \left(\frac{\partial q_y}{\partial x}\frac{dx}{dt} + \frac{\partial q_y}{\partial y}\frac{dy}{dt}\right)\hat{j}$$

$$= \left(q_x\frac{\partial q_x}{\partial x} + q_y\frac{\partial q_x}{\partial y}\right)\hat{i} + \left(q_x\frac{\partial q_y}{\partial x} + q_y\frac{\partial q_y}{\partial y}\right)\hat{j}$$

(4.14b)

In accordance with the maximum modulus principle, an analytical function will assume its maximum modulus on boundary C. Thus both speed and acceleration magnitude will attain maximum values on the boundary.

$$|f(z_0)| \geq |f(z)| ; z_0 \quad \text{on } C$$

Similarly,

$$|f'(z_0)| \geq |f'(z)| ; z_0 \quad \text{on } C$$

$$|f''(z_0)| \geq |f''(z)| ; z_0 \quad \text{on } C$$

4.4 Integration of Complex Functions

The global property of fields can be explored as a field particle moves about from one point to another. This invokes the mathematical operations of the line integral of a complex function along an oriented curve in a complex plane. The fundamental theorems of calculus for contour integrals of real functions are directly extended to complex functions. The physical interpretation of line integral of real functions is the concept of work done when a particle or charge moves around in a field. However, the physical interpretation of complex integrals is elusive and it cannot be considered as an area under a curve. They are defined in terms of line integrals along prescribed plane paths.

Suppose $f(z)$ is continuous and the path is piecewise smooth and single-valued, the complex line integral along an oriented path exists. It allows us to see the difference between any two points in a given field. Consider the line integral of a complex conjugate vector between two arbitrary points 1 and 2 along an arbitrary path as shown in Figure 4.3. The integration process is shown below consisting of a combination of a work integral and a flux integral.

$$\int_1^2 \bar{q}\,dz = \int_1^2 q_x dx + q_y dy + i \int_1^2 q_x dy - q_y dx$$

$$= \int_1^2 q \cdot \hat{l}\,dl + i \int_1^2 q \cdot \hat{b}\,dl \qquad (4.15)$$

where $\qquad \hat{l} = \dfrac{d\vec{r}}{dl} = \left[\dfrac{dx}{dl} \ \dfrac{dy}{dl} \right]; \ \hat{b} = \hat{k} \times \hat{l} = \left[-\dfrac{dy}{dl} \ \dfrac{dx}{dl} \right]$

$\hat{l}$ and $\hat{b}$ are the unit tangent vector and unit normal vector of the path. dl is an elemental arc length along the path. The real and imaginary parts are referred to as:

$$\text{Work integral: } \Delta\phi = \int_1^2 q \cdot \hat{l}\,dl = \int_1^2 \nabla\phi \cdot \hat{l}\,dl = \phi_2 - \phi_1 \qquad (4.16)$$

$$\text{Flux integral: } = \int_1^2 \frac{\partial\phi}{\partial b}\,dl = -\int_1^2 \frac{\partial\psi}{\partial l}\,dl = \psi_1 - \psi_2 \qquad (4.17)$$

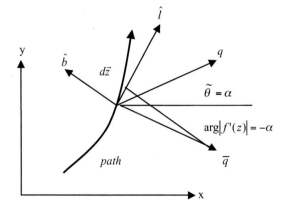

Figure 4.3 Integration of a complex conjugate vector.

The line integral of complex conjugate velocity is path independent and has a unique physical interpretation. The real part is called work integral and is equal to the difference in equipotential values of two end points. It represents work or potential change required in a potential field between two arbitrary points. The imaginary part is referred to as flux integral. It depicts the transport of field property per unit depth between any two points. The flux integral is given by the difference in flux line values. The concepts of flux and potentials are then linked to complex integrals and shed light on the understanding of analytic complex function theory. This also shows the advantage of working with complex conjugate vector as it facilitates the interpretation of complex integrals with physics of the fields.

When the integral path is closed, we obtain complex circulation through Stokes theorem and two-dimensional Gauss's theorem.

$$
\oint_c \bar{q}\,dz = \int_A \omega \cdot \hat{k}\,dA + i \int_A \nabla \cdot q\,dA
$$
$$
= \int_A \omega_z\,dA + i \int_A \rho\,dA
$$

(4.18)

A is an area on the x-y plane enclosed by the path. The above facilitates the assessment of field sources. The real part represents circulation and determines the amount of field lines that have curled up on them in terms of vector source, which generates the whirlpool phenomenon. It is a measure of the strength of a vortex tube. When this integral is not zero, there is a vorticity source inside the closed path, and as a result, field lines tend to form closed loops or extend to infinity at both ends. Whenever the closed real line integral vanishes, the field is said to be irrotational and scalar potential exists in the closed region.

The imaginary part yields net flux generated by a scalar source ρ inside the region per unit depth. It is a measure of dilation rate within a region. If the closed imaginary integral is zero, no field lines originate or terminate, as there is no source in the region. All field lines pass through from one side to other. The field is said to be solenoidal and vector potential exists in the region.

Complex circulation vanishes in harmonic fields. There are situations where vector fields are harmonic everywhere except inside a limited region of space. That is to say, all complex circulations enclosing this region differ from zero and all those not enclosing the region vanish. Complex potential is not analytic on the entire domain. An irrotational vortex is one example. Other field quantities such as forces and moments can be evaluated by Blasiu's theorem through integration.

4.5 Visualization of Complex Potentials

The search for complex potential is a streaming process, which hinges on being able to find the equipotential and flux function. It is necessary that the potentials we search must satisfy the governing equations together with the specified boundary conditions. For harmonic fields, flux function ψ exists and can be found by a pair of first order partial differential equations, each expresses vector field component along mutually orthogonal directions. Scalar potential ϕ can be found likewise. Both are employed to construct complex potential with scalar potential and flux function assuming the role of real and imaginary parts respectively. For a given vector field, solutions for a system of first order partial differential equations can be obtained by integration analytically or numerically. Two graphical methods are presented to visualize the graphs of complex potential. The beauty of graphical methods lies in its simplicity of formulations and the approximate results are quickly displayed.

4.5.1 Trajectory Method

Trajectories provide the basic tools for solving first order differential equations. Consider a trajectory of a general plane system. We think of the system as equation of motion of a particle moving about in a plane. A tangent to the trajectory defines the direction of the trajectory at that point. This simple idea comes from geometric considerations and provides a direct method to visualize the graphs of first order differential equations with the distribution of a set of directed line segments in the concept of directional fields.

4.5.1.1 Tangent Fields and Isoclines

When a set of points in (x, y) plane satisfies a trajectory equation, it can be considered as the graph of respective equation. In this trajectory method, a family of small line segments is drawn on (x, y) plane, such that a line segment that goes through a point (x_0, y_0) has the slope $f(x_0, y_0)$ with respect to positive x-axis. It represents an approximation to the solution of trajectory equation through that point. The line segments indicate tangent directions without the need to solve equations. Motion of a particle at point (x_0, y_0) is indicated by the tangent along which the particle will advance and remain on trajectory while keeping trajectory equation constant. Solution curves (trajectories) are traced out graphically though the field of tangent directions belong to respective curves while critical/singular points are skipped. The procedure is initiated at a given point and proceeds in the tangential direction of line segment through that point to the neighboring segments at each small step. The accuracy of the method increases as the number of line elements increase. As the zigzag length reduces, a smooth curve takes shape. The processes are repeated until a sufficient number of trajectories is traced. In this method, isoclines are introduced to integrate local line segments and to provide information about global characteristics of the vector fields. Isoclines are defined as lines intersecting a system of trajectories at a constant angle equal to the slope of trajectory, that is, the vector field direction. Isoclines are convenient for direction field construction and visualization as they produce many line segments at once. The use of isoclines is advantageous in hodograph mapping. Isoclines can be obtained from its hodograph representation by mapping a straight line passing through origin on hodograph plane. The slope of the straight line (isoclines) corresponds to the constant direction of vector fields on physical plane. The method is simple and effective.

4.5.2 Method of Curvilinear Squares

Curvilinear squares field map is one of the graphical methods for two-dimensional harmonic fields based on geometric considerations and the orthogonal property of a pair of conjugate harmonic functions. The method discretizes the field domain into a network of curvilinear squares formed by flux lines and equipotentials. A curvilinear square means an area that tends to yield true squares as it is subdivided into smaller and smaller areas by going through a successive process of fine-tuning. The resultant field map has unique properties and aids understanding of physical processes through visual effects without the need to struggle with a huge database. The benefits of curvilinear square method include economy and handling problems with complicate boundary conditions. It also facilitates the evaluation of geometric effects of a domain configuration subject to prescribed boundary conditions. A curvilinear field map is independent of field property coefficient such as hydraulic conductivity in hydrodynamics and could be applied directly from one physical field to another sharing the same analogy.

4.5.2.1 Curvilinear Squares Theory and Field Map Construction

Curvilinear squares theory is based on Cauchy-Riemann relation, which ensures that a conjugate pair of harmonic functions satisfies the Laplace equation in any orthogonal coordinate system and that the two systems of contours are orthogonal everywhere. The method makes use of the intrinsic representation of vector fields in streamline coordinates such that vector fields are always tangent to streamlines and depend only on the distributions of streamlines and equipotentials. Consider a typical curvilinear quadrilateral on the physical plane (Figure 4.4) with length Δs, width Δn and depth Δz bounded by the respective equipotential lines and flux lines. The origin of a local flux line coordinate is placed at the center of a cell with flow passing

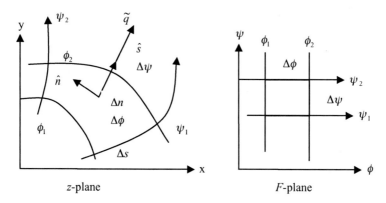

Figure 4.4   Construction of a typical curvilinear square.

through its lateral surfaces, $\Delta n \Delta z$. The average flow velocity for a cell is obtained by linear approximation of the Cauchy-Riemann relation, that is,

$$|\tilde{q}_s| = \frac{\Delta \phi}{\Delta s} = \frac{\Delta \psi}{\Delta n} = |\tilde{q}| \tag{4.19}$$

Hence, $\Delta \phi = |\tilde{q}| \Delta s; \Delta \psi = |\tilde{q}| \Delta n.$

Field maps possess unique properties useful for field visualization and analysis when they are constructed with equal intervals of $\Delta \psi$ and $\Delta \phi$. Such maps are composed of cells of the same kind since each cell has the same potential difference across it. The aspect ratio of any two contiguous sides in the complex F-plane is then equal to that of the corresponding sides in the physical plane. In particular, by choosing aspect ratio of unity, each curvilinear square is a true square in the F-plane with equal size (Figure 4.4).

$$\frac{\Delta \phi}{\Delta \psi} = \frac{\Delta s}{\Delta n} = 1$$

For a given $\Delta \phi$ or $\Delta \psi$, the sides of a curvilinear square are inversely proportional to field strength (or $\tilde{q}_s$). As field intensity changes from point to point, Δn and Δs in the physical plane varies with convergence or divergence of flux lines and equipotential lines. The density of flux lines (or equipotential lines) at a point is proportional to the magnitude of vector field at that point. They give a graphical interpretation of field strength. The distance between contour lines is a geometric measure of the steepness of surface. When flux lines (or equipotentials) are closely packed in the physical plane, field strength is strong and the surface rises or falls rapidly. Conversely, where lines are far apart the field is weak, corresponding to a flat surface. In those places with drastic changes in cell sizes, it indicates rapid changes of field strength. If flux lines are nearly parallel and equidistant from one another the field is uniform. The size of a curvilinear square on the physical plane and its image on the complex F-plane is related by:

$$\Delta\psi\,\Delta\phi = |\bar{q}|^2\,\Delta s\,\Delta n$$

The size of a curvilinear cell on a complex F-plane may bear physical significance depending on the nature of the field being considered. The above expresses the amount of kinetic energy, elastic potential energy or electric energy storage in a cell in a flow field, displacement field or electric field. Field cells of the same kind have the same energy storage regardless of their sizes in the physical plane. A curvilinear square can be scaled up or down in the physical plane as long as the ratio remains unaltered and energy storage in a curvilinear cell is invariant. When contour intervals are made sufficiently small, they tend to form a network of squares on the physical plane. Thus, a curvilinear square is characterized by having four right-angled corners. Furthermore, the diagonals of curvilinear squares intersect each other at right angles. A region formed by two adjacent flux lines is called flux tube. Thus, a field map can be used to estimate flux integrals by counting the number of flux tubes multiplied with interval $\Delta\psi$. Similarly, a region formed by two adjacent equipotential lines is called work tube. It can be used to evaluate line integrals by the number of work tubes multiplied with interval $\Delta\phi$. The tube ratio is a useful parameter in the study of field transport characteristics.

Construction of curvilinear square field maps is a trial and error process involving continuous adjustment and refinement until a satisfactory network is completed. It is usually started with a relatively large network. To improve solution accuracy, the field map is fine-tuned by going through a process of successive subdivisions to form a dense network. In this process, intervals between consecutive values of equipotentials and flux lines are kept equal and made to become sufficiently small. The method does not require analytic expressions for the field but requires practice to obtain good results. The following observations and general guidelines are useful in the construction processes:

- When starting a field map construction, it is good practice to examine the geometry and take advantage of symmetry that may exist. Lines of symmetry separate regions of similar patterns and serve as a boundary with no flux crossing.
- Choose a convenient scale to suit the field domain and identify the necessary boundary conditions whether they are equipotentials or flux lines. Where Direchlet boundary conditions are known, they serve to specify certain potentials. For instance, a conductor boundary serves as an equipotential. Where Neumann type condition is specified, equipotential lines

must be perpendicular at an impermeable boundary while flux lines must be parallel to it and assume constant. For example, an insulated boundary or an impermeable boundary serves as a flux line.

- Orthogonality fails at singular/critical points. Singular points should be identified and located since construction work breaks down there. It is also useful to locate, if any, the positions of critical points (stagnation). This is because field patterns have unique features near these points and they can be deployed to our advantage to speed up the construction process. Dividing flux lines passing through critical points is useful in separating the fields into regions. Once critical points are located and the nearby streamlines and equipotentials are drawn, it becomes a relatively easy matter to sketch the whole fields.

- If the diagonal lines of curvilinear squares in a field map are constructed then the points of intersection of these diagonals are also points on equipotential lines and streamlines. A circle could be inscribed tangent to the four sides in each square to aid the sketching process. Such practice also helps to check/verify the construction work. Accordingly, one can consider starting a freehand sketch by drawing strings of circles of varying sizes to delineate the field domain.

- Curvilinear square field map of flow fields is an approximation of actual flows. It is good practice to observe possible flow separations in real fluids. Field maps are valuable to indicate how boundaries may be streamlined.

4.5.3 Transfer Characteristics and Field Property Evaluation

4.5.3.1 Transfer Number and Shape Factor

Transfer number is introduced to assess the characteristics of a transfer process in a given field map. It is defined as:

$$D = \frac{\Delta\psi_{max}}{\Delta\phi_{max}} = \frac{N_f \Delta\psi}{N_p \Delta\phi} = \frac{N_p}{N_f} = \zeta$$

N_f and N_p are the number of flux tubes and number of work tubes bounded by the global domain configuration. ζ is tube ratio. Both contour intervals are maintained equal and constant, $\Delta\phi = \Delta\psi$.

Shape factor was traditionally introduced in conduction heat transfer study and in potential flows involving Direchlet boundary conditions specified at domain boundaries. It has analogy in other field transport phenomena of harmonic nature. Shape factor describes the geometric characteristics of a configuration and is a measure of the effectiveness of transport processes. It is defined as:

$$S = \frac{|Q|}{k \Delta\varphi_{max}}$$

k is thermal conductivity in heat conduction. Q represents total heat transfer through the configuration driven by thermal potential difference across boundaries, which are equal to $\Delta\varphi_{max} = \Delta T_{max}$ by virtue of maximum principle. Fourier's law gives the average heat flux through a cell with reference to flux line coordinate:

$$|\tilde{q}| = |\tilde{q}_s| = \left| -k\frac{\Delta\varphi}{\Delta s} \right|$$

Hence,

$$|Q| = N_f |\tilde{q}| dA = N_f |\tilde{q}| \Delta n \Delta z.$$

Δz is domain depth normal to (x, y) plane. With $\Delta n = \Delta s$, shape factor reduces to:

$$S = \frac{N_f |\tilde{q}| \Delta n \Delta z}{N_p k \Delta\varphi} = \frac{N_f \Delta z}{N_p} = \zeta \Delta z$$

Since field map is independent of field property coefficient, the shape factor follows suit. Shape factor depends on domain geometry and has dimension in length. It is related to transfer number or tube ratio when it is defined per unit depth.

$$D = \frac{S}{\Delta z} = \zeta$$

Shape factor/transfer number is an important parameter in similarity theory. They are useful indices to compare transfer process characteristics and to assess geometric effects when domain configurations vary. A change in domain geometry causes a change in shape factor. This is because shape factor is a function of geometric shape and is invariant for geometrically similar configurations regardless of its size. Field transmittance of a local cell such as capacitance, electric/thermal conductance/resistance, permeance or permeability is equal to the field medium property independent of physical dimensions. This is because the effects of a change in width cancel the influence of a change in length. The results are applicable to other field problems whenever transport analogy prevails.

In addition to the study of geometric effects, field maps have practical applications in the evaluation of field properties. Once a satisfactory field map has been developed for a given configuration, one can use it in many other ways and extract the information needed. By virtue of analogies, concepts and results can be readily translated from one field into other fields of interests. Table 4.1 presents field analogies including but not limited to electric fields, flow fields and temperature fields.

Table 4.1 Field analogies and correspondence of scalar potential.

	Electric field	Fluid flow field	Temperature field
Potential	V	ϕ	T
Complex potential	$F = V + i\psi$	$F = \phi + i\psi$	$F = T + i\psi$
Field intensity	$E = -\nabla V$	$V = \nabla\phi$	$U = -\nabla T$
Property constant	Dielectric constant ε	Fluid density ρ	Thermal conductivity k
Flux density	$D = \varepsilon E$	$m = \rho V$	$q = kU$
Flux	$Q = \int D \cdot \hat{n}\, dA = \varepsilon(\psi_2 - \psi_1)\Delta z$ $= N_f \varepsilon \Delta\psi \Delta z$	$Q = \int m \cdot \hat{n}\, dA = \rho(\psi_1 - \psi_2)\Delta z$ $= N_f \rho \Delta\psi \Delta z$	$Q = \int q \cdot \hat{n}\, dA = k(\psi_2 - \psi_1)\Delta z$ $= N_f k \Delta\psi \Delta z$
Transfer characteristic index	$S = \dfrac{|Q|}{\varepsilon \Delta V_{max}} = \dfrac{N_f \Delta\psi \Delta z}{N_p \Delta V}$ $= \zeta \Delta z = D\Delta z$	$S = \dfrac{|Q|}{\rho \Delta\phi_{max}} = \dfrac{N_f \Delta\psi \Delta z}{N_p \Delta\phi}$ $= \zeta \Delta z = D\Delta z$	$S = \dfrac{|Q|}{k\Delta T_{max}} = \dfrac{N_f \Delta\psi \Delta z}{N_p \Delta T}$ $= \zeta \Delta z = D\Delta z$
Field transmittance	$C = \dfrac{|Q|}{\Delta V_{max}} = S\varepsilon$	$H = \dfrac{|Q|}{\Delta\phi_{max}} = S\rho$	$K = \dfrac{|Q|}{\Delta T_{max}} = Sk$
Capacitance Capacitance/conductance	$\dfrac{C}{\Delta z} = \zeta\varepsilon$ $\dfrac{C_{cell}}{\Delta z} = \varepsilon$ $\dfrac{G}{C} = \dfrac{\sigma}{\varepsilon}$ $G = \dfrac{\varepsilon}{C}\Delta z = \dfrac{\sigma N_f \Delta E \Delta z}{N_p \Delta V} = S\sigma$ $\dfrac{G}{\Delta z} = \zeta\sigma$ $\dfrac{G_{cell}}{\Delta z} = \sigma$	$\dfrac{H}{\Delta z} = \zeta\rho$ $\dfrac{H_{cell}}{\Delta z} = \rho$	$\dfrac{K}{\Delta z} = \zeta k$ $\dfrac{K_{cell}}{\Delta z} = k$
Conductance Resistance	$R = \dfrac{1}{G} = \dfrac{1}{S\sigma}$	$R = \dfrac{1}{H} = \dfrac{1}{S\rho}$	$R = \dfrac{1}{K} = \dfrac{1}{Sk}$
Energy storage in a cell per unit depth	$W = \dfrac{\varepsilon}{2}\displaystyle\int\!\!\int\!\!\int_{\Delta n\, \Delta s\, \Delta z}|E|^2 dv$	$W = \dfrac{\rho}{2}\displaystyle\int\!\!\int\!\!\int_{\Delta n\, \Delta s\, \Delta z}|V|^2 dv$	$W = \rho c_p(T - T_{ref})\Delta n\Delta s\Delta z$
Energy density	$\Delta W = \dfrac{\varepsilon}{2}\displaystyle\int\!\!\int_{\Delta n\, \Delta s}|E|^2 dA = \dfrac{\varepsilon}{2}\Delta V \Delta\psi$ $K = \dfrac{\Delta W}{\Delta n\Delta s} = \dfrac{\varepsilon}{2}\dfrac{\Delta V}{\Delta s}\dfrac{\Delta\psi}{\Delta n}$	$\Delta W = \dfrac{\rho}{2}\displaystyle\int\!\!\int_{\Delta n\, \Delta s}|V|^2 dA = \dfrac{\rho}{2}\Delta\phi\Delta\psi$ $K = \dfrac{\Delta W}{\Delta n\Delta s} = \dfrac{\rho}{2}\dfrac{\Delta\phi}{\Delta s}\dfrac{\Delta\psi}{\Delta n}$	$\Delta W = \rho c_p(T - T_{ref})\Delta n\Delta s$ $K = \dfrac{\Delta W}{\Delta n\Delta s}$ $= \rho c_p(T - T_{ref})$

4.6 Example 4.1a Visualization of Heat and Fluid Transport in a Corner

This example is used to illustrate different aspects of the problem and is presented in four parts in subsequent chapters. This first part covers scalar potential function, stream function, complex potential, mechanical energy function and heatfunction. Consider an incompressible, inviscid fluid flows around a corner in Figure 4.5. The origin of a Cartesian coordinate is placed at a corner point with its axes coinciding with two orthogonal walls. Each wall has a length of L. The governing equations for a steady state system are:

Continuity equation: $\nabla \cdot V = 0$

Euler equation: $\rho(V \cdot \nabla)V + \nabla p + \rho \nabla U = 0$

Thermal energy equation: $V \cdot \nabla T = \alpha \nabla^2 T = \dfrac{k}{\rho c_p} \nabla^2 T$

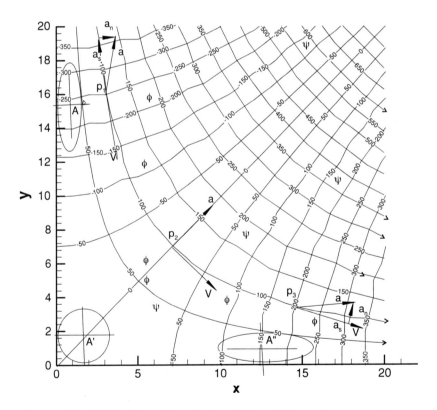

Figure 4.5 Distributions of equipotentials and streamlines on z-plane.

The following model parameters are assumed:

$$0 \leq x \leq L, \ 0 \leq y \leq L.$$

$$\rho = 1\frac{kg}{m^3}, \ c_p = 1\frac{kJ}{kg\,K}, \ k = 0.028e - 3\frac{kW}{m\,K}, \ T_0 = 100^0C,$$

$$T_{ref} = 20^0C, \ L = 20m, \ \nabla U = 0\frac{m}{s^2}$$

$$@V = 0, \ p = p_0\frac{N}{m^2}.$$

Boundary conditions: No flow is allowed to penetrate through two orthogonal walls. This requires:

$$V \cdot \hat{i}\,|_{x=0} = V \cdot \hat{j}\,|_{y=0} = 0$$

$$v_x\Big|_{x=0} = \frac{\partial \psi}{\partial y}\Big|_{x=0} = 0, \ \rightarrow \ \psi = C_1 \text{ along } y\text{-axis.}$$

$$v_y\Big|_{y=0} = -\frac{\partial \psi}{\partial x}\Big|_{y=0} = 0 \rightarrow \psi = C_2 \text{ along } x\text{-axis.}$$

ψ represents stream function. Both coordinate axes coincide with streamlines.

$$@x = 0, y = 0; \ T = T_{ref}$$
$$@x = 20, y = 20; \ T = T_0$$

4.6.1 Complex Velocity Potential

Flows around a corner could be modeled by potential flows subject to conservative forces. In particular, when there are no free surfaces in flow fields, the motion depends on boundaries and is conveniently described by complex velocity potential in a class of boundary value problems. Consider a complex potential, which defines an identity mapping from ζ-plane to F-plane such that the solution in ζ-plane is an analytic function of a complex variable.

$$F\text{-plane: } F = f(\zeta) = \zeta = \phi(\xi, \eta) + i\psi(\xi, \eta) \tag{4.20}$$
$$\zeta\text{-plane: } \zeta = \xi + i\eta$$

To apply conformal mapping, it is necessary to find an analytic function, which maps the geometric arrangement in the z-plane into that in the

ζ-plane to bring about a simplification of the problem. Suppose the ζ-plane is connected to the z-plane by the following conformal mapping defined by a power function:

$$\zeta = \zeta(z) = z^{\pi/\beta} = r^{\pi/\beta} \exp\left(i\frac{\pi}{\beta}\theta\right) \tag{4.21}$$

z-plane: $z = x + iy$

$\zeta(z)$ is an analytic mapping function, which transforms a wedge angle of β with vertex at origin in the z-plane onto a real axis of the ζ-plane, that is, the upper half ζ-plane. A line through origin is mapped from the z-plane to a line passing through origin in the ζ-plane. Such mapping is useful for boundary transformation lying on a line. By setting $\beta = \pi/2$, the wedge angle resembles a right angle corner formed by rectangular coordinate axes. The first quadrant of the z-plane is then mapped onto the upper half of the ζ-plane as the wedge angle doubles. Streamlines in the z-plane are contained inside the right angle corner in the first quadrant and are asymptotic to its two corner walls. They are mapped as straight coordinate lines parallel to the real axis on the upper ζ-plane. By substitution and equating real and imaginary parts, we obtain equipotential and stream function in the z-plane. Both satisfy the Laplace equation and are connected by the Cauchy-Riemann equation.

$$F = f(\zeta) = f[\zeta(z)] = z^2 = (x + iy)^2$$
$$\phi = x^2 - y^2 = C_1 \tag{4.22a}$$
$$\psi = 2xy = C_2 \tag{4.22b}$$
$$\left.\frac{\partial\phi}{\partial x}\right|_{x=0} = \left.\frac{\partial\phi}{\partial y}\right|_{y=0} = 0$$

Or in polar representation,

$$r^2 \cos 2\theta = C_1$$
$$r^2 \sin 2\theta = C_2$$

The complex conjugate velocity is:

$$\overline{V} = v_x - iv_y = \frac{dF}{dz} = 2z$$

$$V = \begin{bmatrix} v_x \\ v_y \end{bmatrix} = \begin{bmatrix} q_x \\ q_y \end{bmatrix} = \begin{bmatrix} 2x \\ -2y \end{bmatrix} = \nabla\phi$$

The flow fields point in the direction of increasing velocity potential. The flow fields are irrotational and satisfy the continuity equation as well as boundary conditions. Circulation does not exist since the line integral for an arbitrary closed path vanishes.

$$\omega = \nabla \times V = 0$$

Or

$$\Gamma = \oint_C V \cdot d\vec{r} = \oint_C (\nabla \times V) \cdot \hat{b} dA = 0$$

Coordinate lines on the complex F-plane representing uniform flows are mapped into flow patterns of conic curves in the z-plane. Figure 4.5 shows the distributions of equipotential contours and streamlines giving rise to an orthogonal field map. Equipotential contours exhibit a system of hyperbola with center at origin and foci at $(\sqrt{2C_1}, 0)$ and vertices at $(\sqrt{C_1}, 0)$ for $C_1 > 0$. It has a conjugate hyperbola $(C_1 < 0)$ with foci at $(0, \sqrt{2C_1})$ and vertices at $(0, \sqrt{C_1})$. Equipotential contours terminate perpendicularly on two corner walls. The contours in the first quadrant are antisymmetric with respect to line $y = x$, which also serves as an oblique asymptote. Streamlines resemble a system of rectangular hyperbola with center at origin and transverse axis coincides with line $y = x$, which is also a line of symmetry in the first quadrant. A pair of streamline asymptotes coincides with coordinate axes. They represent dividing streamlines $\psi = 0$ passing through the critical point (origin) and separate flow fields into first quadrant. Fluids flow along dividing streamline (y-axis) towards the critical point where they change their course of motions abruptly and thereon follow x-axis in the positive direction. The converging/diverging/converging streamlines at upstream and downstream implies high velocity. Upstream fluid velocity starts to decrease until it crosses line $y = x$ after which it increases again towards downstream. To substantiate the above observations with physics of the fields, we display both velocity V and acceleration vector a at a point. Tangents represented by arrow icons at points p_1, p_2, p_3 along a streamline depict the velocity directions. From Equation 4.14a or 4.14b, the acceleration vector is:

$$a = (2x + i2y)2 = 4x + i4y = 4x\hat{i} + 4y\hat{j}$$

The acceleration vectors (arrow icons) at these locations correspond to the tangents of the hodograph at the image points in Figure 5.11 and are mapped onto Figure 4.5. They are displayed at points p_1 to p_3 together with the velocity vectors and are decomposed into streamline tangent and normal directions. The normal component of acceleration vectors always point towards the centers of streamline curvature. It indicates velocity direction changes as motion veers to the left side of streamline (counter-clockwise) on physical plane. At point p_1, the tangent component of acceleration points in the opposite direction of velocity indicating flow is decelerating. Acceleration vector is perpendicular to velocity at point p_2, which lies on $\phi = 0$ equipotential. Speed assumes minimum and constant along equipotential $\phi = 0$, which also serves as an isoclines (part 4.1b). The tangential component at point p_3 points in the same direction of velocity and hence flow is accelerating. The same findings can be obtained by analytical means in part 4.1b. The radius of streamline curvature is determined from Equation 2.41. The radius of curvature decreases towards the critical point and approaches zero as the limiting value. As a result, velocity directions change rapidly in the region close to the critical point. This observation is also reflected in Figure 4.7 or 4.8 by the constant flow direction contours (isoclines) converging towards the corner.

$$R_s = \frac{(x^2 + y^2)^{3/2}}{2xy}$$

Symmetry test shows that the distribution of radius of curvature of streamline exhibits line-symmetry about line $y = x$ in the field domain. The centers of curvature lie on the left sides of streamlines implying that the radii of curvature are positive by convention as flow advances downstream and veers to the left. The change of field potential (or energy per unit mass flow rate) between any two equipotential lines is:

$$\Delta\phi_{12} = \phi_2 - \phi_1$$

Volumetric flow rate per unit depth between two streamlines can be easily determined as the difference in streamline values.

$$\frac{Q}{\Delta z} = \psi_1 - \psi_2$$

The transfer number for the corner section is:

$$D = \frac{\Delta\psi_{max}}{\Delta\phi_{max}} = \frac{800 - 0}{400 - (-400)} = 1$$

The magnification factor and twist angle are:

$$\left| f'(z) \right| = |2z| = 2\sqrt{x^2 + y^2}$$

$$= \sqrt{\left(\frac{\partial\phi}{\partial x} \right)^2 + \left(-\frac{\partial\psi}{\partial x} \right)^2} = |V| \qquad (4.23a)$$

$$\arg\left| f'(z) \right| = \tan^{-1}\left(\frac{\frac{\partial\psi}{\partial x}}{\frac{\partial\phi}{\partial x}} \right) = \tan^{-1}\left(\frac{y}{x} \right) = -\alpha$$

Magnification factor represents field strength (or momentum density per unit mass) and is depicted in Figure 4.6. Distributions of flow field strength contours resemble a system of concentric circles with its center at origin. Field strength increases linearly with distance from origin and corresponds to the decreasing size of curvilinear square in the z-plane (Figure 4.5). The velocity direction could be determined by twist angle reflected about the x-axis.

To supplement the above views, the distributions of kinetic energy are presented in Figure 4.7 given by:

$$K = \frac{|V|^2}{2} = 2(x^2 + y^2) \qquad (4.23b)$$

$$x^2 + y^2 = \left(\sqrt{\frac{K}{2}}\right)^2$$

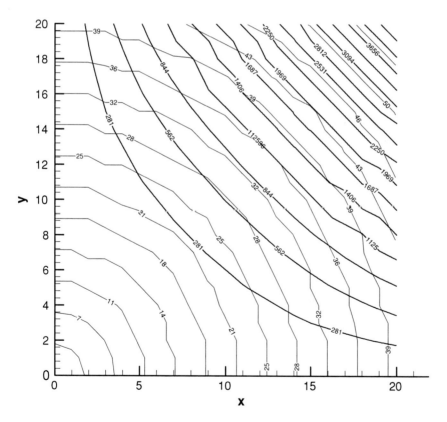

Figure 4.6 Distribution of momentum density (line) and heat flux magnitude contours (dotted).

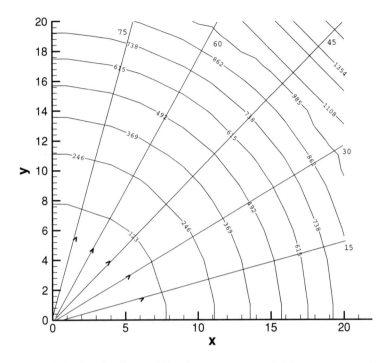

Figure 4.7 Distributions of kinetic energy and twist angle (arrow).

It is observed that distributions of kinetic energy contour, known as isotachs also form concentric circles about origin. A circle of radius $\sqrt{2K}$ on the hodograph plane represents a line of constant kinetic energy K which is mapped to a circle of $\sqrt{\frac{K}{2}}$ on the physical plane. Gradient mapping of kinetic energy is a system of radial vectors from origin giving maximum rate of increase of kinetic energy. Figure 4.7 shows the distributions of twist angle contours, which depict a mirror image of field directions reflected about x-axis. There is a rapid change of flow directions in the corner region as the contours converge towards the origin. The pressure field is obtained from Bernoulli's theorem.

$$\frac{p}{\rho} = \left(B - \frac{V^2}{2} \right) = \frac{p_0}{\rho} - 2(x^2 + y^2)$$

Or

$$v_x^2 + v_y^2 = \frac{2}{\rho}(p_0 - p)$$

Pressure contours are also mapped into concentric circles centered at origin where maximum pressure occurs. The contours are symmetric about origin. Similarly, gradient mapping of pressure gives a system of radial vectors passing through origin. The isobars are also mapped to circles about origin on hodograph plane. Lamb vectors vanish in irrotational flows.

4.6.2 Tangent Fields and Isoclines

The tangent field method is presented to visualize the velocity fields by making use of the following trajectory equations:

$$\text{Streamline trajectory: } \frac{dy}{dx}\Big|_{\psi=C_2} = -\frac{\dfrac{\partial \psi}{\partial x}}{\dfrac{\partial \psi}{\partial y}} = -\frac{y}{x}$$

$$\text{Equipotential trajectory: } \frac{dy}{dx}\Big|_{\phi=C_1} = -\frac{\dfrac{\partial \phi}{\partial x}}{\dfrac{\partial \phi}{\partial y}} = \frac{x}{y}$$

A sufficient number of lineal elements is evaluated at the grid points from the above equations to cover the field domain. Whenever there is a singular/ critical point, it will be skipped and excluded. Figure 4.8 shows the distributions of lineal elements for the streamline trajectories. Accuracy of the method is fair when compared with exact solutions. Similarly, equipotential trajectories can be traced.

The method of isoclines is also illustrated in Figure 4.8. In this method, curves having a constant slope are drawn in. They appear as a system of straight lines passing through origin. Hence, at every point on these lines, line segments have the same slope corresponding to vector field direction. Starting at a point, proceed in the direction corresponding to the isoclines through that point to the neighboring isoclines, and change direction as and when necessary. Repeat the procedures until a sufficient number of curves are traced. As the isoclines converge towards corner, the flow fields change directions rapidly in the region close to the critical point.

4.6.3 Energy Transfer Trajectories

4.6.3.1 Mechanical Energy Function

Mechanical energy function and stream function are within a proportional constant in irrotational flows. Therefore, mechanical energy trajectories and streamlines share the same patterns in Figure 4.5. Mechanical energy transfer per unit depth per unit time bounded between two flux trajectories is given

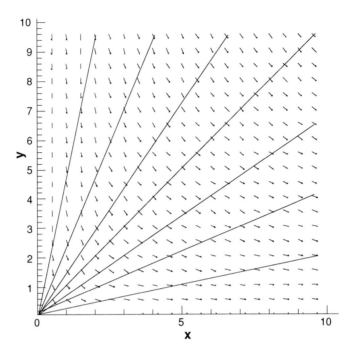

Figure 4.8 Graphical solutions using direction fields (lineal elements) and iso-clines.

by Equation 3.32. Mechanical energy transfer direction coincides with fluid velocity everywhere.

4.6.3.2 Heatfunction

In view of the symmetry with respect to origin, a product function of an odd type in both x and y is constructed to represent temperature field solution, that is,

$$T = c_1 + c_2 xy$$
$$c_1 = T_{ref}, \quad c_2 = \frac{T_0 - T_{ref}}{L^2}$$

The energy governing equation and boundary conditions are satisfied by substitution. The heatfunction introduced by Kimura & Bejan (1983, Bejan

1995) is defined in terms of heat fluxes in x and y-directions. It is obtained by integration.

$$\frac{\partial}{\partial x}\left[\rho c_p v_x(T - T_{ref}) - k\frac{\partial T}{\partial x}\right] + \frac{\partial}{\partial y}\left[\rho c_p v_y(T - T_{ref}) - k\frac{\partial T}{\partial y}\right] = 0$$

$$q_x^* = \left[\rho c_p v_x(T - T_{ref}) - k\frac{\partial T}{\partial x}\right] = \frac{\partial \psi_t}{\partial y}$$

$$q_y^* = \left[\rho c_p v_y(T - T_{ref}) - k\frac{\partial T}{\partial y}\right] = -\frac{\partial \psi_t}{\partial x}$$

$$\psi_t = \rho c_p c_2 x^2 y^2 + \frac{k c_2(x^2 - y^2)}{2} \qquad (4.24)$$

$$\frac{\partial \psi_t}{\partial x}\Big|_{x=0} = \frac{\partial \psi_t}{\partial y}\Big|_{y=0} = 0$$

The first term in the above expression of heatfunction is attributed to convection while the second term is due to conduction. Distributions of temperature (isotherms) and heat flux trajectories are presented in Figure 4.9. The isotherms patterns are similar to those of streamlines showing that the temperature field is intimately affected by flow field in convection heat transfer. The transport processes could be interpreted by displaying the heat flux vector, mechanical energy flux vector and velocity vector at a point. Heat flux vectors are aligned closely in the velocity directions, which coincide with mechanical energy flux vectors indicating energy (thermal and mechanical) transfer is dominated by convection. Conduction heat flux trajectories resemble those of velocity equipotential as heat conducts from outer boundaries towards two corner walls hyperbolically. Conduction is however confined along wall surfaces locally and is given by:

$$\text{Vertical wall } (x = 0): q_x^* = -k\frac{\partial T}{\partial x} = -k c_2 y \qquad (4.25)$$

$$\text{Horizontal wall } (y = 0): q_y^* = -k\frac{\partial T}{\partial y} = -k c_2 x \qquad (4.26)$$

The dividing heat flux trajectory exists only at origin (critical point), $\psi_t(0, 0) = 0$. On the other hand, convection heat flux trajectories resemble families of high order curves similar to flow fields. The convective mode of heat transfer dominates conduction outside wall boundaries. The converging/diverging/converging heat flux lines at upstream and downstream imply heat transfer intensifies at upstream and downstream due to high velocity. As upstream velocity continues to decrease and reaches its minimum at line $y = x$, it causes heat flux lines to diverge towards there. Velocity then

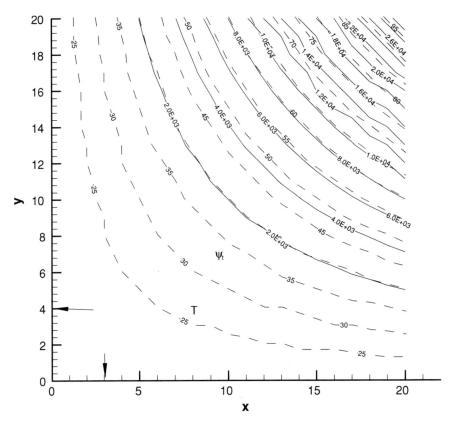

Figure 4.9 Distributions of heat flux trajectories (line) and isotherms (dashed).

starts to increase and heat flux lines converge downstream. The temperature field is rotational as convection fails to vanish. Distributions of heat flux magnitude contours are presented in Figure 4.6. Heat flux vanishes at the corner (critical point) and its magnitude increases away from this point. It assumes maximum at right top corner.

Heat flux trajectories directly depict heat energy transfer paths. The density of heat flux lines drawn at constant contour intervals can be used to visualize heat flux magnitudes similar to streamlines. When they squeeze, it implies transfer process is intensifying with increasing heat flux magnitude. The heat transfer rate per unit depth per unit time between two heat flux lines is given by the difference of heat flux line values.

To visualize interaction between convection mode and conduction mode of heat transfer, we evaluate *RAI* index (Phillips, 1991) below. It can be shown that conduction mode of heat transfer crosses isotherms orthogonally and is everywhere perpendicular to convection mode heat transfer.

$$RAI = V \cdot \nabla T = |V||\nabla T| \cos \gamma = 0$$

Thus,
$$\gamma = \frac{\pi}{2}.$$

4.6.4 Visualization of Flow Field Structure Effects

The fluid system is in global anticlockwise rotations along curved streamlines. To investigate the effects of flow field structure on fluid behaviors in local regions, we call upon Frenet's formula and equation of continuity with reference to streamline coordinates:

$$\nabla \cdot V = \left(\hat{s} \frac{\partial}{\partial s} + \hat{n} \frac{\partial}{\partial n} \right) \cdot (v_s \hat{s} + v_n \hat{n}) = 0$$

$$\rightarrow \frac{\partial v_s}{\partial s} + \frac{v_s}{R_n} = 0$$

In the region bounded by y-axis and line $y = x$, the center of curvature of normal trajectory is located on the left side of it such that the radius of curvature is positive by convention, $R_n > 0$. This implies $\frac{\partial v_s}{\partial s}$ is negative and there is a decrease in velocity as flows advance. When flows arrive at line $y = x$, the normal trajectory becomes a straight line as the radius of curvature approaches infinite. Velocity then assumes local minimum proportional to radius from origin.

$$R_n = \infty; \rightarrow v_s = v_s(n) = \sqrt{v_x^2 + v_y^2} = 2r$$

$$\frac{\partial \alpha}{\partial n} = 0; \rightarrow \alpha = \alpha(s) = \tan^{-1} \left(\frac{v_y}{v_s} \right) |_{y=x} = \tan^{-1}(-1) = -\frac{\pi}{4}$$

In the region bounded by x-axis and line $y = x$, the center of curvature of normal trajectory is located on the right side of it such that the radius of curvature is negative, $R_n < 0$. This means $\frac{\partial v_s}{\partial s}$ is positive and there is an increase in velocity as flows advance downstream.

The equation of motion in streamline normal direction shows that pressure gradient is set up to maintain force balance with centripetal acceleration. For irrotational flows, the velocity gradient is required to satisfy the following condition:

$$\frac{\partial p}{\rho \partial n} = -\frac{v_s^2}{R_s} = -v_s \frac{\partial v_s}{\partial n}; \ \omega_z = 0$$

In the first quadrant, streamline curvature is positive when the center of curvature is located on the left side of streamline. There will be a pressure reduction and velocity increase when moving across curved streamlines towards the center of curvature of streamlines. If streamline is straight such as the walls, then pressure and velocity change in the direction of flow according to Bernoulli's theorem. Velocity is proportional to radius from origin while pressure does the opposite. The streamline tangent angle remains invariant.

$$R_s = \infty; \ \to \ v_s = v_s(s) = \sqrt{v_x^2 + v_y^2} = 2r, \ p = p(s)$$

$$B = \frac{p}{\rho} + \frac{V^2}{2} = C$$

$$\frac{\partial \alpha}{\partial s} = \frac{1}{R_s} = 0$$

Vertical wall: $\alpha = \tan^{-1}\left(\frac{v_y}{v_x}\right)|_{x=0} = -\pi/2$

Horizontal wall: $\alpha = \tan^{-1}\left(\frac{v_y}{v_x}\right)|_{y=0} = 0$

The rate of strain tensor is:

$$\dot{\gamma}_{ij} = \frac{1}{2}\begin{bmatrix} 2\dfrac{\partial v_x}{\partial x} & \dfrac{\partial v_x}{\partial y} + \dfrac{\partial v_y}{\partial x} \\ \dfrac{\partial v_x}{\partial y} + \dfrac{\partial v_y}{\partial x} & 2\dfrac{\partial v_y}{\partial y} \end{bmatrix} = \begin{bmatrix} 2 & 0 \\ 0 & -2 \end{bmatrix}$$

With vanishing off-diagonal components and vanishing trace in the rate of strain tensor, there is no shear deformation and no dilation distortion. The flows in a corner undergo circulatory motions along curved streamlines asymptotic to the orthogonal walls in a state of pure normal strain. At any point, elongation in one coordinate direction is accompanied by contraction of the same in the other coordinate direction. To visualize motions under pure normal strain, Figure 4.5 shows a plane fluid element of conic shape subject to elongation and compression in x-direction and y-direction respectively. The orientation of a local rectangular coordinate attached to the fluid element will be unaltered since there is no vorticity. The shape does not change either due to vanishing shear strain. The fluid element moves along walls from an initial elliptic element A to circle element A' and then elliptic element A'' continuously all in conic shape of same size everywhere, that is, $A = A\prime = A''$. The action of pure normal strain causes the ellipse major axis to become minor axis and vice versa as the fluid element travels from A to A''.

5

Field Mapping and Applications

5.1 Introduction

The general characteristics of a tensor field are that its components depend on the coordinates used. Tensor components change according to a set of rules when there is a change of coordinate systems. However, they represent the same physical phenomena in spite of coordinate systems. The concept of mapping between coordinate systems adds a new dimension to the graphical representation of field phenomena. It extends our visual capability by viewing and studying field structure in different settings. Field patterns may exhibit unique features easily recognized or analyzed conveniently in certain coordinates than others. Information may be hidden implicitly in some views but become explicit in others. In other circumstances, quantitative and accurate information can be obtained in one view while others require interpolation estimates. Mapping could change the field characteristics. Gradient mapping takes a point in the domain of a scalar field onto a gradient vector pointing in the direction of maximum spatial rate of property change at that point. Mapping can be used to emphasize specific details or relocate certain undesirable features/effects. It is a critical step of linking physical quantities to visual primitives. The objective of field mapping is precisely the graphical representation of field structure with analysis capacity in quantitative terms. It is essential to have knowledge about the properties of different transformations so that appropriate mappings could be applied to achieve the objectives or to explore new information. Certain properties that incur loss/retain or remain invariant in a transformation are to be observed.

Visualization of Fields and Applications in Engineering, First Edition. Stephen Tou.
© 2011 John Wiley & Sons, Ltd. Published 2011 by John Wiley & Sons, Ltd.

Geometric transformation has practical applications in arts and sciences from design, cartography and engineering to computer graphics/visions. A vivid example is the mapping of electromagnetic fields in television/radio (Needham, 1997). In this respect, mapping theory is indispensible and has provided us with the necessary tools to carry out transformation in a specific manner.

Deformation/motion in Euclidean space is of great concern and can be expressed by geometric transformation. Coordinate transformation is a geometric mapping process whereby a set of rules assigns a unique member of another. We need to determine the coordinate of a point or components of a tensor in a coordinate that undergoes a transformation with respect to a given coordinate. This will enable us to describe correspondence between various figures/patterns before and after deformation/motion. Affine mapping is commonly applied to translate, rotate or stretch/compress a given object or pattern in Euclidian geometry with co-linearity preservation. Affine reflection maps points not on the line of reflection to the opposite side of it. In non-linear transformation, we could produce twisting, bending or distortion effects in a class of curvilinear mapping. In particular, conformal mapping has the property of preservation of angle and local shapes. It facilitates field visualization and study of complex boundaries using simple ones. When geometries/objects are connected by analytical mapping functions, field patterns can be obtained by conformal mapping from one complex plane to another. Information about field structure could be derived through differentiation/integration of complex analytic functions interpreted as a geometric mapping process. By extending mapping into non-Euclidean geometry, we could explore new information from old.

To visualize the nature and action of mapping processes, it is useful to consider the manner in which the functions map certain points, curves or regions of practical interest. We are interested in mapping certain specific information such as flux lines, equipotentials in general and boundaries/critical/singular points in particular. Singular points, which generally cause problems in solutions, are mapped to points at infinity on some coordinate planes making graphical solution of shape factor possible. Field sources can be removed by mapping the original domain to a model domain with modifications of boundaries and/or variables to render the field solenoidal or irrotational. Field patterns or solutions could be obtained much more easily analytically or graphically on the model domain which is then mapped back to the original by the same mapping relationships. We are also interested in hodograph mapping and representations, which have practical applications in visualization of field characteristics and behaviors. Hodograph representation of trajectories is useful for complimenting the position hodograph on a physical plane, and together they provide complete information about the fields. It allows us to trace the shape of free surface or to obtain field patterns from hodograph plane to physical plane and conversely. We could determine the state of motion by displaying/decomposing the acceleration vector (or tangent of streamline hodograph representation) and velocity (or streamline tangent) at a point on physical plane. We could also visualize the locus of constant vector magnitude or constant direction of vector fields (isoclines) by mapping circles or lines about origin in hodograph plane onto physical plane. The advantage of mapping is that no new or additional field solutions are required except the mapping functions to meet with the prescribed objectives.

5.2 Mapping of Euclidean Geometry

A tensor field is said to be invariant with respect to coordinates if it is reducible from one to another by coordinate transformation. The continuity theorem of composite functions and transformation law of scalars establish the following mapping except at singular points.

$$\Omega(x, y, z) = \Omega[x(x^*, y^*, z^*), y(x^*, y^*, z^*), z(x^*, y^*, z^*)] = \Omega(x^*, y^*, z^*)$$
$$x = x(x^*, y^*, z^*)$$
$$y = y(x^*, y^*, z^*)$$
$$z = z(x^*, y^*, z^*)$$

For one to one correspondence, the inverse mapping takes the image back to the original,

$$x^* = x^*(x, y, z)$$
$$y^* = y^*(x, y, z)$$
$$z^* = z^*(x, y, z)$$

Euclidean geometry has exactly the invariant properties. Proof that contour curves in one coordinate mapped into the same curves in another coordinate follow an observation that along a specific curve identified by a scalar constant is the same for the corresponding image. In composite mapping, a set of independent variables may be thought of as specifying the coordinates of a point in a reference frame. It can be regarded as different coordinates of the same point. The results yield an equivalent model in the chosen coordinate system serving a particular purpose.

5.2.1 Congruent Mapping

Congruent mapping transforms a set of points or an object (original) into another set or object (image) in Euclidean space in an isometry that consists of a single or a combination of rigid-body translation, rotation and/or reflection. The basic property of an isometry is distance-preserving, meaning that distance between any two points will remain invariant before and after transformation. Metric properties such as angle, size and shape are also preserved while orientation/direction may reverse in reflection mappings. The original and image objects are said to be congruent if they have the same shape and size regardless of their position or orientation. Identity mapping is a class of congruent transformation which produces an image copy of the original.

Symmetry is an underlying concept of certain transformation. Symmetry mapping is also a class of congruent transformation that maps a geometric object onto itself. Symmetry calls attention and interest to themselves, not only through patterns in nature, but also a long history of study in geometry. We are keen to search for symmetry in an attempt to get a better understanding of field properties and behaviors, in particular, the interplay between symmetries of field pattern, domain geometry and/or boundary condition. When symmetry exists, we may minimize the visualization effort or field domain. The concept of symmetry evoked intuitively from shapes and patterns has lead to important development and applications in the theory of geometric transformation. The theory establishes quantitative relations between science and the arts of symmetry in terms of mathematical functions with specific properties. The idea of symmetry and mapping goes beyond traditional boundaries. In science, symmetry is more than just the properties of geometric figures, it has the beauty to connect with physical laws like angular momentum.

The intrinsic property of an object (or function) remains the same when plane/line-symmetry or point-symmetry exists. The property of symmetry will be useful for searching and testing symmetry. Symmetry is closely related to reflection mapping in such a way that line-symmetry is the line of reflection, which is the perpendicular bisector of the line segment joining each pairing of points. An object on one side of the line of reflection is a mirror reflection image on the other side of the line at equal distance from the line of symmetry. Reflection mapping could be applied by choosing a line of reflection and by taking each point to a point directly opposite on the other side of the line. Point-symmetry being a form of central inversion coincides with the center of inversion. The reflection of a point about the point of symmetry is a self inverse point such that the point of symmetry is the mid point for every line segment joining the pair of inverse points. Field behaviors and patterns may exhibit one or more types of symmetry. They could have more than one line of symmetry.

Symmetry test helps to determine the possible symmetries of a graph/pattern and could be carried out conveniently for implicit functions. To observe the property of symmetry of graphs of a function, we apply reflection mapping, which takes place about a line or a point through a plane or in space. The property of reflection is that it preserves distance but reverses direction/orientation. To test symmetry for implicitly defined functions in two variables, we replace x with $-x$ and/or y with $-y$ systematically and to observe the invariance transformation property of the functions. A mapping is said to be invariance under a point transformation carrying each point into a new point of the same form in both coordinates. A plane object will have line-symmetry about the y-axis or x-axis under reflection mapping when it folds $180°$ about the coordinate axis of symmetry in space each corresponding point covers each other exactly. Such that,

$$\text{Symmetry with respect to } y - \text{axis}: f(x, y) = f(-x, y)$$
$$\text{Symmetry with respect to } x - \text{axis}: f(x, y) = f(x, -y)$$

A function is classified as symmetric function when it possesses the property of symmetry by permutation of variables in exactly the same identity. Line-symmetry will exist about line $y = x$ when,

$$f(x, y) = f(y, x)$$

Similarly, the graph of a function will have line-symmetry about line $y = -x$ in exactly the same identity when,

$$f(x, y) = f(-y, -x)$$

A graph is said to have point-symmetry with respect to origin when,

$$f(x, y) = f(-x, -y)$$

Point-symmetry corresponds to a form of inversion mapping when the point of symmetry coincides with the center of inversion. The original and image points cover each other exactly but in reverse orientation. Rotation symmetry is an isometry that may exist in many forms depending on the angle and direction of rotation. For example, the following rotation mapping will turn a figure about the origin anticlockwise or clockwise through a right angle:

$$f(x, y) = f(-y, x)$$
$$f(x, y) = f(y, -x)$$

When mapping introduces a change in the sign of a function, it is classified as antisymmetry.

5.2.2 Similitude Mapping

To expand symmetry transformation to some more general transformation involving the action of dilation/shear, we call upon similitude or similarity mapping. Dilation transformation comes with two types of stretching/ shrinking actions: Isotropic dilation is a scaling transformation with equal scaling factor in two mutually orthogonal directions such that angle is preserved but not distance. It allows us to stretch/squeeze an object by changing its size. The transformation takes one figure to another of similar shape and plays an impotent role in topology. Similitude has practical applications in the theory of flow similitude for geometric similar configurations in hydrodynamics. When scaling factors vary with directions, the transformation is said to be anisotropic resulting in distorted figures.

5.2.3 Affine Mapping

To visualize the above transformation from the theory of linear transformation, we consider a general affine mapping in matrix form between two rectangular coordinate systems in two dimensions, that is,

$$\begin{bmatrix} x^* \\ y^* \end{bmatrix} = [B][A][T(\theta)]^{-1} \begin{bmatrix} x \\ y \end{bmatrix} + \begin{bmatrix} x_0 \\ y_0 \end{bmatrix} \tag{5.1}$$

The algebraic operations compose of a multiple sequential transformation, which offers a compact and powerful tool for complex geometric transformation in multi-dimensions. When the order of sequential transformation is altered, it may lead to different results and must be observed. The operation could be interpreted as the mapping of geometric vectors. In the above expression, translation is represented by the displacement vector in the last term. Each point will move the same distance equal to the displacement vector in the same direction. It is then rotated about coordinate origin through an angle θ. When the rotation angle equals to principal angle, the principal axes are aligned to coincide with the coordinate axes in an important class of intrinsic representation obtained from principal state transformation.

The rotation matrix has the property of preserving orientation.

$$\det |T(\theta)| = \det |T(\theta)^{-1}| = 1$$

Reflection mapping in a plane is produced by matrix A defined as:

$$[A] = \begin{bmatrix} \pm 1 & 0 \\ 0 & \mp 1 \end{bmatrix}$$
$$\det |A| = -1$$

Line-reflection will take place about x-axis or y-axis when the diagonal elements of unity have opposite sign. There is a reverse in orientation.

Point-reflection about origin is a form of central inversion and could be accomplished with both diagonal elements set equal to negative unity. Such transformation preserves orientation.

$$[A] = \begin{bmatrix} -1 & 0 \\ 0 & -1 \end{bmatrix}$$
$$\det|A| = 1$$

To produce zooming effects or distorted figures, scaling/shear transformation is introduced. The scaling factors along mutually orthogonal directions are specified by the diagonal elements of dilation/shear transformation matrix:

$$[B] = \begin{bmatrix} k_x & 0 \\ 0 & k_y \end{bmatrix} + \begin{bmatrix} 0 & k_{xy} \\ k_{yx} & 0 \end{bmatrix}$$

where

$$k_{xy} = k_{yx} = 0$$

Dilation mapping preserves parallel lines while size changes. The amount of stretching or contraction is controlled by the scaling factors, which may or may not vary with directions. With $k_x = k_y$, isotropic dilation is a form of similarity transformation that preserves shape and angles but not distance. Transformation is congruent when $k_x = k_y = \det|B| = 1$. For $k_x \neq k_y$, anisotropic dilation produces distortion. It could be applied to map an anisotropic domain onto an isotropic one as illustrated in Example 8.4. In the event that the off-diagonal elements are non-zero, shear deformation takes place resulting in change of angle and shape. Transformation in this case is not conformal unless the deviatorial part of matrix B vanishes.

5.3 Inversion Mapping

To extend mapping beyond Euclidean geometry, inversion mapping is introduced. Circle inversion is a transformation in Euclidean geometry but has connections to non-Euclidean geometry. The mapping operation involves inversion of a point, line or circle with respect to a given circle. This is an exotic, non-linear transformation that does the opposite action or even turns lines into circles and vice versa. This unique property makes it possible to transform linear motion into curvilinear motion and conversely. Inversion mapping opens the door to the world of exploration and makes connections to sources of new information from old. Inversion mapping has the basic invariance of property that angle magnitude is preserved with reverse direction in a class of conformal mapping. Its applications include the arts of pattern design, inverse design, map making and crystal structure study and so on. Inversion in complex variables is the building block of Joukowski transformation when combined with identity mapping. This part will be presented in the next section and an example of image warping in Example 4.1b.

5.3.1 Circle Inversion

Inversion in the plane can be visualized by a cross section of a sphere in three-dimensional graphs that pass through the center of the sphere, which is the point of intersection of all planes. Consider a circle of radius c centered at origin 0 referred to as a circle of inversion and center of inversion respectively. The center of inversion is to ensure a unique image point for each inverse point in the inversion plane in one to one correspondence. The center of inversion, the inverse point and its image point are collinear lying on the same radial line. Figure 5.2a shows the inversion of a point p at a distance r defines its image point p' lying on the same line at a distance $0p'$ relative to origin. An inversion relation is established as follows:

$$\rightarrow r(0p') = c^2$$

c^2 is the geometric mean of the distances. p and p' are measured in the same direction from origin. They are a pair of inverse points with respect to the inversion circle with center at origin. The lengths of position vectors $0p = r$ and $0p'$ are inversely proportional to each other. Points $r < c$ closer to the origin are mapped further away from it (stretching) and conversely (compression). Each point on the circle is mapped to itself being its own inverse point (congruent). The center of inversion and point at infinity are mutually inverse with respect to the inversion circle. Inversion at a point is also called reciprocal radii transformation.

Inversion in real numbers can be extended to complex numbers. Inversion at a complex point is considered a special case of circle inversion in the complex plane. The inversion of a complex point z in the z-plane with respect to circle of inversion c is defined by the following complex function, which fixes its image complex point ζ, such that

$$\zeta = \zeta(z) = \frac{c^2}{z} = \frac{c^2}{r} \exp(-i\theta)$$
$$\zeta - \text{plane} : \zeta = \xi + i\eta \tag{5.2}$$

Above defines inversion in a circle. The invariance property of a circle is that all points have a fixed distance from the center. Figure 5.2b shows that a point p lays on the ray from origin at a distance r is mapped to a point p'' on a ray reflected about the real axis at a distance $\frac{c^2}{r}$ from origin. Thus, points interior to the circle of radius $r = c$ about origin are mapped onto the exterior of the circle about origin of ζ-plane and conversely. It maps circles not through origin onto circles not through origin in reverse sense of direction. From z-plane, vertical/horizontal coordinate axes are mapped into horizontal/vertical axes. Any coordinate lines not passing through origin are mapped into circles through origin in the ζ-plane. Circles corresponding to a pair of rectangular coordinate lines are orthogonal circles. By setting $\theta = 0$, it reduces to reciprocal radii transformation in a form of dilation mapping, that is,

$$\rightarrow |\zeta| = \frac{c^2}{r}$$

By setting $\theta = \pi$ and $r = c$, we have then

$$|\zeta| = -r$$

It corresponds to central inversion as a special form of point-symmetry in reflection mapping. The inverse point is mapped in opposite radial direction to a mirror reflection image with respect to origin.

5.4 Mapping with Complex Functions

When a field is resolved into components with respect to a chosen coordinate system, vector algebraic and vector analysis become ineffective to deal with these scalar quantities and are ready to be taken over by the theory of complex functions and mapping. Mapping processes may invoke the use of complex planes on which complex variables take on physical z-plane and complex functions on complex F-plane representing uniform fields. With the employment of two complex planes (z and F) in a complex function, one overcomes the problem of presenting a pair of complex numbers on four dimensions. Complex planes are linked together by mapping functions.

In a mathematical context, mapping with complex functions is a form of coordinate transformation with a change of variables between complex planes. Mapping functions play a central role in the generation of families of curvilinear coordinates and provide an effective means of extending the totality of coordinate systems. In particular, analytic complex mapping functions offer a simple method of generating orthogonal curvilinear coordinate systems by conformal mapping in two-dimensions. A complex potential transformed by a conformal map from one plane will yield the corresponding complex potential in another plane. The nature of harmonic functions is invariant and both field map contours remain orthogonal. Analytic complex potentials are differentiable and are convenient to produce information about the vector fields on conjugate/hodograph plane. Synthesizing of fields can be accomplished by a combination of complex potentials to yield target fields of complexity. These unique features and properties facilitate field visualization/analysis greatly as we could study fields through synthesizing and mapping.

Consider an analytic complex function (Equation 4.1) in a simply connected region on z-plane. A complex potential represents a Cauchy–Riemann system in terms of two implicit functions. They assume a pair of harmonic conjugate functions and map a set of complex variables into real variables as shown in Figure 5.1. Complex potential describes field phenomena in terms of a numeric pair of potentials (ϕ, ψ) on F-plane corresponding to an image pair of (x, y) on physical plane. By choosing different complex potential, we obtain different field patterns directly. We are particularly interested in patterns that may bear physical significance or represent solutions of a field problem.

To investigate the specific properties of a mapping process, we may consider points, curves or regions defined in z-plane where physical phenomena occur. Interpretation of field behavior and structure is usually based on this plane as the reference source. By simply equating the pair of harmonic real functions to separate constants to assume the role of parameters, contour procedure is invoked to evaluate these constants from a set of grid points on z-plane and presents each of them by a unique level curve, that is,

$$\phi = \phi(x, y) = C_1 \tag{5.3a}$$
$$\psi = \psi(x, y) = C_2 \tag{5.3b}$$

The above then maps systems of rectangular coordinate lines from F-plane to z-plane as the parametric constants vary and conversely. The use of level curves at equal intervals is particularly fruitful and it will strengthen our visual conception of mapping with an insight view of the fields. The distribution contours are preferably drawn at equal intervals for both potentials,

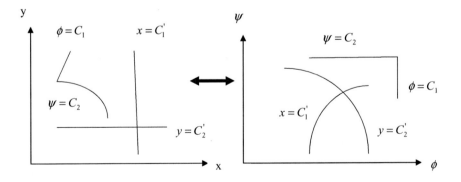

Figure 5.1 Mapping between z-plane and complex F-plane

that is, $\Delta\phi = \Delta\psi$ to result in a network of true squares of the same kind on F-plane. The size of the curvilinear cells in this plane may bear physical significance depending on the nature of the fields.

5.5 Conformal Mapping and Applications

The theory of functions of complex variables is closely related to harmonic fields and has connection with the invariance property of Laplace equation under conformal transformation in two-dimensions. Conformal mapping has important applications in arts/design and physical sciences including but not limited to, hydrodynamics, heat transfer, elasticity, electromagnetic fields and and so on. The main utility of conformal mapping in the context of visualization is that boundary value problems involving the solution of Laplace equation in one plane remains the solution in another plane when complex planes are related by analytic mapping functions. In this respect, conformal mapping is useful to derive unknown vector fields around complex shapes by means of a complex potential from the known vector fields around simple shapes. By the same token, we could study the effects of domain geometry by mapping complex potential from one geometry to another with the introduction of a complex ζ-plane. When complex potential is defined on this plane, we transform it onto another, say z-plane through mapping function $z = z(\zeta)$. The mapping function defines a geometric connection between the shape of a curve/object on ζ-plane and z-plane. This then opens the window of exploration for new complex potentials with a great variety of analytic mapping functions. There is a remarkable source of solutions (field patterns) coming in a pair at a time as the real and imaginary parts of any $f(z)$. Conformal mapping generally transforms continuous lines to continuous lines, circles to circles and curvilinear network to curvilinear network of the same angle. In practice, geometrical transformation may involve mapping one or more planes successively when boundaries are complex. These planes are connected by analytic mapping functions chosen to meet with specific objectives. Applications could be extended to produce visual effects of image distortion using Joukowski transformation (Example 4.1b).

Consider a complex velocity potential F (Equation 4.1) and a change of coordinates between ζ-plane and z-plane. The real axes of both complex planes are aligned in parallel. When complex potential is found in ζ-plane, it is transformed to z-plane by composite mapping to yield the desired complex potential.

$$\zeta = \zeta(z); \; z = z(\zeta)$$
$$\zeta-\text{plane} : \zeta = \xi + i\eta$$
$$z-\text{plane} : z = x + iy \tag{5.4}$$

$$F = f(\zeta) = \phi(\xi, \eta) + i\psi(\xi, \eta)$$
$$= f[\zeta(z)] = f(z) = \phi(x, y) + i\psi(x, y) \tag{5.5}$$

At corresponding points,

$$\phi(\xi, \eta) = \phi[\xi(x, y), \eta(x, y)] = \phi(x, y) \tag{5.6a}$$
$$\psi(\xi, \eta) = \psi[\xi(x, y), \eta(x, y)] = \psi(x, y) \tag{5.6b}$$

$\phi(\xi, \eta)$ and $\psi(\xi, \eta)$ are equipotentials and streamlines in ζ-plane respectively. $\phi(x, y)$ and $\psi(x, y)$ are the corresponding equipotentials and streamlines in z-plane. They all satisfy Laplace equation in both planes, since conformal transformation maps a harmonic function into another harmonic function. The complex analytic function connects the two planes similar to those of z-plane and F-plane connected by the complex velocity potential.

$$\frac{\partial^2 \phi(x, y)}{\partial x^2} + \frac{\partial^2 \phi(x, y)}{\partial y^2} = |\zeta'(z)|^2 \left(\frac{\partial^2 \phi(\xi, \eta)}{\partial \xi^2} + \frac{\partial^2 \phi(\xi, \eta)}{\partial \eta^2} \right) = 0$$

$$\frac{\partial^2 \psi(x, y)}{\partial x^2} + \frac{\partial^2 \psi(x, y)}{\partial y^2} = |\zeta'(z)|^2 \left(\frac{\partial^2 \psi(\xi, \eta)}{\partial \xi^2} + \frac{\partial \psi^2(\xi, \eta)}{\partial \eta^2} \right) = 0$$

$$\frac{\partial \phi(\xi, \eta)}{\partial \xi} = \frac{\partial \psi(\xi, \eta)}{\partial \eta}; \quad \frac{\partial \phi(\xi, \eta)}{\partial \eta} = -\frac{\partial \psi(\xi, \eta)}{\partial \xi}$$

$$\frac{\partial \phi(x, y)}{\partial x} = \frac{\partial \psi(x, y)}{\partial y}; \quad \frac{\partial \phi(x, y)}{\partial y} = -\frac{\partial \psi(x, y)}{\partial x}$$

For illustration purposes, the following elementary mapping functions define some basic mappings with complex functions between z-plane and ζ-plane. It is essential to have knowledge about properties of elementary mapping functions and the results achieved by each of them since they produce a great number of important examples of these mappings.

$$\text{Translation}: \zeta = \zeta(z) = z + z_0 = x + x_0 + i(y + y_0)$$

Translation maps a point to its image point at a distance $|z_0|$ in the direction $\arg |z_0|$.

$$\text{Reflection about the origin}: \zeta = \zeta(z) = -z = -x - iy$$
$$\text{Reflection about the real axis}: \zeta = \zeta(z) = \bar{z} = x - iy$$
$$\text{Rotation and scaling}: \zeta = \zeta(z) = az$$
$$= |a| r \exp(i(\theta + \arg |a|))$$

It transforms an infinitesimal line segment through rotation about origin at an angle equal to $\arg|a|$ and scaled by $|a|$. Circles are mapped to circles in the transformation.

To study the property of conformal mapping, we invoke the theory of complex functions and observe the invariance of field property in a transformation process.

$$\bar{q}_\zeta \equiv \frac{dF}{d\zeta} = \frac{dF}{dz}\frac{dz}{d\zeta} = \bar{q}\,|z'(\zeta)|$$

$$d\zeta = \frac{d\zeta}{dz}dz = |\zeta'(z)|\,dz = \frac{1}{|z'(\zeta)|}dz \qquad (5.7)$$

$\bar{q}_\zeta$ and $\bar{q}$ are complex conjugate velocity in ζ-plane and z-plane respectively. The modulus of velocities is scaled by magnification factor while the modulus of lengths is inversely scaled by magnification factor $|z'(\zeta)|$. Conformal mapping preserves invariance of field property. If c_z denotes the position of a curve in z-plane and c_ζ denotes the mapping of this same curve in ζ-plane, then the line integral of complex conjugate vector is invariant in both planes.

$$\int_{c_z} \bar{q}\,dz = \int_{c_\zeta} \bar{q}_\zeta \frac{|z'(\zeta)|}{|z'(\zeta)|}d\zeta = \int_{c_\zeta} \bar{q}_\zeta\,d\zeta$$

$$F(z) = F[z(\zeta)] = F(\zeta)$$

$$\Delta F = \Delta\phi(x, y) + i\Delta\psi(x, y) = \Delta\phi(\xi, \eta) + i\Delta\psi(\xi, \eta)$$

$$\Delta\phi(x, y) = \Delta\phi(\xi, \eta)$$

$$\Delta\psi(x, y) = \Delta\psi(\xi, \eta)$$

The change of corresponding potentials and transport of flux bounded between corresponding flux lines remain unaltered in both planes.

Kinetic energy storage per unit depth is also invariant in conformal mapping and remains unchanged in corresponding field spaces. Kinetic energy density in two planes is scaled by magnification factor square.

$$\frac{\Delta W}{\Delta z} = \frac{\rho}{2}\iint q \cdot q\,dA$$

$$= \frac{\rho}{2}\iint |q|^2\,dx\,dy = \frac{\rho}{2}\iint |q_\zeta|^2\,|z'(\zeta)|^2\,\frac{d\xi\,d\eta}{|z'(\zeta)|^2} = \frac{\rho}{2}\iint |q_\zeta|^2\,d\xi\,d\eta$$

$$\text{Kinetic energy}: \frac{\rho}{2}q_\zeta^2 = \frac{\rho}{2}q^2\,|z'(\zeta)|^2$$

A constant line source is mapped to another line source of same strength C. The flows bounded by a closed path c around the source are invariant. From Cauchy's integral formula:

$$\oint dF = \frac{C}{2\pi} \oint \frac{dz}{z - z_0}$$

$$= \frac{C}{2\pi} 2\pi i = Ci$$

where

$$F = f(z) = \frac{C}{2\pi} \ln(z - z_0)$$

$$= f[z(\zeta)] = \frac{C}{2\pi} \ln(\zeta - \zeta_0)$$

$$= \phi + i\psi$$

$$\zeta = z$$

$$\oint dF = \oint d\phi + i \oint d\psi = Ci$$

$$\frac{Q}{\Delta z} = \oint_c d\psi(x, y) = \oint_{c_\zeta} d\psi(\xi, \eta) = C$$

By equating the imaginary parts, the same volume flow rate per unit depth is passing through the enclosed path equal to constant source strength. It is shown in Section 7.4 that the nature of point source is also invariant in transformation. Similar analysis leads to the same conclusion for other singularity like vortices or doublets.

The flow field analogy could be applied to other physical fields, provided that it admits complex potential. Field transmittance such as conductance, capacitance or inductance are purely geometric characteristics of the configurations and is invariant in conformal mapping. When they are evaluated in ζ-plane and expressed in terms of z-plane dimensions, they have the same values as if evaluated directly from the field distributions in the physical plane since geometric similarity is preserved in the two planes under conformal transformation. For the same reasons, shape factor per unit depth, transfer number and tube ratio are invariant scalar quantities. Since inverse mapping function is analytic and satisfies the Cauchy–Riemann relation, the inverse mapping is also conformal.

In field visualization, any two-dimensional field patterns or solutions of Laplace equation for a given geometry can be viewed through other geometries by introducing ζ-plane with connection to the physical plane. Consider Joukowski transformation as an application example of conformal mapping,

$$\zeta = z + \frac{c^2}{z} \tag{5.8}$$

Conversely,

$$z = \frac{1}{2}\left[\zeta \pm \sqrt{\zeta^2 - 4c^2}\right]$$

c is a real number referred to as parameter of Joukowski transformation, which defines a conformal mapping between z-plane and ζ-plane in two action parts. The first part is represented by the first term, that is,

$$\zeta = z = r\exp(i\theta)$$
$$F = f(\zeta) = f(z)$$
$$|f'(z)| = \left|\frac{dz}{dz}\right| = 1$$

This term does linear scaling in terms of magnification factor $|f'(z)|$. For unity magnification factor, it defines an identity mapping with no change in the input of the function. Any point in z-plane is transformed to the same point in ζ-plane. The physical interpretation of complex potential is a uniform vector field parallel to real axis in ζ-plane.

The second term represents inversion of a circle of radius c with the center at origin and a reflection about real axis with a reverse in orientation.

$$\zeta = \frac{c^2}{z} = \frac{c^2 x}{x^2 + y^2} - i\frac{c^2 y}{x^2 + y^2}$$

or

$$\xi = \frac{c^2 x}{x^2 + y^2} \tag{5.9a}$$

$$\eta = \frac{-c^2 y}{x^2 + y^2} \tag{5.9b}$$

Or in polar form:

$$\zeta = \frac{c^2}{r \exp i\theta} = \frac{c^2}{r} \exp(-i\theta)$$

This term is of basic importance and has the property of doing exactly the opposite of the original. The properties of inversion are that it does not preserve distance but angle and that the scale of an inverted figure depends on radius of inversion c. A straight radial line at a distance c from origin is mapped into another straight radial line as a mirror image with respect to real axis. This is equivalent to a reflection transformation about real axis. A circle of radius r in z-plane with center at origin is mapped as another circle of radius $\frac{c^2}{r}$ in ζ-plane in reverse direction. Similarly, a circle not through the center of inversion is another circle also not through center of inversion. The magnification factor $|\zeta'|$ shows that small circles are enlarged for $r < c$ whereas they are reduced if $r > c$. Points outside the inversion circle of radius c are then mapped onto points inside the circle and vice versa. There is no change in circle with $r = c$ as the circle of inversion inverts into itself in the reverse sense of direction. It is observed that origin is a singular point, which is mapped to a point at infinity in ζ-plane and they are mutually inverse with respect to the circle.

To visualize lines that do not pass through origin but are mapped to circles, we proceed to map rectangular coordinate lines from ζ-plane to z-plane by writing the mapping functions in conic form:

$$(x^2 + y^2)\xi - c^2 x = 0$$
$$(x^2 + y^2)\eta + c^2 y = 0$$

The first mapping function transforms a set of vertical coordinate lines $\xi = \pm C'_1 \neq 0$ into a pair of circle systems on z-plane symmetric with respect to y-axis. The circle centers are located at $\left(\pm \frac{c^2}{2C'_1}, 0\right)$. The second mapping function transforms a set of horizontal coordinate lines $\eta = \pm C'_2 \neq 0$ into a pair of circle systems symmetric with respect to x-axis. The circle centers are located at $\left(0, \mp \frac{c^2}{2C'_2}\right)$. By varying constants C'_1 and C'_2, the mapping yields two families of orthogonal circles passing through origin on z-plane giving rise to a four-leaf rose petal. Coordinate η-axis and ξ-axis are mapped to y-axis and x-axis respectively. Each of them can be viewed as a circle of infinite radius. The physical interpretation of complex potential is a doublet of strength c^2, that is,

$$F = f(\zeta) = \zeta = \frac{c^2}{z}$$

Figure 5.2b presents a geometric interpretation on the nature and action of Joukowski mapping in point transformation. The modulus of a complex point $|z| = r = 0p$ is scaled to $0p' = \frac{c^2}{r}$ and is then reflected about real axis to yield an image point p'' as the angle changes its sign. The transformation is to be completed by combining the image point with the original point via vector addition. ζ-plane and z-plane coincide at infinity because of identity mapping with magnification factor approaching unity. This implies boundary conditions at infinity are satisfied in both planes. By equating the real parts and imaginary parts of Joukowski transformation, we have,

$$\zeta = z + \frac{c^2}{z}$$

or

$$\xi = x + \frac{c^2 x}{x^2 + y^2}$$

$$\eta = y - \frac{c^2 y}{x^2 + y^2}$$

$$\frac{d\zeta}{dz} = 1 - \frac{c^2}{z^2}$$

It is observed that rectangular coordinate lines from z-plane are mapped as a set of orthogonal coordinate curves in ζ-plane. Conversely, by assigning rectangular coordinate lines ξ and η to different parametric constants, they are mapped as a set of orthogonal curves in z-plane. y-axis is mapped to the

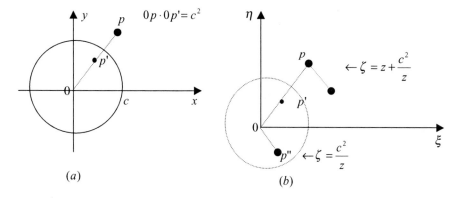

Figure 5.2 Mapping with circle inversion or Joukowski transformation

imaginary η-axis with points $y = \pm c$ to coincide with the origin in ζ-plane. The x-axis is mapped to ξ-axis ($\eta = 0$) and it splits/rejoins at ($\xi = \pm 2c$, $\eta = 0$). These two points correspond to critical points ($x = \pm c$, $y = 0$) on z-plane. The circle $r = c$ represents $\eta = 0$ where conformal mapping breaks down and the magnification factor vanishes. The coordinate lines are distorted most severely in the regions close to critical points when mapped from z-plane.

A different view of transformation can be obtained in the standard form of conics in polar coordinate:

$$\zeta = \left(r + \frac{c^2}{r} \right) \cos\theta + i \left(r - \frac{c^2}{r} \right) \sin\theta$$

$$\text{Or } \left(\frac{\xi}{2c\cos\theta} \right)^2 - \left(\frac{\eta}{2c\sin\theta} \right)^2 = 1 \tag{5.10}$$

$$\left(\frac{\xi}{a} \right)^2 + \left(\frac{\eta}{b} \right)^2 = 1$$

$$\text{where } a = r + \frac{c^2}{r}, \ b = r - \frac{c^2}{r}$$

With $r > c$, it maps radial coordinate lines, $\theta \neq 0$ from z-plane into hyperbolas with foci at $(2c, 0)$, $(-2c, 0)$ in ζ-plane. The other set of coordinate curves r represented by concentric circles is mapped into ellipses in ζ-plane with foci at $(\pm 2c, 0)$. The ellipse is centered at origin with major axis a and minor axis b coincide with ξ-axis and η-axis respectively. The exterior of circle is mapped to the exterior of ellipse. The two systems of coordinate curves remain orthogonal. When $r = c$, magnification factor $|\zeta'(z)|$ vanishes and mapping function reduces to:

$$\zeta = \xi = 2x = 2c\cos\theta$$

$$\eta = y = 0$$

The circle is then mapped into an infinitely thin line segment along real ξ-axis in Figure 5.3, that is, $-2c \leq \xi \leq 2c$, with $c = 5.4$. The locus of points on the circle of radius c is transformed to points on the upper ($0 \leq \theta \leq \pi$) and lower section ($\pi \leq \theta \leq 2\pi$) of the line segment respectively on ζ-plane. Region outside the circle is mapped to entire ζ-plane. In order to have Joukowski mapping single-valued for one complete coverage of ζ-plane, it is necessary to restrict to the exterior of circle $r = c$, which is regarded as a barrier and is mapped to entire ζ-plane outside the thin line segment. The singular point

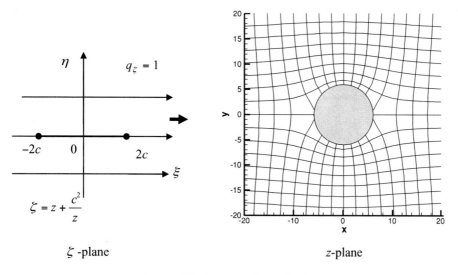

Figure 5.3 Flow past a cylinder

at origin is of course excluded with $r \geq c$. Otherwise it is mapped to a point at infinity.

The physical interpretation of Joukowski transformation is a uniform flow of unit velocity in the positive direction of real axis past a circular cylinder of radius c with center at origin of z-plane in Figure 5.3. Specifically the first term in Joukowski mapping function describes uniform flow parallel to real axis. The second term represents insertion of a circular cylinder of radius c into the stream. Such arrangement is referred to as Milne-Thomson's (1962) circle theorem in hydrodynamics. It describes the superposition of uniform flow and a doublet at origin. On complex F-plane, velocity complex potential describes uniform flows parallel to horizontal real axes, that is,

$$F = f(\zeta) = \zeta = \phi(\xi, \eta) + i\psi(\xi, \eta)$$

The above identity mapping transforms a system of rectangular equipotentials and streamlines crossing each other orthogonally in F-plane onto ζ-plane. They are then mapped by Joukowski transformation to the corresponding equipotentials and streamlines on physical z-plane where flow past a cylinder takes place.

$$F = f(\zeta) = f[\zeta(z)] = \phi(x, y) + i\psi(x, y) = z + \frac{c^2}{z}$$

where

$$\phi = \frac{x(x^2 + y^2) + c^2x}{x^2 + y^2} = x\left(1 + \frac{c^2}{r^2}\right) = \left(1 + \frac{c^2}{r^2}\right)r\cos\theta$$

$$\psi = \frac{y(x^2 + y^2) - c^2y}{x^2 + y^2} = y\left(1 - \frac{c^2}{r^2}\right) = \left(1 - \frac{c^2}{r^2}\right)r\sin\theta$$

$$q_r = \frac{\partial\psi}{r\,\partial\theta} = \left(1 - \left(\frac{c}{r}\right)^2\right)\cos\theta; \quad q_\theta = -\frac{\partial\psi}{\partial r} = \frac{\partial\phi}{r\,\partial\theta} = -\left(1 + \left(\frac{c}{r}\right)^2\right)\sin\theta$$

$$\begin{bmatrix} q_x \\ q_y \end{bmatrix} = \begin{bmatrix} \cos\theta & -\sin\theta \\ \sin\theta & \cos\theta \end{bmatrix}\begin{bmatrix} q_r \\ q_\theta \end{bmatrix}$$

Conjugate velocity is given by:

$$\bar{q} = \frac{dF}{dz} = \frac{dF}{d\zeta}\frac{d\zeta}{dz} = \bar{q}_\zeta\frac{d\zeta}{dz} = 1 - \frac{c^2}{z^2}$$

$$\rightarrow \bar{q}\,|_{r=\infty} = 1$$

where

$$\bar{q}_\zeta = \frac{dF}{d\zeta} = 1 = q_\zeta$$

$$\rightarrow \bar{q}\,|_{r=c} = 0$$

Critical points: $r = c$, $\theta = 0$ and $\theta = \pi$.

Figure 5.3 presents the flow field map showing flow behavior in the neighborhood of cylinder. As flow approaches the cylinder, velocity reduces from unity and assumes zero at critical points. Square cells representing uniform flow in ζ-plane are transformed into curvilinear squares on z-plane as flow approaches the cylinder. Velocity then starts to increase after passing over the cylinder. At sufficiently large distance from it, flow becomes uniformly parallel to x-axis and resumes unity as the effects of second term diminish to nil. Dividing streamline $\psi = 0$ for $\theta = 0$ and $\theta = \pi$ resembles a circle of $r = c$ that goes around circle circumference and splits at critical points to coincide with x-axis in z-plane. It separates the flow fields into lower and upper regions exterior to the circle. This circular streamline is mapped onto ξ-axis in ζ-plane to coincide with the thin line segment. Symmetry test shows that streamlines are symmetric with respect to y-axis and have asymptotes parallel to x-axis respectively. Streamlines are antisymmetric with respect to x-axis as the stream function changes sign. Equipotentials are symmetric and antisymmetric about x-axis and y-axis. They end on circle circumference perpendicularly.

$$\frac{\partial \phi}{\partial r}\Big|_{r=c} = \frac{\partial \psi}{r \partial \theta} = \left(1 - \frac{c^2}{r^2}\right) \cos \theta = 0$$

Joukowski transformation has the property that it causes distortion in the region with small values of z. For large values of z, it reduces to identity mapping. This property is useful to produce warping effects on images at a local region as demonstrated by mapping of rectangular coordinate lines. These visual effects of distortion with angle and local shape preservation differ from those produced by curvilinear mapping, which distorts the relative sizes of a figure in a global sense.

5.6 Hodograph Method and Mapping

Hodograph theory was initiated by Helmholtz and Kirchhoff. W.R. Hamilton developed hodograph method as a mathematical technique in hydrodynamics to solve two-dimensional flows using velocity components as variables. The method is extended as a mapping process to obtain solutions or fields on the physical plane by mapping the field map from hodograph plane. In this respect, hodograph mapping refers to transformation of fields from a given coordinate system onto hodograph plane and conversely. Hodograph theory and complex function theory are closely linked. Complex function theory makes it possible to map regions of flows conformally from F-plane to conjugate/hodograph plane through differentiation whereby information about flow fields could be obtained in terms of velocity components. Conversely, hodograph method is convenient to obtain complex potential relating F-plane and z-plane by mapping conjugate hodograph plane through integration. Hodograph representation of trajectories is essential to provide complete information about vector fields in addition to position trajectories on physical plane. Hodograph plane and physical plane are indeed equally important. Hodograph method enables us to determine the shape of free surface, which is generally not known in the physical plane. In plane fields, hodograph representation of free surface turns out in theory to resemble simple or conic curves on hodograph plane when free surface boundary condition is imposed. By mapping the curve from hodograph plane onto the physical plane, we could visualize the shape of free surface. Each circle about hodograph origin is a hodograph representation of line of constant speed or kinetic energy. Each line through origin defines isocline of constant flow direction. We could then obtain the corresponding position hodograph representation of circles or lines from hodograph plane onto physical plane through mapping. By introducing hodograph plane/space, we are able to make use of hodograph representation of certain field patterns or trajectories with unique features to enhance our visual ability. The origin of hodograph plane corresponds to critical points at which dividing streamlines or equipotentials pass through. Hodograph transformation has the property that an infinitesimal area on physical plane is mapped to finite area on hodograph plane as singular points are mapped to points at infinity. Hodograph representations of some important plane trajectories are then presented in an attempt to visualize the characteristics of vector fields. We are also interested in mapping field phenomena onto hodograph plane from which information is extracted in quantitative terms. The concept and applications of velocity hodograph transformation could be extended to other physical fields such as displacement hodograph, heat flux hodograph or electric field intensity hodograph.

5.6.1 Conjugate Hodograph

Conjugate hodograph plane is obtained by differentiation of complex potential and could be interpreted geometrically as scaling and twisting of an infinitesimal object. In this respect, the interest is focused on magnification factor and twist angle, which control the amount of scaling/twisting between infinitesimal objects. The physical interpretation of magnification factor is flow field strength representing velocity magnitude. The twist angle defines the conjugate velocity direction with respect to the positive horizontal axis. It is convenient to use polar representation in certain situations. We are concerned with the use of complex conjugate velocity and its roles in mapping.

A conjugate hodograph $\bar{q}$-plane is also called inverse hodograph plane being the reflection of hodograph about q_x-axis. It is defined by the derivative of complex potential:

$$\bar{q} \equiv \frac{dF}{dz} = \left(\frac{\partial \phi}{\partial x} + i \frac{\partial \psi}{\partial x} \right) = q_x - iq_y \qquad (5.11)$$

$$= |\bar{q}| \exp(-i\tilde{\theta})$$

$$\bar{q} - \text{plane} : \bar{q} = q_x + iq'_y$$

$$\text{where } q'_y = -q_y; \ \tilde{\theta} = \tan^{-1} \left(\frac{q_y}{q_x} \right)$$

$\tilde{\theta}$ is the inclination angle of modulus measured from positive q_x-axis counterclockwise. $\bar{q}$ defines the mapping functions on $\bar{q}$-plane (or q-plane) on which regions of flow are transformed from z-plane. In essence, it establishes a connection between the conjugate/hodograph plane and z-plane. Mapping of fields onto these planes and the physical plane are essential as they contain all information about fields and provide the venues for field visualization/analysis. In practice, horizontal q_x-axis is aligned parallel to x-axis of z-plane. Velocity components q_x and q_y assume the role of x and y-axes respectively. Complex conjugate velocity at a point is mapped from its image in z-plane as a position vector on conjugate hodograph plane. It is a mirror reflection image of actual velocity vector about q_x-axis. The negative sign on q_y-component has resulted from the use of conjugate vector. The inverse mapping then takes each point back to the z-plane. Complex conjugate velocity $\bar{q}$ is an analytic function and as such possesses conformal transformation properties.

5.6.2 Hodograph

Hodograph plane is similar to conjugate hodograph plane and is also obtained by differentiation of complex potential, that is,

$$q \equiv \overline{\frac{d F(z)}{dz}} = \frac{\partial \phi}{\partial x} - i \frac{\partial \psi}{\partial x} = q_x + i q_y$$

$$= |q| \exp(i\tilde{\theta})$$

$$q - \text{plane}: q = q_x + i q_y$$

Similarly, horizontal q_x-axis is aligned parallel to x-axis of z-plane. The vector components q_y and q_x assume the role of y-axis and x-axis respectively. Hodograph and conjugate hodograph are mirror reflections of one another about real axis and could be obtained from one to the other. The difference is that complex velocity represents actual velocity with no analytic property. In the above polar representation, the equation of hodograph defines vector fields at a point in terms of its strength magnitude and direction. We are concerned about hodograph representations of trajectories and its applications including visualization of boundaries.

5.7 Hodograph Representations and Applications

When an arbitrary curve in a given coordinate is mapped onto hodograph plane/space, it is called hodograph representation of the curve. Without loss of generality, a parametric curve traced by velocity vector defines hodograph representation of the curve. Consider a two-dimensional velocity field in parametric form $q(t)$. The hodograph mapping process is exactly the same as if a plane parametric trajectory is traced by path function $\vec{r}(t)$ on the physical plane. Velocity vector $q(t)$ and position vector $\vec{r}(t)$ are the equation of hodograph representation and equation of position hodograph representation of trajectory in parametric form. For a given t, the respective parametric equation defines a point $q(t)$ on hodograph (or velocity hodograph) plane and a point $\vec{r}(t)$ on the physical plane (or position hodograph plane) in a one to one correspondence. It constitutes a continuous mapping of points from t-axis onto a point-wise trajectory on each plane. Such geometric transformation is a mapping of trajectory (or curve) in four dimensions using two coordinate planes, that is, hodograph representation (q-plane) and coordinate representation (z-plane).

The modulus represents velocity in hodograph plane. Whereas the tangent to a hodograph representation of trajectory defines acceleration vector of a particle in exactly the same manner as the tangent to the position hodograph representation of trajectory defines velocity vector in the physical plane. Acceleration has indeed the same relation to velocity as velocity has to the position vector in geometrical, graphical and mathematical point of views. Acceleration vector and velocity vector always point in the respective directions of hodograph tangent and streamline tangent along which particle advances. They determine the state of motions completely. To visualize the action of these two vector quantities, we translate them like free vectors between image points in physical and hodograph planes in a mapping process (Figure 5.4). When acceleration vector a is mapped from hodograph plane onto physical plane and is displayed with velocity at a point A, it is decomposed into streamline tangential and normal components through the law of parallelogram. The tangential component of acceleration is responsible for speed changes. Speed will increase or decrease when this component acts in the same velocity direction or in the opposite. Speed is constant if the

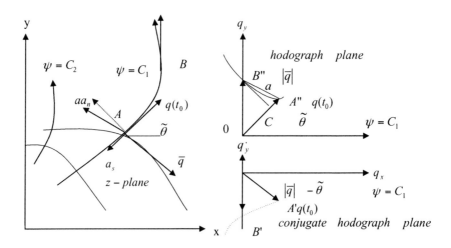

Figure 5.4 Mapping of trajectory on z-plane, $\tilde{q}$-plane and q-plane

tangential component vanishes or the two vectors are perpendicular to each other. The normal component of acceleration always acts towards the center of curvature of trajectory giving rise to centripetal acceleration. This component is responsible for change in motion direction. The velocity tangent $q(t)$ always points in the positive $\hat{s}$-direction of motion while the normal component vanishes. It is mapped to a velocity vector on hodograph plane. The locus of points of velocity vectors (arrowheads) describes hodograph representation of trajectory just like locus of points for the arrowheads of the position vector in physical plane. Hodograph mapping processes could be illustrated by velocity diagram construction using graphical method through the law of parallelogram. Figure 5.4 depicts a velocity vector at an arbitrary point A tangent to a streamline at time t_0 on the physical plane. The motion of a particle at A along a streamline could be traced by first mapping it onto the image point A'' on velocity hodograph. It involves drawing a vector line through origin of hodograph plane parallel to the streamline tangent at point A. The modulus of vector $0A''$ is determined by the velocity magnitude at A. A'' representing actual velocity vector in z-plane is the mirror image of point A' in conjugate hodograph plane reflected about q_x-axis. If point A moves along a streamline to point B at a time interval Δt, the corresponding point A'' will then move to B'' on hodograph plane using the same graphical work. Similarly, the modulus of vector $0B''$ is determined by the velocity magnitude at B. The line segment $A''B''$ represents the change of velocity for time interval Δt. The graphical work leads to the construction of velocity diagram. In general,

$$q(t) = q(t_0) + a(t - t_0)$$

By going through successive positions at small Δt and translating all velocity vectors to start from origin, an array of successive vectors $\Delta q = a\,\Delta t$ then constitutes a continuous mapping of velocity vectors from t-axis giving rise to hodograph representation of trajectory. Velocity vectors of all points along the hodograph representation of trajectory could be drawn from origin like a pencil of rays. We could also obtain hodograph in explicit or implicit form by eliminating parameter t. The relation between velocity components is the equation of hodograph on q-plane. Expansion to hodograph space is straightforward from hodograph plane although complex potential does not exist in three-dimensions.

$$q_x = q_x(t) = q_x[t(q_y)] = f(q_y)$$
$$q_y = q_y(t) = q_y[t(q_x)] = g(q_x)$$
$$h(q_x, q_y) = C$$

Similarly, the position diagram is constructed by the same graphical method. By eliminating parameter t in position trajectory, we obtain coordinate representation of trajectory. The relation between x and y is the equation of trajectory on x-y plane.

$$\vec{r}(t) = \vec{r}(t_0) + q(t - t_0)$$

Given $\bar{q} = \bar{q}(z)$, hodograph representation of streamlines or equipotentials could be obtained by mapping stream function or equipotential from z-plane onto $\bar{q}$-plane or q-plane. Consider mapping of equipotential on $\bar{q}$-plane as an example, the chain rule gives,

$$
\begin{bmatrix} \dfrac{\partial \phi}{\partial x} \\[2mm] \dfrac{\partial \phi}{\partial y} \end{bmatrix} = [J] \begin{bmatrix} \dfrac{\partial \phi}{\partial q_x} \\[2mm] \dfrac{\partial \phi}{\partial q_y'} \end{bmatrix}
\tag{5.12}
$$

$$
\text{where } [J] = \begin{bmatrix} \dfrac{\partial q_x}{\partial x} & \dfrac{\partial q_y'}{\partial x} \\[2mm] \dfrac{\partial q_x}{\partial y} & \dfrac{\partial q_y'}{\partial y} \end{bmatrix}
\tag{5.13}
$$

$$
\text{Conversely, } \begin{bmatrix} \dfrac{\partial \phi}{\partial q_x} \\[2mm] \dfrac{\partial \phi}{\partial q_y'} \end{bmatrix} = \frac{1}{J} \begin{bmatrix} \dfrac{\partial q_y'}{\partial y} & -\dfrac{\partial q_y'}{\partial x} \\[2mm] -\dfrac{\partial q_x}{\partial y} & \dfrac{\partial q_x}{\partial x} \end{bmatrix} \begin{bmatrix} \dfrac{\partial \phi}{\partial x} \\[2mm] \dfrac{\partial \phi}{\partial y} \end{bmatrix}
$$

$$
\text{where } J = \begin{vmatrix} \dfrac{\partial q_x}{\partial x} & \dfrac{\partial q_y'}{\partial x} \\[2mm] \dfrac{\partial q_x}{\partial y} & \dfrac{\partial q_y'}{\partial y} \end{vmatrix}
\tag{5.14}
$$

$$
\text{Hence } \phi(q_x, q_y') = \int \frac{1}{J} \left[\frac{\partial q_y'}{\partial y} \frac{\partial \phi}{\partial x} - \frac{\partial q_y'}{\partial x} \frac{\partial \phi}{\partial y} \right] \partial q_x + f(q_y')
\tag{5.15}
$$

$$
\phi(q_x, q_y') = \int \frac{1}{J} \left[-\frac{\partial q_x}{\partial y} \frac{\partial \phi}{\partial x} + \frac{\partial q_x}{\partial x} \frac{\partial \phi}{\partial y} \right] \partial q_y' + f(q_x)
\tag{5.16}
$$

By substituting equipotential with stream function, transformation of stream function has the exact forms in conjugate hodograph plane, that is,

$$
\psi(q_x, q_y') = \int \frac{1}{J} \left[\frac{\partial q_y'}{\partial y} \frac{\partial \psi}{\partial x} - \frac{\partial q_y'}{\partial x} \frac{\partial \psi}{\partial y} \right] \partial q_x + f(q_y')
\tag{5.17}
$$

$$
\psi(q_x, q_y') = \int \frac{1}{J} \left[-\frac{\partial q_x}{\partial y} \frac{\partial \psi}{\partial x} + \frac{\partial q_x}{\partial x} \frac{\partial \psi}{\partial y} \right] \partial q_y' + f(q_x)
\tag{5.18}
$$

$\phi(q_x, q'_y)$ and $\psi(q_x, q'_y)$ are velocity potential and stream function in conjugate hodograph plane respectively. Streamlines and equipotentials remain orthogonal in the transformation process. Reflection mapping then takes conjugate hodograph plane onto hodograph plane or by replacing q'_y with q_y from Equations 5.15 to 5.18. Alternatively, hodograph representation of these curves on hodograph plane could be traced directly by composite mapping when mapping function $z = z(q)$ is known.

$$\psi(x, y) = \psi[(x(q_x, q_y), y(q_x, q_y)] = \psi(q_x, q_y) = C_2 \qquad (5.19a)$$

$$\phi(x, y) = \phi[(x(q_x, q_y), y(q_x, q_y)] = \phi(q_x, q_y) = C_1 \qquad (5.19b)$$

$$x = x(q_x, q_y)$$

$$y = y(q_x, q_y)$$

The relations between F-plane and conjugate hodograph plane are essential in the application of hodograph method. Knowledge of hodograph mapping is useful because it is often simpler than that of physical plane and mapping is conformal. When complex conjugate velocity is known on F-plane, hodograph method provides an alternative means of getting mapping from F-plane to z-plane and conversely.

$$z = \int \frac{dF}{\bar{q}} + C = \int \frac{dF}{\bar{q}(F)} + C = z(F) \qquad (5.20a)$$

Above solution can be obtained by graphical or numerical method of integration. In the graphical mapping process, the expressions are written approximately along a streamline and an equipotential as follows:

$$\text{Along streamline}: \ x = \int_{\phi_0}^{\phi} \frac{\cos \alpha}{|\bar{q}|} d\phi \approx \frac{\cos \alpha}{|\bar{q}|}(\phi - \phi_0);$$

$$y = \int_{\phi_0}^{\phi} \frac{\sin \alpha}{|\bar{q}|} d\phi \approx \frac{\sin \alpha}{|\bar{q}|}(\phi - \phi_0) \qquad (5.20b)$$

$$\text{Equipotential}: \ x = -\int_{\psi_0}^{\psi} \frac{\sin \alpha}{|\bar{q}|} d\psi \approx -\frac{\sin \alpha}{|\bar{q}|}(\psi - \psi_0);$$

$$y = \int_{\psi_0}^{\psi} \frac{\cos \alpha}{|\bar{q}|} d\psi \approx \frac{\cos \alpha}{|\bar{q}|}(\psi - \psi_0) \qquad (5.20c)$$

where

$$\alpha = \tan^{-1}\left(\frac{q_y}{q_x}\right) = \tilde{\theta}$$

To initiate the graphical method of integration using the field map on F-plane, it is preferable to use critical point as the starting point (x_0, y_0) since its location is generally known in physical plane. Critical points correspond to the origin of hodograph plane. Curves passing through origin are dividing streamlines or equipotentials. For good accuracy, small intervals are used between ϕ_0 and ϕ_1 along ψ_0 curve and between ψ_0 and ψ_1 along ϕ_0 curve. Using the mean values of $\cos\alpha/|\bar{q}|$ and $\sin\alpha/|\bar{q}|$, we could determine point (x_1, y_1) lying on each respective curve. Repeating the graphical work until sufficient points are located to complete the field map network. The inverse then takes it back to z-plane.

$$F = f(z)$$

Similarly, if function $F(\bar{q})$ is known, the method could be applied to visualize the vector fields by mapping from conjugate hodograph plane to z-plane and inversely.

$$z = \int \frac{dF}{\bar{q}} + C = \int \frac{dF}{d\bar{q}} \frac{d\bar{q}}{\bar{q}} + C = z(\bar{q}) \qquad (5.20d)$$
$$\bar{q} = \bar{q}(z)$$

As an extension of hodograph method, a fictitious plane $\bar{q}_H$ is introduced into above integral written as:

$$z = \int \bar{q}_H \frac{d\bar{q}}{\bar{q}} + C = z(\bar{q}) \qquad (5.21)$$

$$\bar{q}_H - \text{plane} : \bar{q}_H \equiv \frac{dF}{d\bar{q}} = \frac{dF}{dz}\frac{dz}{d\bar{q}} = (q_x - iq_y)/|\bar{q}'(z)| = q_{xH} - iq_{yH} \qquad (5.22)$$

$$q_H \equiv \overline{\frac{dF}{d\bar{q}}} = (q_x + iq_y)/|\bar{q}'(z)| = q_{xH} + iq_{yH} \qquad (5.23)$$

$$|\bar{q}'(z)| = \left|\frac{\partial q_x}{\partial x} + i\frac{\partial q_y'}{\partial x}\right| = \sqrt{\left(\frac{\partial q_x}{\partial x}\right)^2 + \left(\frac{\partial q_y'}{\partial x}\right)^2} \qquad (5.24)$$

$$|\bar{q}_H| = \frac{|q|}{|\bar{q}'(z)|}$$

$\bar{q}_H$ represents a fictitious flow field and serves as an ancillary plane. It is regarded as a version of modified conjugate hodograph plane. A vector drawn from origin to any point in this plane then agrees in direction with velocity vector at corresponding point in conjugate hodograph plane whose magnitude is modified by a factor $|\bar{q}'(z)|$. Critical points and singular points are mapped to origin and infinity respectively on $\bar{q}_H$-plane.

Hodograph representations have practical applications. We could determine the relative velocity between two arbitrary points A and B in z-plane by mapping them onto hodograph plane. The line segment joining these two image points in hodograph plane is the relative velocity between them. If the time required to travel from A to B is Δt, then the acceleration of particle at A could be determined from the velocity diagram and is represented by the tangent to the hodograph of trajectory. The direction of acceleration is parallel to $A''B''$ in hodograph in Figure 5.4 with a magnitude equal to:

$$|a| = \sqrt{\left(\frac{B''C}{\Delta t}\right)^2 + \left(\frac{CA''}{\Delta t}\right)^2} \tag{5.25}$$

where

$$a = a_s\hat{s} + a_n\hat{n} = \frac{dV}{dt} = \frac{V_B - V_A}{\Delta t} = \frac{0B'' - 0A''}{\Delta t} = \frac{A''B''}{\Delta t} = \frac{B''C}{\Delta t} + \frac{CA''}{\Delta t}$$

$$a_s = \left|\frac{CA''}{\Delta t}\right| = \left|\frac{dV_A}{dt}\right| = \frac{V_A \cdot a}{|V_A|} = \hat{s} \cdot a$$

$$a_n = \left|\frac{B''C}{\Delta t}\right| = \left|\frac{B''C}{\Delta\alpha}\right| \frac{\Delta\alpha}{\Delta t} \frac{\Delta s}{\Delta s}$$

$$= |0A''| \frac{\Delta\alpha}{\Delta s} \frac{\Delta s}{\Delta t}$$

$$= |V_A| \frac{|V_A|}{R_s} = \frac{V_A^2}{R_s}$$

$$B''C \perp 0A''; \Delta t \to 0; \Delta s \to 0; \Delta\alpha \to 0$$

Or

$$a_n = \frac{|V_A \times a|}{|V_A|} = \frac{|V_A \times a|}{|V_A|^3}|V_A|^2 = \frac{V_A^2}{R_s} = \hat{n} \cdot a$$

Δs denotes elemental arc length along streamline in the physical plane. R_s is radius of curvature of streamline. It shows that tangent to the hodograph trajectory represents acceleration vector. The tangential component of acceleration vector a_s equals to the time rate of change of speed. The normal component a_n is centripetal acceleration responsible for the change in velocity direction and always points towards center of curvature of streamlines. Alternatively, acceleration vector could be determined from the complex potential and Equation 4.14a or 4.14b analytically. It is possible to trace the distributions of constant kinetic energy (isotachs) or speed contours on physical plane by mapping concentric circles about origin on hodograph plane. Each circle represents a contour of constant kinetic energy or speed. If hodograph representation of trajectories is circles about origin on hodograph plane, then the kinetic energy of a particle moving at constant speed is conserved in the motion. Examples of such flow phenomena include vortex motions, objects moving in circular planet orbits in central force fields or charged particles in uniform magnetic fields. Hodograph plane is useful to trace/visualize certain boundaries in general and free surfaces of unknown shapes in particular. These include straight boundaries (streamline/equipotential/seepage), free surfaces and some special trajectories.

5.7.1 Straight Boundaries

Straight boundary is a special class of boundaries in simple regular shape, which could be described by equation of a straight line. Hodograph representation of boundary of this type could be easily found when the straight boundary assumes a streamline, an equipotential or seepage. The hodograph representation method based on kinematics transforms a straight boundary from physical plane onto the hodograph plane by another straight line of arbitrary length. Unlike point mapping, points along the boundary/line may not be identified. The hodograph representation is then extended from motions along a straight streamline to a system of straight streamlines in parallel flow/uniform flow by connecting the physical plane and hodograph plane with the trajectory equation.

5.7.1.1 Impervious Boundary

Consider a straight impervious boundary or a streamline $\psi = C$. The tangent at a point along the streamline is inclined at an angle α with positive x-axis in physical plane. The boundary condition requires that the directional derivative of stream function vanishes along its flow path (streamline). It leads to the following equation of hodograph for a straight line on hodograph plane for any homogeneous medium:

$$\frac{\partial \psi}{\partial s} = \frac{\partial \psi}{\partial x}\frac{\partial x}{\partial s} + \frac{\partial \psi}{\partial y}\frac{\partial y}{\partial s} = \frac{\partial C}{\partial s} = 0$$
$$\rightarrow q_y = q_x \tan \alpha$$
$$= q_x \tan \bar{\theta}$$

s denotes arc length along the boundary. Hodograph representation of a straight impervious boundary/streamline is a straight line passing through origin and makes the same angle with positive q_x-axis on hodograph plane. This is because flow is tangent to a streamline everywhere. Because angle α will vary from point to point along a curved boundary, hodograph representation is ideal for straight boundaries with constant angle α such that above expression defines a straight line in hodograph plane corresponding to isocline at a velocity angle equal to tangent angle. It simply maps points along an impervious boundary onto a line segment of arbitrary length in parallel and through the origin on hodograph plane. Each point has the same direction defined by the streamline direction in terms of the tangent angle with respect to positive x-axis. Dividing impervious boundary/streamline will pass through the origin corresponding to a critical point.

5.7.1.2 Equipotential Boundary

The boundary of a reservoir is modeled by an equipotential line of constant value. A straight equipotential represented by a line of arbitrary length extends in two opposite directions and is referred to as a dividing equipotential when it passes through a critical point. The boundary condition shows that a straight equipotential inclined at an angle γ in z-plane is mapped to a straight line through origin making a velocity angle equal to $(\gamma - \pi/2)$ with respect to positive q_x-axis in hodograph plane, that is,

$$\phi = C$$
$$\frac{\partial \phi}{\partial s} = \frac{\partial \phi}{\partial x} \cos \gamma + \frac{\partial \phi}{\partial y} \sin \gamma = 0$$
$$q_y = q_x \tan(\gamma - \pi/2)$$

where

$$q_x = -\frac{\partial \phi}{\partial x}; \ q_y = -\frac{\partial \phi}{\partial y}$$

Above hodograph representation of a straight equipotential boundary is a straight line through origin and perpendicular to the image equipotential in physical plane.

5.7.1.3 Seepage Surface

Both equipotential and stream function may vary continuously along a seepage surface. A seepage surface is defined as an interface where pressure assumes constant and fluids may flow across the boundary. It expresses the piezometric head as a function of water elevation explicitly. The boundary condition of constant pressure for a straight seepage surface in an anisotropic medium yields the following equation of hodograph representation for a straight seepage boundary:

$$\varphi = y + \frac{p}{\rho g}$$

$$\frac{\partial \varphi}{\partial s} = \frac{\partial \varphi}{\partial x} \cos \gamma + \frac{\partial \varphi}{\partial y} \sin \gamma = \frac{\partial y}{\partial s} = \sin \gamma$$

$$q_y = \frac{k_v}{k_h} q_x \tan(\gamma - \pi/2) - k_v$$

k_h and k_v denote hydraulic conductivity in the x and y-direction respectively. s denotes arc length along the boundary. The straight seepage boundary is mapped to a straight line through point $(0, -k_v)$ having a slope equal to $\frac{k_v}{k_h} \tan(\gamma - \pi/2)$ with respect to positive q_x-axis in hodograph plane. It is normal to the seepage surface in physical plane. For isotropic medium, it reduces to:

$$q_y = q_x \tan(\gamma - \pi/2) - 1$$

Above is the hodograph representation of a straight seepage boundary. It is a straight line through point $(0, -1)$ on hodograph plane and is perpendicular to the seepage boundary in physical plane. Because fluids may flow across a seepage boundary at some finite velocities, a seepage boundary will not pass through the origin, which corresponds to a critical point.

5.7.2 Free Surface

The shape of free surface is generally unknown in physical plane. Like seepage surface, both equipotential and stream function may vary continuously along a free surface. A free surface is defined as an interface where pressure assumes constant. We consider it as a material surface expressed in Euler description in terms of an implicit function:

$$f(x_i, t) = \varphi - y = \frac{p}{\rho g} = C$$

Suppose S denotes a region enclosed by a boundary of a free surface on which fluid is in motion. The function maps each particle in the region from one position to another as the bonding surface moves with fluid flows. Although the material surface may change shape, the particles bounded by S are invariant in time. Because velocity of a free surface at any point is normal to the surface and vanishes, the free (material) surface moving with fluids in steady flow is conserved in the motion. It is necessary that the material rate of change of quantity f vanishes in the direction of flow.

$$\frac{Df}{Dt} = \frac{\partial f}{\partial t} + q \cdot \nabla f = 0$$

The function is constant on a material surface which contains the same set of particles moving together with the stream and provides the basic kinematic condition for free surfaces. This invariance of property means a material line/surface does not change when evaluated at particles on a streamline/surface. It is useful for deriving boundary conditions for any material surfaces.

$$(\nabla \varphi - \nabla y) \cdot q = 0$$

$$q_x \left(\frac{\partial \varphi}{\partial x} \right) + q_y \left(\frac{\partial \varphi}{\partial y} \right) - q_y = 0$$

$$-q_x \left(-\frac{k_h}{k_h} \frac{\partial \varphi}{\partial x} \right) - q_y \left(-\frac{k_v}{k_v} \frac{\partial \varphi}{\partial y} \right) - q_y = 0$$

Or in conic expression:

$$\frac{q_x^2}{(0.5\sqrt{k_h k_v})^2} + \frac{(q_y + 0.5k_v)^2}{(0.5k_v)^2} = 1 \qquad (5.26)$$

The above equation of hodograph for a plane free surface traces the locus of velocity vectors from origin like a pencil of rays that end on an ellipse boundary giving rise to the hodograph representation of the free surface. It shows that the graph of a plane free surface (a curve of unknown shape/length) is mapped to an ellipse image on hodograph plane. A free surface may take up the whole or just a part of ellipse boundary. Points on the free surface may be identified by applying the kinematic condition. The ellipse center is located at $(0, -0.5k_v)$ with major axis, $\sqrt{k_h k_v}$ and minor axis, k_v parallel to q_x-axis and q_y-axis respectively. By mapping locus of points on the ellipse boundary back to the physical plane, we obtain the shape of free surface. Thus hodograph representation of free surface is essential to visualize its shape in the physical plane. It is observed that velocity vector from origin to any point on the ellipse boundary has a negative y-component everywhere. This suggests the free boundary is sloping downwards continuously. The maximum velocity components are represented by the ellipse axes and depend on hydraulic conductivity. The velocity vector at any point along the free surface can be determined by graphical construction. A line segment drew between origin o and a point p' on the ellipse boundary is used to define the velocity at the image point on physical plane. The line o-p' is then translated in parallel until it becomes a tangent of the free surface at the point in question p on physical plane. The tangent line o-p' defines the direction of flow at point p measured with respect to positive x-axis and the modulus gives the magnitude of velocity. With velocity perpendicular to the free surface normal defined by the gradient of f, the kinematic condition is satisfied for a free surface. When sufficient points are located by mapping from image points along ellipse boundary, the shape of free surface could be traced on physical plane. For isotropic medium, $k_h = k_v = k$, the ellipse degenerates into a circle of radius $0.5k$ centered at $(0, -0.5k)$. The mapping process could be taken over by mapping functions connecting the physical plane and hodograph plane.

5.7.3 Special Field Patterns

Hodograph representation is useful for interpretation of field phenomena because certain field patterns have unique features on hodograph plane. These special field patterns include critical/singular points, unidirectional flows, isoclines, uniform flows and potential flows. The origin of hodograph plane corresponds to critical point and curves passing through origin are dividing streamlines or equipotentials. Singular points are mapped to points at infinity. For unidirectional flow fields, the hodograph representation is a straight line through origin inclined at the velocity angle $\tilde{\theta}$ from q_x-axis counterclockwise as per Equation 5.27a. The velocity angle defines unidirectional flow fields along straight streamline with constant tangent angle $\alpha = \tilde{\theta}$ from x-axis. The speed magnitude at a point defines the length of line segment representing the modulus on hodogaph plane. In this process, the trajectory equation connects the hodograph representation of unidirectional flow and the position hodograph of straight streamlines on physical plane.

$$\frac{q_y}{q_x} = \tan\tilde{\theta} = \tan\alpha = \frac{dy}{dx} \tag{5.27a}$$

In polar representation:

$$q = |q|\exp(i\tilde{\theta}) = |q|\exp(i\alpha)$$

Furthermore, the captioned hodograph representation of unidirectional flow corresponds to isocline along which each point has the same direction equal to constant velocity angle $\tilde{\theta}$. We could then trace isoclines contours (flows of constant direction) implicitly by mapping the straight isoclines from hodograph onto physical plane with the velocity angle assuming the role of parametric constant.

$$\frac{q_y}{q_x} = f(q_x, q_y) = f(x, y) = \tan\tilde{\theta} = C'$$

where

$$q_x = q_x(x, y)$$
$$q_y = q_y(x, y)$$

Above is the equation of position hodograph representation of isoclines in implicit form. By varying the parameter at constant intervals, a system of isoclines contours is generated on physical plane. When the isoclines contours squeeze, it indicates high isoclines density and a rapid change of flow directions in the region.

Consider steady unidirectional flow of incompressible fluids parallel to x-axis. The continuity equation in Cartesian coordinate requires:

$$\frac{\partial q_x}{\partial x} = 0; \text{ with } \tilde{\theta} = \alpha = 0; q_y = 0$$

Hence,

$$(q \cdot \nabla)q = 0$$
$$q_x = \frac{dx}{dt} = f(y)$$
$$a = \frac{dq}{dt} = \frac{dq_x}{dx}q_x = q_x\frac{df(y)}{dx} = 0$$
$$a \times q = 0$$
$$a \cdot q = 0$$
$$\psi = \int q_x \partial y = g(y)$$

The conditions for unidirectional flow are satisfied such that velocity remains constant in the direction of flow along a streamline with vanishing acceleration. The integration of trajectory equation yields the position hodograph representation of trajectory.

$$y = x \tan \alpha + C$$
$$= C$$

From Equation 5.27a,

$$q_y = q_x \tan \tilde{\theta} = 0$$
$$|q| = \sqrt{q_x^2 + q_y^2} = q_x = f(y)$$

The stream function ψ depends only on y variable. It describes motions in straight streamlines parallel to the x-axis while speed may vary in the y-direction from one streamline to another. The x-component velocity profile to be obtained from the governing equation may be linear or non-linear depending on the forces in the system and boundary conditions. The same streamline pattern is obtained from the trajectory equation with the integration constant assuming the role of parameter. It shows that the position hodograph representation of unidirectional flow is a family of streamlines parallel to x-axis. The hodograph representation of unidirectional flow is a system of line segments of finite length specified by the x-component velocity magnitude and coincides with the q_x-axis on q-plane. It is convenient to use the polar representation in certain situation.

For uniform flow fields, flow speed is required to assume a constant magnitude in addition to a specific direction. The hodograph representation of uniform flows is a straight line segment inclined at angle $\tilde{\theta}$ through origin with a finite length equal to constant speed $|q|$.

By setting each velocity component constant, the conditions for uniform flow at a given magnitude and inclination angle are satisfied.

$$q_x = |q_0| \cos\alpha; \quad q_y = |q_0| \sin\alpha$$
$$|q| = \sqrt{q_x^2 + q_y^2} = |q_0|$$
$$\psi = |q_0| (y \cos\alpha - x \sin\alpha)$$

The stream function depicts a system of parallel streamlines inclined at the velocity angle $\tilde{\theta} = \alpha$. The velocity profile is uniform across streamlines and the magnitude is constant everywhere.

For potential flows, the Bernoulli's theorem gives,

$$B - \left(\frac{p}{\rho} + U \right) = \frac{q^2}{2}$$

or

$$q_x^2 + q_y^2 = C(p)$$

Hodograph representation of pressure isobar is a system of circles of radius $\sqrt{C(p)}$ about origin when potential energy is invariant or negligible. Each circle corresponds to constant speed $\sqrt{C(p)}$ or constant kinetic energy $C(p)/2$. Similarly, hodograph representation of constant potential energy is also a system of circles about origin when pressure is invariant.

In general, the equation of hodograph for a straight line of arbitrary length in point-slope form or parametric form is given by:

$$q_y - q_{y0} = (q_x - q_{x0}) \tan \tilde{\theta} \qquad (5.27b)$$

or

$$q_x = q_{x0} + mt; q_y = q_{y0} + nt \qquad (5.27c)$$

Above is the parametric equation of a straight line through point (q_{xo}, q_{y0}) with slope equal to $\tan \tilde{\theta} = n/m$ and t serving as the parameter. When the point coincides with origin $(q_{xo} = 0, q_{yo} = 0)$, the straight line through origin corresponds to isoclines which define the constant direction of vector fields equal to $\tilde{\theta}$. Equation 5.27b reduces to 5.27a representing the equation of hodograph for constant flow direction $\tilde{\theta}$ in unidirectional flows. Unidirectional flow will become uniform flow when speed assumes a constant magnitude. Hodograph representation of other special trajectories includes motions in constant force fields, central force fields and circular motions at constant speed.

5.7.4 Projectile Trajectory in Constant Force Fields

Hodograph representation of projectile is presented since it is one of important trajectories of practical interest in constant force fields. When altitude changes are negligible, acceleration due to gravity is considered constant in magnitude as well as direction. For constant force fields, rectangular coordinate is convenient for motion analysis. Consider a particle of mass m moves frictionless from Cartesian coordinate origin 0, with an initial velocity V_0 at time $t_0 = 0$. The vertical y-axis points upwards as shown in Figure 5.5a. The vector representation of motion in physical plane is also known as position hodograph, which depicts the trajectory position as a function of time. The particle is subjected to a constant gravitational acceleration pointing through its center downwards. The solution of equation of motion determines the particle velocity, acceleration and the shape of trajectory.

Horizontal x − direction : $m\dfrac{dv_x}{dt} = 0$

Vertical y − direction : $m\dfrac{dv_y}{dt} = -mg$

Initial conditions : $t = 0$; $\vec{r}_0 = \begin{bmatrix} x_0 & y_0 \end{bmatrix} = \begin{bmatrix} 0 & 0 \end{bmatrix}$; $V_0 = \begin{bmatrix} v_{x0} & v_{y0} \end{bmatrix}$; $a = \begin{bmatrix} 0 & -g \end{bmatrix}$

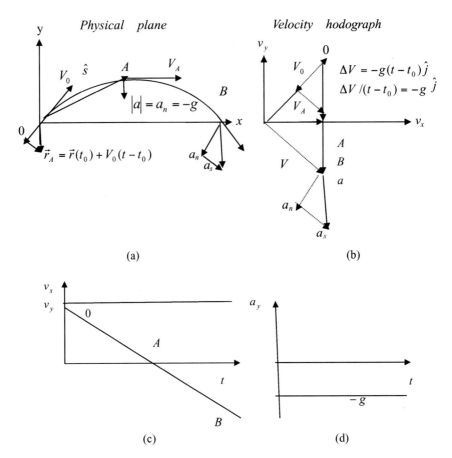

Figure 5.5 (a) Projectile trajectory; (b) Mapping of projectile on hodograph; (c) Particle velocity vs. time; (d) Particle acceleration vs. time.

The solution yields the following parametric equation of hodograph for the projectile trajectory:

$$v_x = v_{x0}$$

$$v_y = -gt + v_{y0}$$

or

$$V = v_{x0}\hat{i} + (-gt + v_{y0})\hat{j} = V_0 + \Delta V$$

$$\Delta V = V - V_0 = -gt\hat{j}$$

$$|V| = \sqrt{v_{x0}^2 + v_{y0}^2 + (gt)^2 - 2gtv_{y0}}$$

With $m = 0$; $n = -g$ in Equation 5.27c, the equation of motion is satisfied. Such that,

$$\tan \alpha = \frac{n}{m} = \infty; \quad \rightarrow \alpha = \frac{\pi}{2}$$

At time $t_0 = 0$, the initial point corresponds to point 0 at origin in physical plane with velocity V_0. Hodograph representation of projectile trajectory is a parametric equation of a line of infinite slope with t as a parameter. It is a straight vertical line passing through point (v_{x0}, v_{y0}) on velocity hodograph in Figure 5.5b. In this figure, velocity diagram could be constructed by translating the initial velocity vectors of a particle along its trajectory in Figure 5.5a to start from the origin of hodograph plane and terminate at point (v_{x0}, v_{y0}). This velocity vector represents the initial velocity of particle at origin. The particle velocity at time t is obtained by adding to the initial velocity a vertical vector $\Delta V = -g(t - t_0)\hat{j}$ and is represented by a vector from origin to the end point. In general,

$$V = V_0 + \Delta V$$

For examples, at point A and B, velocity will assume $(v_{x0}, 0)$ and $(v_{x0}, -v_{y0})$ respectively. Hence every vector representing instantaneous velocity at a given time has its end point lie on the vertical line from the end of initial velocity and the head of velocity vector traces the hodograph representation of trajectory like a pencil of rays from origin. This is attributed to the fact that horizontal velocity component is constant, $v_x = v_{x0}$. As time t varies, particle velocity is changing continuously in the course of motion. The position vector obtained by integration of velocity fields traces the parametric trajectory as a function of time.

$$\vec{r} = \int_{t_0}^{t} V \, dt$$

or

$$x = v_{x0} t$$

$$y = v_{y0} t - \frac{gt^2}{2} + y_0 = v_{y0} t - \frac{gt^2}{2}$$

By eliminating t, above trajectory equation is obtained in explicit form on x-y plane:

$$\rightarrow y = -\frac{gx^2}{2v_{x0}^2} + \frac{v_{y0}}{v_{x0}} x$$

The graph of the trajectory depicted by the above position hodograph is a parabolic curve concave downwards in Figure 5.5a. The results seem to suggest that hodograph representation may be a straight line in a constant force field although the position hodograph of trajectories are not closed in physical plane. Figure 5.5c and d show the variation of velocity and acceleration as a function of time. In launching a projectile, acceleration vector due to gravity points downwards while velocity always points in positive $\hat{s}$-direction of motion. y-component velocity decreases and eventually vanishes when particle reaches the maximum height at point A. Velocity then becomes negative and increases in the downward motion. The particle accelerates and arrives at point B of the same initial level some time later and traces a parabolic trajectory.

Figure 5.5a and b represent a mapping of projectile trajectory from t-axis onto physical plane and hodograph plane. All information about projectile motion is contained in these two graphs. Modulus and tangents to hodograph representation of trajectory on hodograph plane are useful to derive qualitative information about the motion of particle. They define velocity vector and acceleration vector of particle at a point respectively. To determine the state of motion, we display/translate them like free vectors between image points in physical and hodograph planes in a mapping process. Tangent arrow icons to the trajectory at points 0, A and B on physical plane represent velocity directions. Acceleration vector represented by the tangent arrow icons along hodograph is a straight vertical line depicting the characteristics of constant force fields. The force fields describing constant body force always point downwards and maintain force balance with particle's acceleration in accordance with Newton's law of motion. To visualize the action of acceleration vector in the context of velocity, both acceleration and velocity vectors are mapped to points 0, A and B on the physical plane in Figure 5.5a and decomposed into trajectory tangential and normal directions. The tangential component of acceleration vector acts in the opposite direction of velocity at point 0 indicating speed decreases in the upward motion. It is perpendicular to velocity at point A implying speed is constant. The tangential component is in the same velocity direction at B indicating speed increases in the downward motion. The same results are obtained on hodograph plane in Figure 5.5b using the modulus and acceleration vector at these points. Thus the particle experiences deceleration from 0 to A and acceleration from A to B. It assumes constant velocity at point A where the two vectors are mutually orthogonal. In Figure 5.5a, the normal component of acceleration representing centripetal acceleration always points towards the center of trajectory curvature and indicates motion direction changes veering to the right side of trajectory on the physical plane. Velocity always points in positive $\hat{s}$-direction of motion along which particle advances while the normal component vanishes.

$$\text{Trajectory tangent}: V = \frac{d\vec{r}}{dt} = |V|\hat{s} = \begin{bmatrix} v_{x0} & -gt + v_{y0} \end{bmatrix}$$

Acceleration (tangent to hodograph): $a = \dfrac{dV}{dt} = \dfrac{\Delta V}{t - t_0} = -g\,\hat{j}$

$$\rightarrow a_x = \dfrac{dv_x}{dt} = 0;\ a_y = \dfrac{dv_y}{dt} = -g$$

Tangent slope of trajectory: $\tan \alpha = \dfrac{v_y}{v_x} = \dfrac{dy}{dx} = -\dfrac{gx}{v_{x0}^2} + \dfrac{v_{y0}}{v_{x0}}$

Maximum height at point A: $y_{\max} = \dfrac{v_{y0}^2}{2g};\ \alpha = \dfrac{dy}{dx} = 0;$

$$x = \dfrac{v_{x0}v_{y0}}{g};\ v_y = 0;\ v_x = V_A = v_{x0}$$

Flight time to reach maximum height at A: $t = \dfrac{v_{y0}}{g}$

Trajectory curvature at A:

$$\frac{1}{R_s} = -\frac{v_{x0}g}{v_{x0}^3} = -\frac{g}{v_{x0}^2}$$

where

Trajectory curvature : $\dfrac{1}{R_s} = \dfrac{ds/dt}{d\alpha/dt} = \dfrac{(dx/dt)(d^2y/dt^2) - (dy/dt)(d^2x/dt^2)}{[(dx/dt)^2 + (dy/dt)^2]^{3/2}}$

At this point, curvature assumes maximum. Centripetal acceleration responsible for change of velocity direction also attains maximum equal to gravitational acceleration and maintains force balance between them. The trajectory tangent turns clockwise most rapidly as the particle begins to move downwards.

Horizontal range from the launch point 0 to point B: $R = \dfrac{2v_{x0}v_{y0}}{g}$

Kinetic energy per unit mass:

$$\frac{V^2}{2} = \frac{1}{2}\left(v_x^2 + v_y^2\right) = \frac{1}{2}(v_{x0}^2 + v_{y0}^2 - 2v_{y0}gt + (gt)^2)$$

$$= \frac{1}{2}\left(v_{x0}^2 + v_{y0}^2\right) - g(y - y_0) = \frac{1}{2}\left(v_{x0}^2 + v_{y0}^2\right) - gy$$

It follows that total mechanical energy equals to the sum of kinetic energy and potential energy and is conserved:

$$\frac{V^2}{2} + gy = \frac{1}{2}\left(v_{x0}^2 + v_{y0}^2\right) = \text{Constant}$$

$$\rightarrow y = \frac{1}{2g}\left(v_{x0}^2 + v_{y0}^2 - V^2\right)$$

Above is a mapping of kinetic energy into a system of horizontal coordinate lines on the physical plane. The elevation of horizontal coordinate lines represents particle's potential energy level with respect to origin. The expression describes the exchange of energy between potential and kinetic energy in a conservative system. There is a loss/gain of kinetic energy when particle is moving upwards/downwards. This corresponds to a gain/loss of potential energy of particle in accordance with energy conservation principle. We could visualize the state of constant kinetic energy or constant speed or constant potential energy on physical plane as locus of points when horizontal coordinate lines intersect the parabolic trajectory. For instance, x-axis is one of such horizontal coordinate lines represents zero potential energy and intersects the projectile at point 0 and B at which the speed equals to $|V| = |V_0|$.

The energy exchange processes could be interpreted from the concept of work done. Work done by the particle moving in a force field is equal to the change of kinetic energy between two points, that is, 0 and A:

$$W_\Phi = \int_0^A f \cdot d\bar{r} = \int_0^A m \frac{dV}{dt} \cdot \frac{d\bar{r}}{dt} dt = \int_0^A \frac{m \, dV}{dt} \cdot V dt$$

$$= \frac{mV_A^2}{2} - \frac{mV_0^2}{2} = \Delta \frac{mV^2}{2} \qquad (5.28)$$

Work done by the gravitational force field is equal to the change in potential energy of the particle between two points, that is, 0 and A.

$$\text{Work done}: W_U = \int_0^A -f \cdot d\bar{r} = \int_0^A mg \, dy = mg \int_0^A dy = mg(y_A - y_0) = \Delta U$$

Hence, the energy conservation principle is written for the change of energy between two points:

$$\Delta \frac{mV^2}{2} + \Delta U = 0$$

$$\text{Distance travelled}: \frac{ds}{dt} = |V| = \sqrt{v_{x0}^2 + v_{y0}^2 - 2gy}$$

$$\rightarrow s = C^* \pm \int \sqrt{v_{x0}^2 + v_{y0}^2 - 2gy} \, dt$$

The results for projectile motions are applicable to a charged particle traveling in a uniform constant electric field parallel to y-axis, $E = [0 \; -E_y]$, which also constitutes a constant force field.

5.7.5 Motion Trajectory in Central Force Fields

Unlike constant force fields, force may change its direction and/or magnitude continuously in time and space. An important example is central force fields. Paths traveled by planets and satellites are conic sections driven by central force fields. Kepler has laid the foundation work for the study of planet's orbits. We are concerned about hodograph representation of trajectories to provide another source of information about motions in an attempt to interpret the characteristics of force fields. Polar coordinate is convenient for central force fields and is adopted as the reference coordinate. By placing the fixed point to coincide with polar coordinate origin and (r, θ) plane of motion perpendicular to z-axis, particles are confined to move on the plane of motion containing both velocity and position vectors. The cross product of these two vectors is by definition, the angular momentum of a particle per unit mass m about origin. There is only one nonzero z-component of angular momentum Y_z.

$$Y \equiv m(\vec{r} \times V) = m \begin{vmatrix} \hat{r} & \hat{\theta} & \hat{z} \\ r & 0 & 0 \\ \dfrac{dr}{dt} & r\dot{\Omega}_z & 0 \end{vmatrix}$$

$$Y_z = mr^2\dot{\Omega}_z; \; Y_r = Y_\theta = 0$$

$$Y \cdot V = \begin{bmatrix} 0 & 0 & mr^2\dot{\Omega}_z \end{bmatrix} \cdot \begin{bmatrix} \dfrac{dr}{dt} & r\dot{\Omega}_z & 0 \end{bmatrix} = 0$$

For central force fields, the conservation of angular momentum states that:

$$\sum \Gamma = \frac{dY}{dt} = 0; \; \rightarrow Y = \text{constant}$$

where

$$V = \frac{d\vec{r}}{dt} = \frac{d(r\hat{r})}{dt} = \begin{bmatrix} \dfrac{dr}{dt} & r\dot{\Omega}_z \end{bmatrix}$$

$$\dot{\Omega} = \begin{bmatrix} 0 & 0 & \dot{\Omega}_z \end{bmatrix}; \; \dot{\Omega}_z = \frac{d\theta}{dt}; \; v_\theta = r\dot{\Omega}_z$$

When potential energy of a system in motion depends only on distance from a fixed point and the force acting on a particle is directed towards or away from the fixed point (center of attracting body), it is classified as motions in a central force field. This implies central force fields are spherical symmetry and conservative in nature and can always be derived from a potential. Force of this type depends on distance and produces no torque and hence angular momentum relative to the fixed point is conserved in central force fields as stated by angular momentum principle. Furthermore the direction of angular momentum must be perpendicular to the plane of motion. The motion depends on angular momentum and mechanical energy (kinetic plus effective potential) of the system. The trajectory model is described by the vector equation of motion:

$$m\frac{dV}{dt} = f(r)$$

Or in component form:

$$\hat{r}\text{-direction:} \, m\left(\frac{d^2r}{dt^2} - r\dot{\Omega}_z^2\right) = f(r)$$

$$\hat{\theta}\text{-direction:} \, m\left(r\frac{d^2\theta}{dt^2} + 2\frac{dr}{dt}\dot{\Omega}_z\right) = 0$$

where

$$f(r) = -\frac{dU(r)}{dr} = -\frac{GMm}{r^2}$$

$$U(r) = -\frac{GMm}{r}$$

G, M and m are the universal gravitational constant and masses of the objects. Above expresses an attractive inverse-square law of force derived from potential energy $U(r)$. Hence, the force depends on distance and always points towards the fixed point opposite to that of position vector.

Initial conditions: $t = 0$, at point B (perigee)

$$r = r_{min}; v_r = \frac{dr}{dt} = 0; \theta = \theta_0 = 0; v_\theta = r\frac{d\theta}{dt} = r\dot{\Omega}_z = \frac{h}{r} = \frac{h}{r_{min}} = v_{max}$$

Taking the cross product of position vector and vector equation of motion, we obtain the following solution:

$$m\frac{dV}{dt} = f(r) = -\frac{GMm\hat{r}}{r^2}$$

$$m\vec{r} \times \frac{dV}{dt} = -\vec{r} \times \frac{GMm\hat{r}}{r^2} = 0$$

$$m\frac{d(\vec{r} \times V)}{dt} = 0$$

$$Y = m(\vec{r} \times V) = \text{Constant}$$

Hence,

$$Y_z = mr^2\dot{\Omega}_z = mr\,v_\theta = mr_{\min}v_{\max} = mh$$

$$Y_r = Y_\theta = 0$$

Thus, angular momentum is conserved in central force fields. The transverse velocity component and distance are inversely related. The solution satisfies $\hat{\theta}$-direction equation of motion by substitution. Similarly,

$$-Y \times \frac{dV}{dt} = -Y \times \left(-\frac{GM\hat{r}}{r^2}\right)$$

$$\frac{d(V \times Y)}{dt} = r\hat{r} \times m\frac{d(r\hat{r})}{dt} \times \left(\frac{GM\hat{r}}{r^2}\right)$$

$$= GMm\left[r\left(\frac{d\hat{r}}{dt}\right)\left(r\hat{r} \cdot \frac{\hat{r}}{r^2}\right) - \frac{\hat{r}}{r^2}\left(r\hat{r} \cdot \frac{r\,d\hat{r}}{dt}\right)\right]$$

$$= GMm\left[\frac{d\hat{r}}{dt} - \hat{r}\left(\frac{\hat{r} \cdot d\hat{r}}{dt}\right)\right] = GMm\frac{d\hat{r}}{dt}$$

Integration yields the following solution for the trajectories:

$$V \times Y = GMm\hat{r} + \vec{p} = mr^3\dot{\Omega}_z^2\hat{r}$$

$$m\vec{r} \cdot (GMm\hat{r} + |p|\,\hat{p}) = m\vec{r} \cdot mr^3\dot{\Omega}_z^2\hat{r} = Y_z^2$$

or

$$GMr + \frac{r}{m}|p|\cos(\theta - \vartheta) = h^2$$

where

$$\hat{r} = \begin{bmatrix} \cos\theta & \sin\theta \end{bmatrix}$$
$$\hat{p} = \begin{bmatrix} \cos\vartheta & \sin\vartheta \end{bmatrix}$$

$\vec{p}$ is an arbitrary constant vector for the indefinite integral. ϑ is the angle between unit vector $\hat{p}$ and positive x-axis counterclockwise. The unit vector $\hat{p}$ is arbitrary and is placed to coincide with x-axis, $\vartheta = 0$. θ is the angle between vector $\vec{r}$ and x-axis. $|p|$ and h are determined by the initial conditions, which are chosen at perigee for convenience.

$$|p| = \frac{m(h^2 - GMr_{min})}{r_{min}}$$

where

$$h = r\,v_\theta = r_{min}v_{max} = \text{constant}$$

The trajectory reduces to polar equation of a conic section similar to Kepler's first law.

$$r = \frac{h^2}{GM(1 + e\cos\theta)} \tag{5.29}$$

Or in Cartesian form:

$$\frac{(x + ae)^2}{a^2} + \frac{y^2}{b^2} = 1$$

where

$$\text{Ellipse semi-major axis:} a = \frac{h^2}{GM(1 - e^2)} = \frac{r_{min}}{1 - e} = \frac{r_{max}}{1 + e} = \frac{b}{\sqrt{1 - e^2}}$$

$$\text{Ellipse semi-minor axis:} b = \frac{h^2}{GM\sqrt{1 - e^2}} = \sqrt{r_{min}r_{max}}$$

$$r_{min} = \frac{h}{v_{max}} = \frac{h^2}{GM(1 + e)}; \quad r_{max} = \frac{h}{v_{min}} = \frac{h^2}{GM(1 - e)}$$

$$r_m = \frac{r_{min} + r_{max}}{2} = \frac{a(1 - e) + a(1 + e)}{2} = a = \frac{h}{v_m} = \frac{h^2}{GM}; \quad \frac{h^2}{GM} = \frac{b^2}{a}$$

$$r_{min} + r_{max} = 2a$$

Ellipse center: $(-ae, 0)$

Foci with respect to ellipse center: $(c, 0)$; $c = \pm\sqrt{a^2 - b^2} = \pm ae$

Foci with respect to center of attracting body (polar origin or Cartesian origin):

$$(c, 0) + (-ae, 0) = \begin{cases} (0, 0) \\ (-2ae, 0) \end{cases}$$

$$\begin{bmatrix} v_x \\ v_y \end{bmatrix} = \begin{bmatrix} \cos\theta & -\sin\theta \\ \sin\theta & \cos\theta \end{bmatrix} \begin{bmatrix} v_r \\ v_\theta \end{bmatrix}$$

The eccentricity of trajectory e is related to physical parameters given by:

$$e = \frac{|p|}{GMm} = \frac{h^2}{GMr_{min}} - 1 = \frac{h v_{max}}{GM} - 1 = \frac{r_{min} v_{max}^2}{GM} - 1$$

$$v_{max} = \sqrt{\frac{GM(1+e)}{r_{min}}} = \frac{GM}{h}(1+e)$$

$$v_{min} = \frac{h}{r_{max}} = \sqrt{\frac{GM(1-e)}{r_{max}}} = \frac{GM}{h}(1-e)$$

$$v_m = \frac{v_{max} + v_{min}}{2} = \frac{GM}{h}$$

The results show that trajectories driven by central forces are conic sections. The shapes include circle $e = 0$, ellipse $e < 1$, parabola $e = 1$ or hyperbola $e > 1$.

The velocity components are:

$$v_r = \frac{dr}{dt} = \frac{h^3 e \sin\theta}{GM(1 + e\cos\theta)^2 r^2} = \frac{GMe\sin\theta}{h} \tag{5.30a}$$

$$v_\theta = r\frac{d\theta}{dt} = \frac{h}{r} = \frac{GM}{h}(1 + e\cos\theta) \tag{5.30b}$$

By applying initial conditions, the results are the same as above and are consistent with the invariance of angular momentum.

$$|V|_{t=0} = \sqrt{v_x^2 + v_y^2} = \sqrt{v_r^2 + v_\theta^2}\,|_{\theta_0=0} = \sqrt{\left(\frac{GM}{h}(1+e)\right)^2} = \frac{GM}{h}(1+e) = v_{max}$$

Hence,

$$v_{max}r_{min} = \frac{GM}{h}(1+e)\frac{h^2}{GM(1+e)} = h = v_{min}r_{max} = v_m r_m$$

The hodograph equation of trajectory in parametric form with polar angle as a parameter is expressed in Cartesian coordinate as follows:

$$v_x \cos\theta + v_y \sin\theta = \frac{GMe}{h}\sin\theta$$

$$-v_x \sin\theta + v_y \cos\theta = \frac{GM}{h}(1 + e\cos\theta)$$

By grouping terms and taking squares of both sides, polar angle can be eliminated after adding above equations to yield standard conic form:

$$v_x^2 + \left(v_y - \frac{GMe}{h}\right)^2 = \left(\frac{GM}{h}\right)^2 = v_m^2$$

Above hodograph representation of trajectory in central force fields is a circle with center $0'$ at $\left(0, \frac{eGM}{h}\right)$ and radius equal to $\frac{GM}{h}$. It shows that conic trajectories on physical plane are mapped to circles on hodograph plane. For the special case of circular trajectories, $(e = 0)$ the circle center $0'$ coincides with coordinate origin 0 (center of attracting body) in hodograph plane meaning that kinetic energy is conserved in circular motions. Velocity of moving body is perpendicular to acceleration vector everywhere. As a result, both tangential component of acceleration vector and tangential component of attracting force vanish. The motion is at constant speed. The normal component of acceleration gives rise to centripetal force and maintains force balance with the central attracting force.

$$V \cdot a = V \cdot \frac{dV}{dt} = \frac{dV^2}{2dt} = \frac{d}{2dt}\left(\frac{GM}{h}\right)^2 = 0$$

$$V \cdot \vec{r} = \left[\frac{dr}{dt} \quad r\frac{d\theta}{dt}\right] \cdot \left[r \quad 0\right] = 0$$

where

$$\frac{dr}{dt} = 0; \quad V_{e=0}^2 = \left(\frac{GM}{h}\right)^2$$

$$a = \left[-r\left(\frac{d\theta}{dt}\right)^2 \quad r\frac{d^2\theta}{dt^2}\right] = \left[-\frac{v_\theta^2}{r} \quad 0\right]$$

Furthermore, transverse velocity component and angular speed are required by angular momentum principle to assume constant since there is no change in radius distance. Potential energy is constant as well. The total mechanical energy is invariant at every point on the trajectory as stated by conservation of energy principle.

Figure 5.6 depicts the general case of an elliptic trajectory ($e < 1$) for an object in slow motions on physical plane and its hodograph representation (circle) on hodograph plane. The ellipse's center is at ($-ae$, 0) and one of its focuses coincides with the center of attracting body as shown. The center of attracting body at (0,0) is the fixed point at polar origin coinciding with Cartesian coordinate origin 0. For a given polar angle, we can determine the moving object's position on each plane in Figure 5.6 from the position hodograph representation of trajectory and hence its velocity can be obtained directly. Each point is mapped from θ-axis in positive (counter-clockwise) direction onto its image points along trajectories on each plane. At an arbitrary point p, tangents along trajectories on physical plane and hodograph plane define velocity and acceleration vector of the moving object respectively. To visualize the action of these two vectors, it is best to display both tangent arrow icons at a point by translation and then decompose them into the trajectory tangential and normal components in the same way as central force does. The tangential component of velocity points in the direction of motion while normal component vanishes. The normal component of acceleration vector always points towards the center of curvature giving rise to centripetal acceleration as motion direction changes along curved trajectories. The tangential component of acceleration determines speed changes. In elliptic orbits, the attracting force always pointing towards the fixed point is no longer perpendicular to the direction of motion and both velocity and distance vary continuously. Nonetheless it maintains force balance with the acceleration of the object everywhere in accordance with Newton's law of motion. In

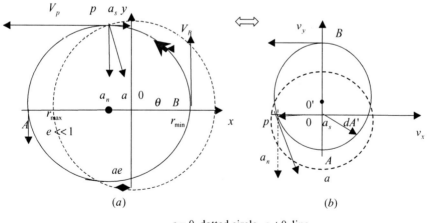

(a)   (b)

$e = 0$ dotted circle, $e \neq 0$ line

Figure 5.6 (a) Conic trajectories; (b) Hodograph representation of trajectories

Figure 5.6, point B is located at perigee such that distance is minimum and velocity is maximum and perpendicular to acceleration vector. Hence, there is no tangential component of acceleration vector acting in the direction of motion and speed is constant equal to v_{max} of moving object. At point p, the tangential component of acceleration vector acts opposite to velocity direction and the object is slowing down. In Figure 5.6a, the normal component (centripetal acceleration) acts towards the center of curvature of trajectory indicating velocity direction changes to the left side of trajectory on physical plane. When the object travels in anticlockwise from point B to point p then A, (apogee) kinetic energy decreases continuously and assumes minimum at A. In this process, distance increases steadily. At point A potential energy assumes maximum. As the object continues to travel from point A to point B in anticlockwise direction, the kinetic energy and potential energy processes reverse in the opposite manner and thereafter repeats the cycle at point B. The total mechanical energy is constant at every point on the trajectory. Points B and A are mapped to the points of intersection between hodograph circumference of circle $0'$ and v_y-axis. They correspond to maximum and minimum speed (or kinetic energy). The shifting of circle center from the coordinate origin reflects speed changes along the path.

The areal velocity defined as the time rate of area swept out by the radius vector $|\vec{r}|$ on physical plane is found equal to constant given by Kepler's second law.

$$\frac{dA}{dt} = \frac{1}{2}\left|\vec{r} \times \frac{d\vec{r}}{dt}\right| = \frac{1}{2}|\vec{r} \times V| = \frac{r^2 \, d\theta}{2 \, dt} = \left(\frac{r^2 \dot{\Omega}_z}{2}\right) = \frac{h}{2} = \text{Constant}$$

where

$$dA = \frac{1}{2}|\vec{r} \times d\vec{r}|$$

Hence, angular speed varies from point to point as a function of polar angle. It is constant only for circular orbits.

$$\frac{d\theta}{dt} = \frac{h}{r^2} = \frac{(GM)^2}{h^3}(1 + e \cos \theta)^2$$

$$\text{At } B \text{ perigee, } \frac{d\theta}{dt}\Big|_{\theta=0} = \frac{(GM)^2}{h^3}(1 + e \cos \theta)^2 = \frac{(GM)^2}{h^3}(1 + e)^2$$

$$= \dot{\Omega}_{z\,max}; \ r = \frac{h^2}{GM(1 + e)} = r_{min}$$

$$\text{At } A \text{ apogee, } \frac{d\theta}{dt}\Big|_{\theta=\pi} = \frac{(GM)^2}{h^3}(1 - e)^2 = \dot{\Omega}_{z\,min}; \ r = \frac{h^2}{GM(1 - e)} = r_{max}$$

An attempt is then made to demonstrate the use of hodograph plane in the context of physical plane. Applying the idea of areal velocity and its geometric interpretation, the areal velocity on hodograph plane is defined as the time rate of area swept out by the velocity radius $|V|$. It is a function of polar angle.

$$\frac{dA'}{dt} = \frac{1}{2}\left|V \times \frac{dV}{dt}\right| = \frac{1}{2}|V \times a| = \frac{|V|^2}{2}\frac{d\theta}{dt} = \frac{|V|^2 h}{2r^2}$$

$$= \frac{(GM)^4}{2h^5}(1 + 2e\cos\theta + e^2)(1 + e\cos\theta)^2$$

5.7.5.1 For Circular Trajectory e= 0

$$\frac{dA'}{dt} = \frac{|V|^2}{2}\frac{d\theta}{dt} = \frac{|V|^2 h}{2r^2} = \frac{(GM)^4}{2h^5} = \text{constant}$$

5.7.5.2 Elliptical Orbits: 1 > e> 0

At perigee, $\dfrac{dA'}{dt}\Big|_{\theta=0} = \dfrac{|V|^2}{2}\dfrac{d\theta}{dt} = \dfrac{|V|^2 h}{2r^2}$

$$= \frac{(GM)^4}{2h^5}(1 + 2e + e^2)(1 + e)^2 = \frac{dA'}{dt}\Big|_{\max}$$

At apogee, $\dfrac{dA'}{dt}\Big|_{\theta=\pi} = \dfrac{|V|^2}{2}\dfrac{d\theta}{dt} = \dfrac{|V|^2 h}{2r^2}$

$$= \frac{(GM)^4}{2h^5}(1 - 2e + e^2)(1 - e)^2 = \frac{dA'}{dt}\Big|_{\min}$$

The time it takes to go around the attracting body is the orbital period T. On hodograph plane, the area swept out by the velocity radius vector in one period is equal to the circle area of constant value with circle radius equal to constant mean speed. On the physical plane, the area swept out by the position radius vector in one period is equal to the area of an ellipse.

$$A' = \int_0^T \frac{dA'}{dt}dt = \int_0^T \frac{|V|^2}{2}\frac{d\theta}{dt}dt = \int_0^T \frac{(\pi |V|^2)_{circle}h}{2(\pi r^2)_{ellipse}}dt$$

$$= \int_0^T \frac{v_m^2 h}{2ab}dt = \frac{v_m^2 hT}{2ab} = \pi v_m^2$$

or

$$\rightarrow T = \frac{2\pi ab}{h}$$

$$T^2 = \left(\frac{2\pi ab}{h}\right)^2 = \frac{4\pi^2 a^3 h^2}{h^2 GM} = \frac{4\pi^2 a^3}{GM}$$

The orbital period and distance ratio obtained from hodograph plane is identical to Kepler's third law derived from the physical plane. The fixed point at ellipse focus is a singular point and the critical point is at infinity. They are mapped to point at infinity and origin respectively on hodograph plane.

By taking the product of elemental displacement with $\hat{r}$-direction equation of motion and then integrate to yield the mechanical energy equation:

$$\int \left[m\left(\frac{d^2 r}{dt^2} - r\dot{\Omega}_z^2\right) - f(r) \right] dr = W_1 + W_2 + W_3 = E$$

where

$$W_1 = \int m\frac{dV}{dt}\frac{dr}{dt}dt = \int mVdV = \frac{mV^2}{2}$$

$$W_2 = -\int mr\dot{\Omega}_z^2 dr = -mh^2 \int \frac{dr}{r^3} = \frac{mh^2}{2r^2}$$

$$W_3 = \int_{\infty}^{r} -f(r)dr = \int_{\infty}^{r} \frac{GMm}{r^2}dr = -\frac{GMm}{r}\Big|_{\infty}^{r} = -\frac{GMm}{r} = U(r)$$

E is the integration constant and assumes the role of total mechanical energy. Its value could be negative, zero or positive and determines the state of the system. The work done W_1 associated with an object moving in a force field is equal to the change of kinetic energy between two points. Work done W_2 is referred to as the centripetal potential energy since the force derived from the negative gradient of this potential is interpreted as centripetal force. Centripetal potential energy constitutes a portion of kinetic energy and accounts for the change in motion directions.

$$f_c = -\frac{d}{dr}\left(\frac{mh^2}{2r^2}\right) = \frac{mh^2}{r^3} = \frac{mv_\theta^2}{r}$$

The work done W_3 is required to bring an object from an infinite distance to a point r and is equal to the negative of potential, which together with the fields will vanish at infinity. The sum of W_2 and W_3 is the effective potential. The total mechanical energy of a moving object is the statement of energy conservation of the equivalent one-dimensional system. For circular motions, each form of energy is conserved by itself.

$$E = \frac{mV^2}{2} + \frac{mh^2}{2r^2} + U(r)$$

In summing up the study on central force fields, trajectories are conic curves and the shapes depend on eccentricity (Kepler's first law). Hodograph representation of conic trajectories is a circle of radius equal to GM/h. The circle center coincides with origin only for circular orbits. It is interesting to take note of Galileo's investigation when planet's orbits were mapped onto hodograph plane taking the shape of circular trajectories. Areal velocity based on hodograph plane generally varies as a function of polar angle. It is constant only for circular orbits. Otherwise, it assumes maximum and minimum values at perigee and apogee respectively. The orbital period and distance ratio derived on hodograph plane is identical to Kepler's third law.

5.7.6 Trajectory of Charged Particles in Uniform Magnetic Fields

Consider a charged particle q of mass m moving on x-y plane in a uniform and constant magnetic field (B-field) perpendicular to the plane of motion. The charged particle is released at a point (x_0, y_0) and has an initial velocity V_0. It is subject to a deflection magnetic force perpendicular to its motion. The equation of motion for a charged particle with negligible body force is:

$$\frac{md V}{dt} - q(V \times B)$$

where

$$B = \begin{bmatrix} 0 & 0 & B_z \end{bmatrix}$$

For convenience, one of Cartesian coordinate axis say x-axis is aligned parallel to the initial velocity. The above is written in component form with respect to Cartesian coordinate:

$$\frac{dv_x}{dt} = \frac{q B_z v_y}{m}$$

$$\frac{dv_y}{dt} = -\frac{q B_z v_x}{m}$$

$$\frac{dv_z}{dt} = 0; \rightarrow v_z = C = 0$$

Initial conditions : $t = 0; x = x_0; y = y_0; v_x = V_0; v_y = 0$

By eliminating y-velocity component, the above can be written to describe simple harmonic motion:

$$\frac{d^2 v_x}{dt^2} + k v_x = 0; \; k \equiv \left(\frac{q B_z}{m}\right)^2$$

Solution:

$$v_x = V_0 \cos(\sqrt{k}t)$$

$$v_y = -\frac{q B_z}{m} \int v_x dt + C = -V_0 \sin(\sqrt{k}t)$$

By eliminating t, the relation between velocity components is the equation of hodograph for the charged particle's trajectory in magnetic fields:

$$\rightarrow v_x^2 + v_y^2 = V_0^2 = |V|^2$$

$$a_x = \frac{dv_x}{dt} = -V_0 \sqrt{k} \sin(\sqrt{k}t)$$

$$a_y = \frac{dv_y}{dt} = V_0 \sqrt{k} \cos(\sqrt{k}t)$$

$$a \cdot V = \begin{bmatrix} a_x & a_y \end{bmatrix} \cdot \begin{bmatrix} v_x & v_y \end{bmatrix} = 0$$

$$|a| = \sqrt{a_x^2 + a_y^2} = V_0 \sqrt{k} = \frac{V_0^2}{r}$$

The hodograph representation of trajectory is a circle about origin with a radius V_0 equal to initial speed. Kinetic energy is conserved in the motion. The equation for particle trajectory on physical plane is the position hodograph representation of trajectory obtained by integration of velocity.

$$x = \int v_x dt + C = x_0 + \frac{V_0}{\sqrt{k}} \sin(\sqrt{k}t)$$

$$= x_0 - \frac{v_y}{\sqrt{k}}$$

$$y = \int v_y dt + C = y_0 - \frac{m V_0}{q B_z} + \frac{V_0}{\sqrt{k}} \cos(\sqrt{k}t)$$

$$= y_0 - \frac{V_0}{\sqrt{k}} + \frac{v_x}{\sqrt{k}}$$

The magnitude of magnetic force per unit mass is equal to the centripetal force per unit mass:

$$F = \frac{q B_z}{m}\sqrt{v_y^2 + (-v_x)^2} = \frac{q B_z V_0}{m} = V_0\sqrt{k} = \frac{V_0^2}{r}$$

$$\rightarrow (x - x_0)^2 + \left(y - \left(y_0 - \frac{V_0}{\sqrt{k}}\right)\right)^2 = \frac{V_0^2}{k} = r^2$$

It can be seen that the position hodograph representation of particle trajectory describes circular motion about point $\left(x_0, y_0 - \frac{V_0}{\sqrt{k}}\right)$ with a radius $r = \frac{V_0}{\sqrt{k}} = \frac{m V_0}{q B_z}$ lying on the plane of motion. Since radius corresponds to the radius of curvature R_s in circular trajectory, the initial velocity is determined by the model parameters, $V_0 = \frac{R_s q B_z}{m} = |V|$. The circular trajectory is mapped from physical plane to another circle of radius equal to initial speed about origin on hodograph plane and conversely. The radius is a measure of momentum possessed by the particle. It is proportional to particle momentum and inversely to magnetic effects. Figure 5.7 depicts the trajectory on physical plane and its hodograph representation on hodograph plane. The circle centers are located at $\left(x_0, y_0 - \frac{V_0}{\sqrt{k}}\right)$ and origin in the physical plane and hodograph plane respectively. Tangents at an arbitrary point p define velocity and acceleration vector of the particle at p on the physical plane and hodograph plane. By displaying both acceleration vector and velocity at point p, we could determine the state of motion. The circular motion is at constant speed and acceleration is perpendicular to velocity everywhere. The tangential component of acceleration

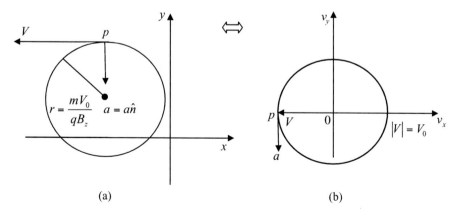

(a)   (b)

Figure 5.7 (a) Trajectory on physical plane; (b) Hodograph representation of trajectory

vanishes while the normal component represents centripetal acceleration acts towards circle center indicating velocity direction changes on the physical plane (Figure 5.7a). The z-component angular momentum is constant and is directed perpendicularly to the plane of motion.

$$Y \equiv \vec{r} \times Vm = m \begin{vmatrix} \hat{i} & \hat{j} & \hat{k} \\ x - x_0 & y - \left(y_0 - \frac{V_0}{\sqrt{k}}\right) & 0 \\ V_0 \cos(\sqrt{k}t) & -V_0 \sin(\sqrt{k}t) & 0 \end{vmatrix}$$

$$Y_z = \frac{m}{\sqrt{k}} V_0^2 (\cos^2 \sqrt{k}t + \sin^2 \sqrt{k}t) = \frac{m V_0^2}{\sqrt{k}} = \frac{(m V_0)^2}{q B_z}$$

$$Y_r = Y_\theta = 0$$

In a different approach, the equation of motion is written with reference to streamline coordinate.

$$\hat{s} - \text{direction} : m \left(\frac{\partial v_s}{\partial t} + v_s \frac{\partial v_s}{\partial s} - v_n \omega_z \right) = -q B_z v_n$$

$$\text{or } \frac{d v_s}{dt} = a_s = 0$$

$$\hat{n} - \text{direction} : m \left(\frac{\partial v_n}{\partial t} + v_n \frac{\partial v_n}{\partial n} + \frac{v_s^2}{R_s} \right) = q B_z v_s$$

$$\text{or } \frac{v_s^2}{R_s} = \frac{q B_z v_s}{m} = a_n; \text{ with } v_n = 0$$

$$\rightarrow R_s = \frac{m v_s}{q B_z} = \frac{m V_0}{q B_z} = r$$

$$v_s = \frac{R_s q B_z}{m} = \text{constant}$$

Hence,

$$a \cdot V = \begin{bmatrix} 0 & \dfrac{v_s^2}{R_s} \end{bmatrix} \cdot \begin{bmatrix} v_s & 0 \end{bmatrix} = 0; \; |V| = v_s$$

$$\hat{z} - \text{direction} : \frac{d v_z}{dt} = 0; \rightarrow v_z = C = 0$$

Centripetal acceleration is constant equal to magnetic force per unit mass to keep particles on circular tracks. The results are the same as those obtained from the solutions using Cartesian coordinate. If the initial z-component velocity has a constant magnitude other than zero, the trajectory will be a helix. The axis of the winding path is aligned with magnetic field in the same z-direction.

5.8 Example 4.1b Mapping of Field Patterns and Image Warping

This second part illustrates the specific properties of mapping by complex functions. It involves images of curves in various planes. Warping of images by Joukowski transformation is also presented.

5.8.1 Mapping of Field Patterns

In part 4.1a, the map from z to z^2 demonstrates the simplification of Laplace equation in the wedge using a power function. Conformal mapping allows us to visualize field patterns in various wedge angles by choosing different power of z. The complex potential in part 4.1a maps streamlines and equipotentials from z-plane as a system of straight horizontal lines and straight vertical lines on complex F-plane in Figure 5.8. Maximum/minimum

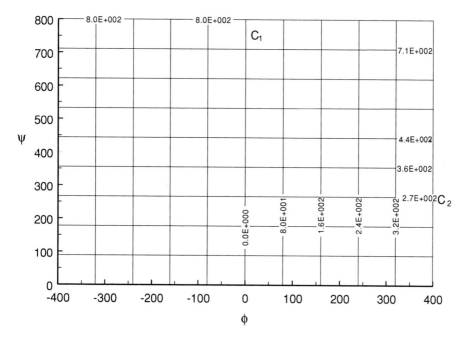

Figure 5.8 Mapping of equipotentials and streamlines from z-plane to complex F-plane

values of streamlines or equipotentials are on the boundaries of domain. In this plane, all true squares have the same size and identical kinetic energy storage since contour intervals are maintained as equal and constant. Kinetic energy density distributions in flow fields can be interpreted with reference to curvilinear square map on z-plane and its images on F-plane. The smallest curvilinear square on physical plane indicates the greatest kinetic energy density. In this plane, the transfer number is obtained from tube ratio close to unity in part 4.1a, that is,

$$\frac{S}{\Delta z} = \zeta = \frac{N_p}{N_f} = \frac{9}{10} = 0.9$$

To map rectangular coordinate lines from Cartesian coordinate to potential-stream function coordinate, the mapping functions in part 4.1a are inverted as follows:

$$x = \pm \left[\frac{1}{2} \left(\phi + \sqrt{\phi^2 + \psi^2} \right) \right]^{1/2} \qquad (5.31a)$$

$$y = \pm \psi \left[2 \left(\phi + \sqrt{\phi^2 + \psi^2} \right) \right]^{-1/2} \qquad (5.31b)$$

By setting $x = C'_1$ or $y = C'_2$ as a parameter, above is expressed in conic forms.

$$\psi^2 = 4C'^2_1(C'^2_1 - \phi)$$
$$\psi^2 = 4C'^2_2(C'^2_2 + \phi)$$

As the parameter C'_1 varies, a family of parabola is generated with focus at the origin and vertex at $(C'^2_1, 0)$ on the right. By varying C'_2, a system of parabola is obtained with focus at the origin and vertex at $(-C'^2_2, 0)$ on the left. The results are presented in Figure 5.9. The flow fields are transformed into uniform flow parallel to the horizontal axis of potential-stream function coordinate as well as on F-plane. The first quadrant of z-plane is mapped to the upper half of potential-stream function coordinate as the right angle doubles. In other words, the vertical wall y-axis and horizontal x-axis are mapped to the entire horizontal ϕ-axis. The orthogonal families of rectangular coordinate lines are transformed into two co focal families of parabolas having ϕ-axis as the axis of parabola. The ϕ-axis is also an axis of symmetry. The two sets of curves remain orthogonal since inverse mapping is conformal. Furthermore, both x and y contours terminate perpendicularly on ϕ-axis.

$$\frac{\partial x}{\partial \psi} \Big|_{\psi=0} = \frac{\partial y}{\partial \psi} \Big|_{\psi=0} = 0$$

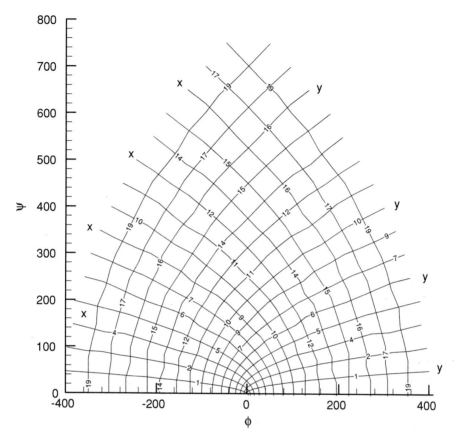

Figure 5.9 Mapping of coordinate lines from Cartesian frame to potential-stream function coordinate

$\psi = 0$ and $\phi = 0$ are mapped to origin on (ϕ, ψ) plane. Every pair of streamline $\psi = C_2$ and equipotential $\phi = C_1$ that satisfies $|f(z)| = \sqrt{\phi^2 + \psi^2} = C$ is mapped to a circle of radius C about origin on (ϕ, ψ) plane. For example, given a boundary point at $x = 20$, $y = 0$, $|z|_{boundary} = 20$; $\phi = 400$, $\psi = 0$, we have then,

$$|f(z)|_{boundary} = \sqrt{\phi^2 + \psi^2} = C_{max} = 400$$
$$q_x = 2x = 40; \quad q_y = -2y = 0; \quad |V|_{max} = 40$$

The maximum modulus principle ensures that both complex potential and complex velocity assume maximum values on the boundary.

It is possible to determine the direction of vector fields in the physical plane from β angle on potential-stream function coordinate. Consider for example at the point of intersection $\phi = 0$, $\psi = 200$ and its image point($x = y = 10$) in the physical plane. The velocity direction αangle in the physical plane is determined from β angle by measuring the tangent slope along $y = 10$ curve on potential-stream function coordinate:

$$\tan \beta = \left. \frac{d\psi}{d\phi} \right|_{\substack{\phi=0 \\ \psi=200}} = 1 \tag{5.32}$$

$$\chi \equiv \tan^{-1} \left. \frac{dy}{dx} \right|_{y=10} = 0$$

$$\beta = \chi - \alpha = \frac{\pi}{4}$$

$$\rightarrow \alpha = -\frac{\pi}{4} @x = 10, \, y = 10$$

Or using $x = 10$ curve,

$$\tan \beta = \left. \frac{d\psi}{d\phi} \right|_{\substack{\phi=0 \\ \psi=200}} = -1$$

$$\chi = \tan^{-1} \left. \frac{dy}{dx} \right|_{x=10} = \frac{\pi}{2}$$

$$\beta = \frac{\pi}{2} - \alpha = \frac{3\pi}{4}$$

$$\rightarrow \alpha = -\frac{\pi}{4} @x = 10, \, y = 10$$

In polar coordinates, circles are mapped to circles and rays are mapped to double angle.

$$f(z) = z^2 = r^2(\cos 2\theta + i \sin 2\theta) = R(\cos \varphi + i \sin \varphi)$$
$$\rightarrow r^2 = R$$
$$2\theta = \varphi$$

Mapping of equipotentials and stream function from the physical plane to conjugate hodograph plane is accomplished with the aid of Equations 5.15 to 5.18.

$$\phi = \int \frac{1}{J} \left[-\frac{\partial q_x}{\partial y} \frac{\partial \phi}{\partial x} + \frac{\partial q_x}{\partial x} \frac{\partial \phi}{\partial y} \right] \partial q_y' + f(q_x)$$

$$= -\frac{q_y'^2}{4} + f(q_x)$$

$$\phi = \int \frac{1}{J}[2q_x]\partial q_x + f(q_{\dot{y}})$$

$$= \frac{q_x^2}{4} + f(q_{\dot{y}})$$

where $q_x = \dfrac{\partial \phi}{\partial x} = v_x = 2x$ (5.33a)

$$q_{\dot{y}} = -q_y = -\frac{\partial \phi}{\partial y} = -v_y = 2y \quad\quad\quad (5.33b)$$

$$J = \frac{\partial(q_x, q_y')}{\partial(x, y)} = \begin{vmatrix} \dfrac{\partial q_x}{\partial x} & \dfrac{\partial q_y'}{\partial x} \\[2mm] \dfrac{\partial q_x}{\partial y} & \dfrac{\partial q_y'}{\partial y} \end{vmatrix} = 4$$

$$\phi = \frac{q_x^2 - q_{\dot{y}}^2}{4}$$

$$\psi = \int \frac{q_y'}{2}\partial q_x + f(q_y')$$

$$= \frac{q_x q_y'}{2} + f(q_y')$$

$$\psi = \int \frac{1}{J}\left[-\frac{\partial q_x}{\partial y}\frac{\partial \psi}{\partial x} + \frac{\partial q_x}{\partial x}\frac{\partial \psi}{\partial y}\right]\partial q_y' + f(q_x)$$

$$= \frac{q_x q_y'}{2} + f(q_x)$$

$$\psi = \frac{q_x q_y'}{2}$$

The same results could be obtained by composite mapping.

$$\psi(x, y) = \psi[(x(q_x, q_y'), y(q_x, q_y')] = \psi(q_x, q_y')$$
$$\phi(x, y) = \phi[(x(q_x, q_y'), y(q_x, q_y')] = \phi(q_x, q_y')$$

where

$$x = x(q_x, q_y') = \frac{q_x}{2}; \quad y = y(q_x, q_y') = \frac{q_y'}{2}$$

Hence,

$$F = f(\bar{q}) = \phi(q_x, q_y') + i\psi(q_x, q_y') \quad\quad\quad (5.34)$$

Figure 5.10 presents mapping of streamlines and equipotential contours from z-plane to conjugate velocity hodograph plane. Each contour pattern

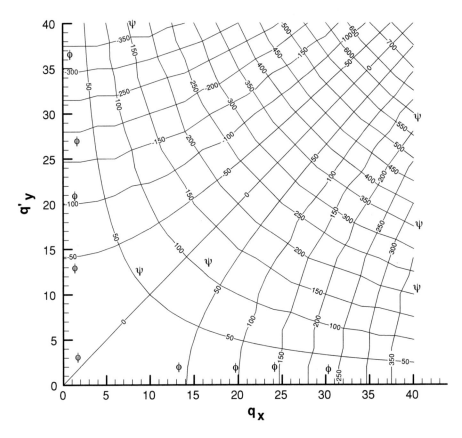

Figure 5.10 Mapping of streamlines and equipotential from z-plane to conjugate velocity hodograph plane

resembles a system of hyperbolas. In this figure, vertical and horizontal wall are mapped to q'_y-axis and q_x-axis respectively. The stagnation point is mapped to origin. The two systems of contours remain orthogonal when mapped conformally onto velocity hodograph. As an application of hodograph method, the inverse mapping from hodograph plane to z-plane is obtained below:

$$z = \int \frac{dF}{d\bar{q}} \frac{d\bar{q}}{\bar{q}} + C$$

$$= \int \frac{(q_x - iq_y)}{2\bar{q}}(dq_x - idq_y) + C$$

$$x = q_x/2; \; y = -q_y/2$$

Mapping of equipotentials and stream function from the physical plane to hodograph plane can be accomplished by substituting $q'_y = -q_y$. The hodograph representation of these curves is a pair of conic curves:

$$\phi(x, y) = x^2 - y^2 = \frac{q_x^2 - q_y^2}{4} = \phi(q_x, q_y)$$

$$\psi(x, y) = 2xy = -\frac{q_x q_y}{2} = \psi(q_x, q_y)$$

Equipotential contours and streamlines are presented on velocity hodograph plane in Figure 5.11. Equipotentials exhibit a system of hyperbola with center at origin whereas streamlines are families of rectangular hyperbola with center at origin. The two systems of contours remain orthogonal on velocity hodograph. When following the path of a streamline, velocity vector swings

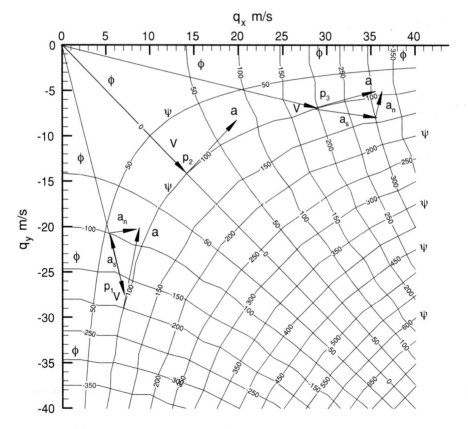

Figure 5.11 Hodograph representations of equipotentials and streamlines

from one end to another as velocity direction changes steadily and continuously within a moderate range. Tangents (arrow icons) along streamline on hodograph plane define the acceleration vectors and are depicted at points p_1, p_2, p_3. They are decomposed into streamline tangential and normal direction. The tangential component of acceleration vector is in the opposite velocity direction at p_1 but in the same direction at point p_3. It vanishes at point p_2 where the two vectors are mutually perpendicular. Flow speed is decreasing in approaching p_2 after which it starts to increase. The same findings are obtained when acceleration vectors are mapped to the physical plane in Figure 4.5 and then decomposed in the like manner. In this graph, the normal component of acceleration vector always points towards the center of streamline curvature indicating flow direction veering to the left side of trajectories. Alternatively, one may come to the same conclusion analytically with the aid of Equation 4.14a or 4.14b.

$$\text{At point } p_1 = (3, 16), \; a \cdot \frac{V}{|V|} < 0$$

$$\text{At point } p_2 = (7, 7), \; a \cdot \frac{V}{|V|} = 0$$

$$\text{At point } p_3 = (14, 3), \; a \cdot \frac{V}{|V|} > 0$$

Velocity at a point defined by a pair of (ϕ, ψ) can be obtained directly from its image position vector on hodograph plane. It is convenient to use this plane to describe the state of flow fields along an arbitrary streamline by the position vectors as pencil of rays from origin. Speed at a point is proportional to the distance from origin of hodograph, which corresponds to critical point. The two corner walls coincide with coordinate axes are straight dividing streamlines. They are mapped in parallel to coincide with hodograph coordinate axes as a consequence of hodograph representation theory. By the same token, straight equipotential represented by line $y = x$ (or $\gamma = \pi/4$) is mapped to a straight line passing through origin inclined at an angle $\gamma - \pi/2 = -\pi/4$. Circles of radius C with center at origin representing lines of constant speed (or constant kinetic energy) are mapped to a circle of radius $C/2$ about origin on physical plane in Figure 4.6. The equation of hodograph for constant speed is:

$$|V| = \sqrt{q_x^2 + q_y^2} = C$$

$$= 2\sqrt{x^2 + y^2} = C$$

$$r = \frac{C}{2}$$

At this point, we recall hodograph representation of isoclines is a straight line through origin inclined at an angle equal to constant flow direction of isoclines (Equation 5.27a). For corner flows, each straight isocline in Figure 5.11 represented by a radial line is mapped to the respective straight line through origin in Figure 4.8 and reflected about the horizontal x-axis following the change of sign of angle.

$$q_y = q_x \tan \tilde{\theta}$$
$$-2y = 2x \tan \tilde{\theta}$$

or

$$y = x \tan(-\tilde{\theta})$$

On the physical plane, each streamline crosses the captioned straight line at a point at which the streamline tangent is equal to the constant direction of isocline inclination angle. We could trace isoclines contours on physical plane by varying the inclination angle to assume the role of parametric constant. It gives rise to a system of straight isoclines contours in Figure 4.8 squeezing towards the corner and indicating a rapid change of flow directions and streamline curvatures in the corner region. This could be anticipated since the corner is a velocity critical point where streamline curvature is found to assume infinite in Example 4.1a.

By substitution, hodograph representation of an arbitrary pressure isobar is a circle about origin.

$$\frac{p(x, y)}{\rho} = \frac{p_0}{\rho} - 2(x^2 + y^2) = \frac{p_0}{\rho} \quad \frac{1}{2}(q_x^2 + q_y^2)$$

or

$$q_x^2 + q_y^2 = C(p)$$
$$x^2 + y^2 = C''(p)$$

where

$$C(p) = 2\left(\frac{p_0 - p}{\rho}\right)$$
$$C''(p) = \frac{p_0 - p}{2\rho}$$

The pressure isobars are also mapped to another system of circles on physical plane. In both cases, pressure assumes the role of parametric constant of the implicit functions and defines the circle radii.

Hodograph representations of heat flux trajectories and streamlines are presented in Figure 5.12 via composite mapping:

$$\psi_t(x, y) = \psi_t[x(q_x^*, q_y^*), y(q_x^*, q_y^*)] = \psi_t(q_x^*, q_y^*) \tag{5.35}$$

$$\psi(x, y) = \psi[x(q_x^*, q_y^*), y(q_x^*, q_y^*)] = \psi(q_x^*, q_y^*) \tag{5.36}$$

The heat flux critical point at z-plane origin is mapped to the origin in heat flux hodograph plane and conversely. Dividing heat flux trajectory exists only at origin since no heat flux trajectory passes through the critical point.

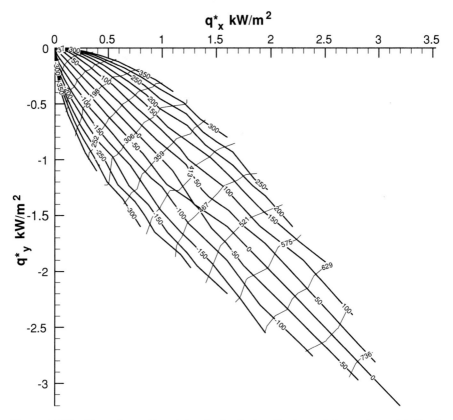

Figure 5.12 Mapping of heat flux trajectories (line) and streamlines (dotted) from z-plane to heat flux hodograph plane

Heat flux increases away from this point and assumes maximum at right top corner in z-plane. The directions of heat flux change moderately when following a streamline of parabolic shape from upstream to downstream. The y-component heat flux decreases while x-component heat flux increases. From this pattern of variation of heat flux components, one may observe the heat transfer characteristics. In addition, the heat transfer has a linear relation with distance from origin following the path of dividing streamline (corner walls). Heat flux diminishes to nil along the vertical wall towards origin thereafter it increases along horizontal wall in a linear fashion with distance. The heat flux trajectories close to origin indicate heat flux changes direction rapidly in the corner region around the heat flux critical point. Heat flux directions change steadily and continuously within a moderate range when following a heat flux trajectory away from the origin. This is reflected by the heat flux vector swinging from one end to another about origin. We could trace the locus of constant heat flux magnitudes C on physical plane by mapping circles about the origin in heat flux hodograph as parameter C varies. The equation of hodograph for constant heat flux magnitude is given by:

$$|q^*| = \sqrt{q_x^{*2} + q_y^{*2}} = C$$

where

$$q_x^* = q_x^*(x, y) = \frac{\partial \psi_t}{\partial y}$$

$$q_y^* = q_y^*(x, y) = -\frac{\partial \psi_t}{\partial x}$$

Mapping of heat flux trajectories and isotherms from the physical plane to potential-stream function coordinate could be accomplished by composite mapping.

$$\psi_t(x, y) = \psi_t[x(\phi, \psi), y(\phi, \psi)] = \psi_t(\phi, \psi) \tag{5.37}$$

$$T(x, y) = T[x(\phi, \psi), y(\phi, \psi)] = T(\phi, \psi) \tag{5.38}$$

The results are presented in Figure 5.13. Vertical wall and horizontal wall are mapped to left side and right side of horizontal ϕ-axis respectively such that boundaries of physical domain coincide with horizontal axis. Furthermore, the horizontal axis corresponds to a dividing streamline and the coordinate origin is a multiple critical point. Heat flux trajectories show that heat transfers from left side to right side. This corresponds to the directions

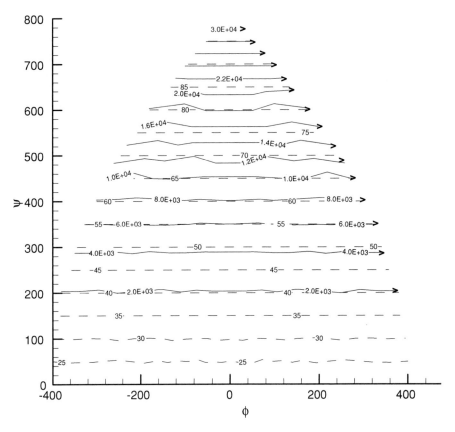

Figure 5.13 Mapping of heat flux trajectories (arrows) and isotherms (dashed) from z-plane to potential-stream function coordinate

from top to right side in the physical domain and tallies with the pattern of variation of heat flux components. The transfer process is most intensifying at right top corner where flux density is the greatest. Heat flux trajectories are nearly parallel to stream function coordinate lines. This implies heat paths and streamlines are about to coincide with each other. Tangents along heat flux trajectories and isotherms are practically horizontal, such that the β angle vanishes on (ϕ, ψ) plane.

$$\tan \beta = \frac{d\psi}{d\phi}\Big|_{\psi \approx \psi_t + C} \approx 0, \text{ or } \beta \approx 0$$

$$\rightarrow \chi\Big|_{\psi \approx \psi_t + C} \approx \alpha$$

Thus, the tangents along heat flux trajectory in physical plane correspond to the directions of flow fields approximately. In a different approach, we could apply the same idea of *RAI* index to examine if relation/correlation exist between the directions of heat flux vector and velocity from the angles between them. This may be visualized by the displays of both vectors at a point. It is also observed that isotherms are symmetric with respect to line $y = x$. Figures 4.9 and 5.13 depict the temperature distributions similar to those of streamlines and heat flux trajectories. Temperature appears to have no significant change along flow paths. Maximum temperature occurs at right top corner specified by temperature condition. We could visualize conduction heat transfer paths by taking gradient mapping of isotherms onto gradient vectors pointing in the direction of maximum spatial rate of temperature decrease. The gradient vector fields are perpendicular to velocity everywhere as indicated by the *RAI* index in part 4.1a.

5.8.2 Image Warping

To produce flow fields in a corner with warping distortion at the region near corner on physical plane, we apply conformal mapping using composite functions. This is the technique of successive transformation, which provides solutions to many problems. Conformal mapping method transforms region on ζ-plane (an auxiliary plane) onto regions in both z-plane and F-plane and hence connects these two planes with the composite functions. In this process, we first establish a mapping between F-plane and ζ-plane using Joukowski transformation, that is,

$$F = f(\zeta) = \phi(\xi, \eta) + i\psi(\xi, \eta) = \zeta + \frac{c^2}{\zeta} \tag{5.39}$$

Or

$$\phi(\xi, \eta) = \frac{\xi(\xi^2 + \eta^2) + c^2\xi}{\xi^2 + \eta^2}$$

$$\psi(\xi, \eta) = \frac{\eta(\xi^2 + \eta^2) - c^2\eta}{\xi^2 + \eta^2}$$

On F-plane, the vertical axis ($\phi = 0$) and horizontal axis ($\psi = 0$) are mapped to η-axis and ξ-axis respectively on ζ-plane. A double straight-line (thin strip) from $-2c \leq \phi \leq 2c$ along ϕ-axis is real value everywhere and is mapped to a circle of radius c with center at origin in ζ-plane. Points exterior of the circle in ζ-plane are mapped to the entire F-plane. The complex potential maps uniform flows on F-plane onto flows past a cylinder of radius c in ζ-plane with center at origin (Section 5.5). We then establish a mapping between

ζ-plane and z-plane. The complex potential for corner flows in the first quadrant on z-plane with corner at origin is given by Example 4.1a using a real power function, that is,

$$\zeta = \zeta(z) = \xi(x, y) + i\eta(x, y) = z^2 \qquad (5.40)$$
$$0 \le \theta \le \pi/2$$

Or

$$\xi(x, y) = x^2 - y^2$$
$$\eta(x, y) = 2xy$$

With power of two, the upper half ζ-plane is mapped to the first quadrant on z-plane representing a right angle corner. The cylinder of radius c is mapped to a circle of radius equal to square root of c on z-plane. The complex potential maps uniform flows parallel to the real axis of ζ-plane to a flow field in a corner on z-plane. The flow fields resemble a system of hyperbola. The F-plane and z-plane are connected through composite mapping as follows:

$$F = f(\zeta) = f[\zeta(z)] = f(z) = \phi(x, y) + i\psi(x, y)$$
$$= \zeta + \frac{c^2}{\zeta} = z^2 + \frac{c^2}{z^2}$$
$$= (x^2 - y^2)(1 + \frac{c^2}{(x^2 + y^2)^2}) + i2xy(1 - \frac{c^2}{(x^2 + y^2)^2}) \qquad (5.41)$$

$$\text{Hence, } \phi(x, y) = (x^2 - y^2)(1 + \frac{c^2}{(x^2 + y^2)^2})$$

$$\psi(x, y) = 2xy(1 - \frac{c^2}{(x^2 + y^2)^2})$$

Dividing streamline: $\psi = 0$; when $x = 0$ or $y = 0$ or $r = |z| = \sqrt{x^2 + y^2} = \sqrt{c}$

$$\bar{q} = \frac{dF}{dz} = 2z - \frac{2c^2}{z^3}$$
$$= 2r \exp(i\theta) - \frac{2c^2}{r^3} \exp(-3i\theta)$$
$$\bar{q}\,|_{r=\sqrt{c}} = 8\sqrt{c}\sin^2\theta\cos\theta + i\frac{8c^2}{(\sqrt{c})^3}\cos^2\theta\sin\theta$$

Critical points: $\theta = 0$, $z = \sqrt{c}$ and $\theta = \pi/2$, $z = i\sqrt{c}$
Flow speed around the circular arc: $|\bar{q}|_{r=\sqrt{c}} = 4\sqrt{c}\sin 2\theta$
Maximum speed around circular arc: $|\bar{q}|_{r=\sqrt{c}} = 4\sqrt{c}$ @ $\theta = \pi/4$

Consider the first quadrant in physical plane. Above describes flow around a right angle corner with one quarter of a circle of radius $z = \sqrt{\zeta} = \sqrt{c}$ center at origin (corner). The quarter circular arc connects the vertical wall ($x = 0$) and horizontal wall ($y = 0$) giving rise to the dividing streamline $\psi = 0$ since there are critical points lie on this streamline. One of them is located at the intersections of circular arc and y-coordinate axis the other at the intersection

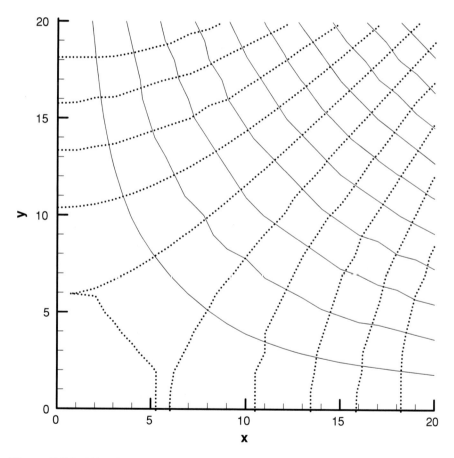

Figure 5.14 Warping of an image (ϕ dotted, ψ line) by Joukowski transformation

of x-coordinate axis. The dividing streamline splits abruptly at the critical point ($x = 0$, $y = \sqrt{c}$) going around the circular arc and then rejoins x-axis at the other critical point ($x = \sqrt{c}$, $y = 0$). It separates the fields into first quadrant and serves as asymptotes for nearby streamlines. Indeed, on the circle, the imaginary part of the complex potential vanishes. Therefore $\psi = 0$ represents a dividing streamline around the circle. The composite function maps the region in the first quadrant and those exterior to the circle of radius Z conformally onto the upper half F-plane as the angle doubles. The results are presented in Figure 5.14 with $c = 30$. The composite function (Equation 5.5) transforms equipotentials into corresponding equipotentials and streamlines into corresponding streamlines between F-plane, ζ-plane and z-plane. The quarter circle acts as an obstacle that causes flow distortion in its neighborhood. The resultant field patterns appear to be warped from those without the circle at corner. The curvilinear squares network preserves angles and local shapes except at critical points. This gives a new way of looking at flows in a corner with a circle centered at corner. Warping can also be produced using different conic shapes. It is observed that,

$$z \rightarrow \infty; \quad F = z^2 + \frac{c^2}{z^2} = z^2$$

Uniform flow on F-plane is mapped onto flow around a corner on z-plane and uniform flow in ζ-plane as distance increases from the corner.

$$\zeta \rightarrow \infty; \quad F = \zeta + \frac{c^2}{\zeta} = \zeta$$

6

Tensor Representation, Contraction and Visualization

6.1 Introduction

Tensors are multivariate data embedded with complex information in time and space. A second order tensor is a mathematical notion used to describe a physical field where vector descriptions fail to do so. It consists of a system of multiple vectors and occurs in many different phenomena. Tensors of practical interests may conveniently be represented in matrix form with respect to a reference coordinate system. These include for example strain tensors, stress tensors and convective momentum flux tensors. Tensor of this sort is a quantity having physical significance and satisfying certain transformation laws. Visualization of tensor fields is complicated and as such it has been lagging behind scalar fields and vector fields. A complete diagrammatic representation of tensors remains as challenging as ever.

The level of difficulty in field visualization escalates rapidly as the order of tensors or domain dimensions increase. The number of components (or degrees of freedom) required to describe a tensor field is:

$$N = D^n$$

D is domain dimensions and n is tensor orders. There is a need to focus on extracting relevant information from a large volume of multivariate data and to present it in visual forms. In the present study, tensor rank reduction

Visualization of Fields and Applications in Engineering, First Edition. Stephen Tou.
© 2011 John Wiley & Sons, Ltd. Published 2011 by John Wiley & Sons, Ltd.

techniques are applied. These include: contractions, tensor invariants, principal states and principal axes trajectories combined with evolving tensor ellipses.

6.2 Development of Tensor Visualization Techniques

Study in strain/stress analysis has contributed significantly to the development of tensor visualization techniques both experimentally and theoretically. For example, the two-dimensional photo-elasticity technique in stress analysis has provided experimental means of tensor visualization. Several graphical plots have been developed from tensor/matrix theory to aid visualization of tensor fields. They are summarized below:

6.2.1 Mohr's Circle

Traditionally, the state of stress acting on elements rotated about the coordinate axes can be visualized by Mohr's circles. They provide a geometric interpretation of stress fields. In the case of plane stress (x-y), only one Mohr's circle is required for an element rotated about z-axis. Mohr's circle is basically a stress diagram representing shear stress and normal stress along two orthogonal axes. The circle center is defined by mean normal stress along normal stress (x-axis) at a location equal to the mean principal stress. Mean normal stress acts on the plane of maximum shear stress. The circle radius representing maximum shear stress is presented on y-axis and is equal to half of principal stress difference. The state of stress at a point is then described by points on the circumference of Mohr's circle. Such stress diagram is a useful means for the understanding of stress patterns. A series of Mohr's circles could be constructed at selected points.

6.2.2 Tensor Field Line Trajectories (Lines of Principal Stress)

The concept of field lines in vector fields could be applied to describe stress tensors in terms of tensor field line trajectories. The trajectory tangents define the directions of principal stresses aligned with principal vectors (principal directions). They are regarded as the intrinsic representation of tensors independent of coordinates. When principal directions are known at every point in the field, one may trace out a system of orthogonal curves (lines of principal stress Woods (1903)) which have these directions as tangents. Each pair of orthogonal curves defined by a pair of principal directions at a point is associated with the corresponding pair of principal stresses. Principal stress magnitudes vary as we travel along a tensor field line trajectory. An additional graph is required to show the magnitudes of each principal stress. This is usually presented as stress contours just like other scalar fields. Stress contours are lines of constant principal stresses. One set is required for each principal stress. Tensor field line trajectories and stress contours provide an effective means of capturing a tensor field in visual forms.

6.2.3 Isochromatics

They are curves along which the principal stress-difference is constant.

6.2.4 Isoclines

These are a set of curves on which the vector fields or tensor principal axes make a constant angle of inclination with respect to a reference direction. The field direction at each point on isoclines does not change. The hodograph representation of an isocline is a straight line through origin inclined at an angle equal to the field direction with respect to positive horizontal axis.

6.2.5 Stress Trajectories

Under certain conditions such as pure shear or plane stress, the stress tensor is conveniently contracted to yield a two-dimensional tensor vector on a reference plane or at a point. The concept of field lines or streamlines can then be applied to describe planar vector in terms of stress trajectories whose tangents define the directions of the resultant stresses. For example, plane stress trajectories described by Prandtl stress function in the study of torsion are analogous to streamlines described by stream function.

6.2.6 Slip Lines

In most problems, great interest centers on the determination of slip lines along which the tangent gives the maximum shear direction at a point. In these shear directions the relative movement between adjacent parallel planes is the maximum giving rise to the phenomenon of slip at the point. They form an orthogonal family of curves at all points bisecting the angles between the directions of principal stresses.

6.2.7 Isopachs

These are curves of constant first invariant, that is, mean normal stress or mean normal strain.

6.3 Tensor Description and Representation

Tensor order ranking is a measure of the amount of information at a point in space. By adopting right-handed coordinate systems in Figure 6.1, a second order tensor is defined with respect to three sets of base vectors, each on a mutually orthogonal plane passing through a given point. They are best represented as three independent column (or row) vectors in a three by three matrix. Thus, a second order tensor has nine components, each being a

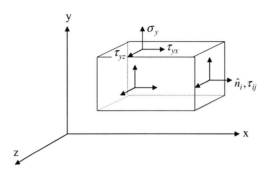

Figure 6.1 Specification of a second order tensor field at a point and convention

function of coordinate variables. This allows us to describe conveniently the state of static tensor fields in terms of tensor components depending on position and not on the orientations of the orthogonal planes at that point. Tensor components are independent of each other and they vary continuously throughout field domain. This prompts the idea of representing a second order tensor in terms of a single scalar function following the concept of potential for vector field descriptions. In potential theory, the natural choice of generating a first order tensor such as a velocity vector field is the gradient of a scalar potential. Similarly, a second order tensor could be derived from the second derivatives of a scalar potential. In this theoretical development, Fung (1965) presents Finzi's work on the generation of a stress tensor from a potential, that is,

$$\tau_{ij} = \varepsilon_{imr}\varepsilon_{jns}\varphi_{rs,mn} = \varepsilon_{imr}\varepsilon_{jns}\frac{\partial^2 \varphi_{rs}}{\partial x_m \partial x_n}$$

Above is a generalization of the well-known Airy stress function, Maxwell's stress function and Morera's stress function by setting certain components to vanish using the index r and s. For instance the two dimensional Airy stress function introduced in elasticity defines the stress components as follows:

$$\sigma_x = \tau_{xx} = \frac{\partial^2 \varphi}{\partial y^2}; \quad \sigma_y = \tau_{yy} = \frac{\partial^2 \varphi}{\partial x^2}; \quad \tau_{xy} = \tau_{yx} = -\frac{\partial^2 \varphi}{\partial x \partial y} \quad (6.1)$$

Stress functions generate a system of equilibrating stress components corresponding to the actual states of stress. It can be verified by direct substitutions that the stress components satisfy the equilibrium equation.

Stress tensor components acting on a reference plane are referred to by dual subscripts. The first subscript i denotes unit normal vector of the reference plane corresponding to and parallel to i-coordinate axis. The second subscript

j is the stress force direction parallel to j-coordinate axis. Thus, stress acting on a given plane is a vector but cannot be operated with other stress vectors on different planes. In matrix representation, the first and second subscripts refer to i-row and j-column position respectively. In Figure 6.1, a positive surface (or face) of a control volume is defined by an outward surface normal pointing in the positive coordinate axis direction. To bring into coincidence the sign of stress tensor components, stress direction conventions are introduced based on the definition of strain and Hooke's law. In elasticity, positive directions of stresses acting on each positive surface are in the respective positive coordinate axis directions with reference to right-handed coordinate systems. In Figure 6.1, all three surfaces of the cubic control volume are positive surfaces and all stress directions shown are positive directions. The remaining three faces in obscured views are designated as negative surfaces whose outward surface normal are in the direction of negative coordinate axes. On these negative surfaces, all positive directions of stresses (not shown) point towards the negative directions of coordinate axes. At principal states, eigenvalues represent extreme normal stresses acting on respective principal planes. Positive normal stresses are classified as tension whereas negative normal stresses are taken as compression.

In hydrodynamics, interests are focused on the transport of property fluxes other than deformation. It is more convenient to reverse the conventions in elasticity. This is because pressure (normal stress) is taken positive in compression and the negative sign is associated with Newton's law of viscosity in hydrodynamics. The conventions are then consistent with convective momentum flux, viscous momentum flux and pressure tensor in thermodynamics of the same sign affixed. It is also consistent with other transport laws such as Fourier's law and Fick's law. Such that diffusive fluxes point towards the directions of decreasing property concentration in transport processes. At principal states, eigenvalues represent extreme fluxes acting on respective principal planes. The plus or minus sign represents out-going flux or in-coming flux on the outward surfaces of a control volume respectively. Attention should be drawn to the adoption of convention and the sign in transport laws so that a correct interpretation of results could be made.

Potential functions are useful to describe tensor fields with reduction of variables. They are of little value in visualization of the structure of tensors since they merely produce tensor components. A common and basic technique of representing tensor fields is to display its individual components as scalar quantities in contours separately. As directional information is lost, correlations between components present a difficult task. Furthermore, tensor topological structure is ruined, the results serve little purpose and such technique is far from satisfaction. Methods for direct graphical representation of second order tensors have existed for some time. In particular, tensor icons have laid the groundwork for tensor field representation and visualization.

6.3.1 Tensor Icons and Classification

6.3.1.1 Tensor Point Icons

Lame's stress ellipsoid/ellipse and Cauchy's stress quadrics are regarded as tensor point icons displaying tensor fields at a point discretely. These methods suffer a major setback of failing to display information, which varies continuously throughout field domains. Tensor glyphs are then developed for tensor visualization. It is a geometric point icon that represents multivariate or higher dimensional information at a given point. Data values are mapped to various physical attributes of the icon in terms of geometry, color and so on. A vector arrow or a probe is an example of its kind. The method has been modified continuously and sophisticated icon such as three-dimensional streamball was developed. Tensor point icons generally consume a finite amount of display space and cause visual cluster.

6.3.1.2 Tensor Line Icons

Tensor field line trajectories introduced formally by Woods (1903), Dickinson (1989, 1991) in stress analysis are useful line icons to depict the lines of principal stress along which forces are transmitted in the system continuously. Hyperstreamlines were then presented by Delmarcelle (1993) as an extension of tensor field lines similar to the concept of streamlines in vector fields. Hyperstreamlines provide a continuous representation of tensor fields along principal trajectories. Similar to streamlines, hyperstreamlines never cross with each other except at critical/singular points.

6.4 Tensor Decomposition and Tensor Rank Reduction

The insight into the characteristics of a tensor is expressed by the so-called Cauchy-Stokes decomposition theorem. It provides one of the important techniques to separate a second order tensor into distinct components for subsequent visualization and analysis. In tensor theory, a tensor can be decomposed into two parts: a symmetric part and an antisymmetric part with unique mathematical properties. The symmetric matrix possesses real eigenvalues and orthogonal bases exist. The antisymmetric matrix yields pure imaginary eigenvalues and orthogonal bases do not exist. The decomposition process also separates the scalar and vector sources into the former and latter parts respectively. Each part assumes specific roles and has unique properties. Both are invariant to a rotation of coordinate system and can occur simultaneously in mixed problems.

In elasticity, interests are focused on deformation. Consider an instantaneous state of deformation in a medium represented by a cubic control volume. We interpret the deformation process from the point of affine mapping by decomposing it into a rotation component, dilation component, shear deformation component and a translation component. The orientation of control volume is defined with respect to a rectangular coordinate system with each axis parallel to the unit normal of each cube surface. When initial equilibrium condition is disturbed, the cube will respond by undergoing a rigid body rotation given by the principal angle with respect to the reference coordinate. It is distorted along the principal axes as the cube elongates/contracts with or without volumetric change. The cube may also change in shape under shearing action, which assumes maximum along slip line. The control volume then translates into a new position until equilibrium condition is restored in the system. Deformation in solids is completely described by the displacement vector of a particle as it moves from its initial position before deformation to that occupied by the same material point after deformation. To visualize deformation phenomena, we expand the displacement fields in terms of Taylor series by linear approximation for small displacements:

$$U = \sum_{n=0} \frac{(dr)^n \cdot [\nabla^{(n)} U]}{n!} \approx U_0 + dr \cdot [\nabla U]|_0$$

U_0 represents the linear component of local displacement (translation). It gives rise to change of linear positions of a particle. The second term in brackets is the gradients of displacement field derived from ∇-operator in terms of first derivative of displacement vector with respect to spatial variables x_i. In Cartesian coordinate system, it is written as:

$$\nabla U \equiv \sum_{i=1} \sum_{j=1} \frac{\hat{e}_i \partial (u_j \hat{e}_j)}{h_i \partial x_i} = \sum_i \sum_j \frac{\partial u_j}{h_i \partial x_i} \hat{e}_i \hat{e}_j = \begin{bmatrix} \dfrac{\partial u_x}{\partial x} & \dfrac{\partial u_y}{\partial x} & \dfrac{\partial u_z}{\partial x} \\[2mm] \dfrac{\partial u_x}{\partial y} & \dfrac{\partial u_y}{\partial y} & \dfrac{\partial u_z}{\partial y} \\[2mm] \dfrac{\partial u_x}{\partial z} & \dfrac{\partial u_y}{\partial z} & \dfrac{\partial u_z}{\partial z} \end{bmatrix} \quad (6.2)$$

The gradient operation yields a second order tensor formed by dyadic product. It contains information about the spatial rate of change of displacement fields. The tensor describes the state of deformation and is useful to reveal the structure of displacement fields. In tensor analysis, it is decomposed into two distinct parts, that is,

$$[\nabla U] = [\gamma] + [\tilde{\Omega}] \quad (6.3)$$

The symmetric part and antisymmetric part are given by:

$$\gamma_{ij} = \frac{1}{2}[[\nabla U] + [\nabla U]^T] = \frac{1}{2}\left[\frac{\partial u_j}{\partial x_i} + \frac{\partial u_i}{\partial x_j}\right]$$

$$= \frac{1}{2}\begin{bmatrix} 2\dfrac{\partial u_x}{\partial x} & \dfrac{\partial u_x}{\partial y} + \dfrac{\partial u_y}{\partial x} & \dfrac{\partial u_x}{\partial z} + \dfrac{\partial u_z}{\partial x} \\[2ex] \dfrac{\partial u_y}{\partial x} + \dfrac{\partial u_x}{\partial y} & 2\dfrac{\partial u_y}{\partial y} & \dfrac{\partial u_y}{\partial z} + \dfrac{\partial u_z}{\partial y} \\[2ex] \dfrac{\partial u_z}{\partial x} + \dfrac{\partial u_x}{\partial z} & \dfrac{\partial u_z}{\partial y} + \dfrac{\partial y_y}{\partial z} & 2\dfrac{\partial u_z}{\partial z} \end{bmatrix} = \gamma_{ji} \qquad (6.4)$$

$$\tilde{\Omega}_{ij} = \frac{1}{2}[[\nabla U] - [\nabla U]^T] = \frac{1}{2}\left[\frac{\partial u_j}{\partial x_i} - \frac{\partial u_i}{\partial x_j}\right]$$

$$= \frac{1}{2}\begin{bmatrix} 0 & -\dfrac{\partial u_x}{\partial y} + \dfrac{\partial u_y}{\partial x} & -\dfrac{\partial u_x}{\partial z} + \dfrac{\partial u_z}{\partial x} \\[2ex] -\dfrac{\partial u_y}{\partial x} + \dfrac{\partial u_x}{\partial y} & 0 & -\dfrac{\partial u_y}{\partial z} + \dfrac{\partial u_z}{\partial y} \\[2ex] -\dfrac{\partial u_z}{\partial x} + \dfrac{\partial u_x}{\partial z} & -\dfrac{\partial u_z}{\partial y} + \dfrac{\partial u_y}{\partial z} & 0 \end{bmatrix}$$

For $i = j$, $\tilde{\Omega}_{ij} = 0$

$$i \neq j, \ \tilde{\Omega}_{ij} = -\tilde{\Omega}_{ji}$$

6.4.1 Strain Tensor and Stress Tensor

The symmetric part is referred to as strain tensor, which describes two types of deformation: Shear deformation and dilation distortion represent shape changes and volume changes respectively. Strain tensor component γ_{ij} accounts for strain flux on the i-face of a control volume generated by linear displacement gradients in j-direction and is equal to one-half (or average) angular displacements associated with two mutually orthogonal lines. The diagonal components represent normal strain parallel to coordinate axes and characterize dilation distortion due to elongation/contraction. The degree of dilation distortion is measured by its trace (strain first invariant or mean normal strain) in terms of percentage of volumetric change and assumes the role of a scalar source.

$$\sum_{i=1}^{3} \gamma_{ii} = \nabla \cdot U$$

If the trace vanishes or is under pure shear, then volume/density will be invariant and deformation is said to be isochronic. Such fields are the inherent characteristics of solenoidal fields. It is convenient to split a strain tensor into an isotropic part with vanishing off-diagonal components and a deviatorial part with vanishing diagonal components. The isotropic part and deviatorial part represent normal and shear strain. If the deviatorial part vanishes, it is in a state of pure normal strain. When isotropic part vanishes, it is said to be in a state of pure shear strain.

Strain tensor is intimately connected to stress tensor via Hooke's law.

$$\tau_{ij} = \lambda I_1 \delta_{ij} + 2\mu \gamma_{ij}$$

$$\text{Or } \gamma_{ij} = \frac{1}{2\mu} \left[\tau_{ij} - \frac{\lambda}{3\lambda + 2\mu} I_1 \delta_{ij} \right]$$

The elastic constants are defined as:

$$\lambda = \frac{E\nu}{(1+\nu)(1-2\nu)} \tag{6.5a}$$

$$\mu = \frac{E}{2(1+\nu)} \tag{6.5b}$$

ν is the Poisson's ratio, E is the modulus of elasticity. τ_{ij} is a symmetric tensor representing stress acting on i-face of a control volume driven by strain fluxes in j-direction. The diagonal and off-diagonal components are measures of normal stress and shear stress respectively. Both stress and strain tensor fields are unified by deformation phenomena and occur simultaneously. Strain tensors depict external characteristics of deformation as shape/volume changes. Stress tensors describe internal characteristics of deformation, as stress fields develop to maintain the system in equilibrium. This symmetric part does work on deformation. If deformation tensor vanishes, it corresponds to pure translation in a class of rigid body motion. All lines connecting particles of the body will retain their original lengths, directions and senses. The distance between any two points remains constant in time.

6.4.2 Rotation Tensor

The antisymmetric part referred to as rotation tensor represents local angular displacement. It is interpreted as rigid body rotation of a particle about its own axis perpendicular to the plane of rotation. For instance, tensor component $\tilde{\Omega}_{xy}$ describes z-component of angular rotation Ω_z generated by the net average angular displacements between two mutually orthogonal lines on x-y plane of rotation about z-axis. Rotation is positive counterclockwise as motion veers to the left side of trajectory following right hand rule such that the rotational axis points towards the positive direction of z-axis (Figure 8.1). It causes change of angular positions of particles while shape/volume remain invariant. This part does not contribute to work on deformation.

In tensor analysis, every antisymmetric tensor can be associated with a vector referred to as axial vector of three independent components. Axial vectors describe directions of rotation differ from regular vectors, which depict vector field directions. Axial vectors have the nature of flux vectors produced by vector cross product and have the origin of an antisymmetric tensor.

$$\Omega = \begin{bmatrix} \Omega_x & \Omega_y & \Omega_z \end{bmatrix} = \frac{1}{2}\nabla \times U = \frac{\varpi}{2} \tag{6.6}$$

where $\varpi_x = \frac{\partial u_z}{\partial y} - \frac{\partial u_y}{\partial z} = 2\Omega_x$; $\varpi_y = \frac{\partial u_x}{\partial z} - \frac{\partial u_z}{\partial x} = 2\Omega_y$; $\varpi_z = \frac{\partial u_y}{\partial x} - \frac{\partial u_x}{\partial y} = 2\Omega_z$

$$\tilde{\Omega}_{ij} = \varepsilon_{ijk}\Omega_k = \frac{1}{2}\varepsilon_{ijk}\varpi_k = [\tilde{\Omega}] \tag{6.7}$$

$$= \begin{bmatrix} 0 & \Omega_z & -\Omega_y \\ -\Omega_z & 0 & \Omega_x \\ \Omega_y & -\Omega_x & 0 \end{bmatrix}$$

The implication is that an antisymmetric tensor is treated as a rotation vector and is connected to a vector source derived from the curl of vector fields. Each rotation vector component has a magnitude equal to half of vector source and both point in the direction normal to the plane of displacement. The resultant rotation vector defines the rotational axis about which rotation takes place. For two-dimensional fields on x-y plane, rotation vector is parallel to z-axis with a single nonzero z-component, which is regarded as a scalar quantity. The displacement vectors lie on x-y plane.

$$\varpi = \begin{bmatrix} 0 & 0 & \varpi_z \end{bmatrix} = \varpi_z\hat{k}$$
$$\varpi \cdot U = 0 \tag{6.8}$$

If this antisymmetric part vanishes, the displacement fields are irrotational.

6.4.3 Rate of Strain Tensor and Viscous Stress Tensor

In fluid systems, interests are focused on motions in terms of velocity, that is, the rate of change of displacement. To extend the analysis from solids to fluids, the rate of change of field property is introduced to relate fluid deformation to the applied stress. Because fluids cannot withstand shear stress, they continue to deform without a definite shape. Fluid elements moving along streamlines undergo local deformation due to stress induced by velocity gradients. By expressing velocity in Taylor series with linear approximation, we obtain information about the fields through the gradient of velocity fields. The gradient of velocity fields is a second order tensor written in a similar way to the gradient of displacement fields.

$$V = \sum_{n=0} \frac{(dr)^n \cdot [\nabla^{(n)} V]}{n!} \approx V_0 + dr \cdot [\nabla V]|_0$$

where

$$[\nabla V] \equiv \sum_{i=1} \sum_{j=1} \frac{\hat{e}_i \partial v_j \hat{e}_j}{h_i \partial x_i} = \sum_{i=1} \sum_{j=1} \frac{\partial v_j}{h_i \partial x_i} \hat{e}_i \hat{e}_j = \begin{bmatrix} \dfrac{\partial v_x}{\partial x} & \dfrac{\partial v_y}{\partial x} & \dfrac{\partial v_z}{\partial x} \\[2mm] \dfrac{\partial v_x}{\partial y} & \dfrac{\partial v_y}{\partial y} & \dfrac{\partial v_z}{\partial y} \\[2mm] \dfrac{\partial v_x}{\partial z} & \dfrac{\partial v_y}{\partial z} & \dfrac{\partial v_z}{\partial z} \end{bmatrix} \quad (6.9)$$

V_0 represents the rate of local displacement, that is, particle velocity in translation. The tensor describes the state of fluid deformation at a point in terms of local velocity gradients. If this tensor vanishes, the translational motion is isochronic with changes in linear positions. All lines connecting particles of the body retain their original directions, lengths and senses and remain invariant in time. Gradient of velocity fields is also decomposed into a symmetric tensor $[\dot{\gamma}]$ and an antisymmetric tensor $[\Omega]$, that is,

$$[\nabla V] = [\dot{\gamma}] + [\Omega]$$

The symmetric part is called rate of strain tensor and is given by:

$$\dot{\gamma}_{ij} = \frac{1}{2}[[\nabla V] + [\nabla V]^T] = \frac{1}{2}\left[\frac{\partial v_j}{\partial x_i} + \frac{\partial v_i}{\partial x_j}\right]$$

$$= \frac{1}{2}\begin{bmatrix} 2\dfrac{\partial v_x}{\partial x} & \dfrac{\partial v_x}{\partial y} + \dfrac{\partial v_y}{\partial x} & \dfrac{\partial v_x}{\partial z} + \dfrac{\partial v_z}{\partial x} \\[2mm] \dfrac{\partial v_y}{\partial x} + \dfrac{\partial v_x}{\partial y} & 2\dfrac{\partial v_y}{\partial y} & \dfrac{\partial v_y}{\partial z} + \dfrac{\partial v_z}{\partial y} \\[2mm] \dfrac{\partial v_z}{\partial x} + \dfrac{\partial v_x}{\partial z} & \dfrac{\partial v_z}{\partial y} + \dfrac{\partial v_y}{\partial z} & 2\dfrac{\partial v_z}{\partial z} \end{bmatrix} = \dot{\gamma}_{ji} \quad (6.10)$$

This part describes two types of deformation: Shear deformation and dilation distortion. Tensor component $\dot{\gamma}_{ij}$ accounts for the i-component of linear motion contributed by the rate of strain flux in j-direction and is equal to one-half (or average) of rotational speeds associated with two mutually orthogonal lines. The off-diagonal components give rise to the rate of shear strain and are responsible for shape changes. The diagonal components determine the rate of normal strain, which causes dilation distortion due to elongation/contraction. The tensor trace is a measure of dilation rate or mean rate of normal strains and assumes the role of a scalar source,

$$\sum_{i=1}^{3} \dot{\gamma}_{ii} = \nabla \cdot V \tag{6.11}$$

If the tensor trace vanishes, fluid is considered incompressible and motions are isochronic. Such fields are the inherent characteristics of solenoidal fields. The rate of strain tensor can be further split into an isotropic part with vanishing off-diagonal components and a deviatorial part with vanishing diagonal elements. The isotropic part and deviatorial part represent normal and shear rate of strain. If the deviatorial part vanishes, it is in a state of pure normal strain. When isotropic part vanishes, it is said in a state of pure shear strain. In the event that the rate of strain tensor vanishes, it is said rigid body motion.

The rate of strain tensor is related to viscous stress tensor in accordance with the Newton's law of viscosity, that is,

$$\tau_{ij} = -\mu[-\frac{2}{3}(\nabla \cdot V)\delta_{ij} + [\nabla V] + [\nabla V]^{T}]$$
$$= -[\lambda(\nabla \cdot V)\delta_{ij} + 2\mu\dot{\gamma}_{ij}] \tag{6.12}$$

Newton's law of viscosity describes viscous momentum flux driven by velocity gradients in terms of a tensor of second order. The tensor components are interpreted as momentum fluxes at a point giving rise to stresses. The diagonal and off-diagonal components represent normal stresses and shear stresses respectively. Viscous stresses are generated due to molecular diffusion of momentum driven by momentum concentration (or velocity) gradients from high to low concentration regions in accordance with the negative sign. Each stress tensor component has unique physical significance. For example, τ_{xy} represents viscous momentum flux per unit area on x-face of a control volume associated with the x-momentum in y-direction and is equivalent to shear stress. This symmetric part does work on deformation.

6.4.4 Vorticity Tensor

The antisymmetric tensor represents the local angular component of rigid body rotation referred to as vorticity. Vorticity (rotation) is a natural component of fluid motions given by:

$$\Omega_{ij} = \frac{1}{2}[[\nabla V] - [\nabla V]^{T}] = \frac{1}{2}\left[\frac{\partial v_{j}}{\partial x_{i}} - \frac{\partial v_{i}}{\partial x_{j}}\right]$$

$$= \frac{1}{2} \begin{bmatrix} 0 & -\dfrac{\partial v_x}{\partial y} + \dfrac{\partial v_y}{\partial x} & -\dfrac{\partial v_x}{\partial z} + \dfrac{\partial v_z}{\partial x} \\[3mm] -\dfrac{\partial v_y}{\partial x} + \dfrac{\partial v_x}{\partial y} & 0 & -\dfrac{\partial v_y}{\partial z} + \dfrac{\partial v_z}{\partial y} \\[3mm] -\dfrac{\partial v_z}{\partial x} + \dfrac{\partial v_x}{\partial z} & -\dfrac{\partial v_z}{\partial y} + \dfrac{\partial v_y}{\partial z} & 0 \end{bmatrix} \qquad (6.13)$$

For $i = j,\ \Omega_{ij} = 0$

$$i \neq j,\ \Omega_{ij} = -\Omega_{ji}$$

Tensor component Ω_{xy} describes z-component of angular speed $\dot{\Omega}_z$ produced by the net average angular speeds between two mutually orthogonal lines on x-y plane of rotation about z-axis. The convention of vorticity rotation direction follows those of rotation vector Ω_z. Particles undergo angular displacement while shape/volume remain invariant. Similarly, this antisymmetric tensor is treated as an axial vector of rotation equal to half of vorticity vector derived from the curl of velocity fields, that is,

$$\dot{\Omega} = \begin{bmatrix} \dot{\Omega}_x & \dot{\Omega}_y & \dot{\Omega}_z \end{bmatrix} = \frac{1}{2}\nabla \times V = \frac{\omega}{2} \qquad (6.14)$$

Or $\omega_x = \dfrac{\partial v_z}{\partial y} - \dfrac{\partial v_y}{\partial z} = 2\dot{\Omega}_x,\ \omega_y = \dfrac{\partial v_x}{\partial z} - \dfrac{\partial v_z}{\partial x} = 2\dot{\Omega}_y,\ \omega_z = \dfrac{\partial v_y}{\partial x} - \dfrac{\partial v_x}{\partial y} = 2\dot{\Omega}_z$

$$(6.15a)$$

$$\Omega_{ij} = \varepsilon_{ijk}\dot{\Omega}_k = \frac{1}{2}\varepsilon_{ijk}\omega_k$$

$$= \begin{bmatrix} 0 & \dot{\Omega}_z & -\dot{\Omega}_y \\ -\dot{\Omega}_z & 0 & \dot{\Omega}_x \\ \dot{\Omega}_y & -\dot{\Omega}_x & 0 \end{bmatrix}$$

In orthogonal curvilinear coordinates:

$$2\dot{\Omega} = \nabla \times V = \frac{1}{h_1 h_2 h_3} \begin{vmatrix} h_1\hat{\xi}_1 & h_2\hat{\xi}_2 & h_3\hat{\xi}_3 \\ \partial/\partial\xi_1 & \partial/\partial\xi_2 & \partial/\partial\xi_3 \\ h_1 v_1 & h_2 v_2 & h_3 v_3 \end{vmatrix} = \omega \qquad (6.15b)$$

Cylindrical coordinate : $v_1 = v_r,\ v_2 = v_\theta,\ v_3 = v_z,\ h_1 = h_3 = 1, h_2 = r$

$$\omega_r = \frac{\partial v_z}{r\,\partial\theta} - \frac{\partial v_\theta}{\partial z} = 2\dot{\Omega}_r;\ \omega_\theta = \frac{\partial v_r}{\partial z} - \frac{\partial v_z}{\partial r} = 2\dot{\Omega}_\theta;\ \omega_z = \frac{\partial(r v_\theta)}{r\,\partial r} - \frac{\partial v_r}{r\,\partial\theta} = 2\dot{\Omega}_z$$

Streamline coordinate: $v_1 = v_s$, $v_2 = v_n$, $v_3 = v_z$, $h_1 = h_2 = h_3 = 1$

$$v_n = 0; \; R_s = r = -n; \; \frac{\partial h_i}{\partial n} = \frac{h_i}{n}$$

$$\omega_s = \frac{\partial v_z}{\partial n} - \frac{v_z}{R_s}$$

$$\omega_n = \frac{\partial v_s}{\partial z} - \frac{\partial v_z}{\partial s}$$

$$\begin{aligned}
\omega_z &= \frac{h_3}{h_1 h_2 h_3} \left(\frac{\partial(h_2 v_n)}{\partial s} - \frac{\partial(h_1 v_s)}{\partial n} \right) \\
&= -\frac{h_1 \partial v_s}{h_1 h_2 \partial n} - \frac{v_s \partial h_1}{h_1 h_2 \partial n} \\
&= -\frac{1}{h_2} \frac{\partial v_s}{\partial n} - \frac{v_s h_1}{h_1 h_2 n} \\
&= -\frac{\partial v_s}{\partial n} + \frac{v_s}{R_s}
\end{aligned} \qquad (6.15c)$$

Above expresses kinematic balance of rotation driven by velocity gradients between two mutually orthogonal axes. The vorticity vector is normal to the plane of rotation and defines the rotational axis through particle center about which rotation takes place. It is a measure of vector source and is related to the concept of circulation density of velocity fields. For two-dimensional $(x\text{-}y)$ fields, vorticity vector is perpendicular to the local velocity vector parallel to z-axis everywhere with a single non-zero component ω_z. Then only the x-y plane contains circulation. If vorticity vanishes, the flow field is irrotational.

$$\omega = \begin{bmatrix} 0 & 0 & \omega_z \end{bmatrix} = \omega_z \hat{k}$$
$$\omega \cdot V = 0$$

Other tensors of engineering interest include momentum-stress flux tensors, Maxwell stress tensors and so on. In flow fields, the dyadic product of velocity gives convective momentum flux tensor:

$$m_{ij} = \rho \, [VV] = \rho v_i v_j \hat{e}_i \otimes \hat{e}_j = \rho \sum_i \sum_j v_i v_j \hat{e}_i \hat{e}_j$$

$$= \rho \begin{bmatrix} v_x v_x & v_x v_y & v_x v_z \\ v_y v_x & v_y v_y & v_y v_z \\ v_z v_x & v_z v_y & v_z v_z \end{bmatrix} = m_{ji}$$

Tensor component $\rho v_i v_j$ is interpreted as the convective transport of i-component of momentum concentration ρv_i on the i-face of a control volume by particle v_j moving in j-direction. It gives rise to normal stress acting on i-coordinate plane ($i = j$) or shear stress ($i \neq j$). The diagonal components are always positive implying that normal stresses are compressive in the same manner as pressure tensor. The above shows that momentum transports whether by diffusion or convection is more complicated than heat or mass since it is a second order tensor. Note however that tensor vectors of convective momentum are not linearly independent since the determinant vanishes. The second order tensor possesses only one eigenvalue equal to kinetic energy density, that is, $\lambda_1 = \rho V^2$ while the other two eigenvalues assume zero and vanish. The eigenvalues could also be interpreted as principal stresses in exactly the same manner as the eigenvalues of stress tensors. The combined momentum-stress flux tensor is:

$$q_{ij} = m_{ij} + \pi_{ij}$$
$$\text{where } \pi_{ij} = p\delta_{ij} + \tau_{ij} = \pi_{ji} \qquad (6.16)$$

π_{ij} represents the total stress tensor in hydrodynamics consisting of static pressure tensor and viscous stress tensor.

Maxwell stress tensor describes the state of stress in electromagnetic fields associated with the transfer of electromagnetic momentum flux in a manner similar to convective momentum tensors. They share the same physical interpretation carrying energy and momentum in different media.

$$M_{ij} = \varepsilon \left(\frac{\delta_{ij} E^2}{2} - E_i E_j \right) + \mu \left(\frac{\delta_{ij} H^2}{2} - H_i H_j \right) = M_{ji} \qquad (6.17)$$

E and H represent electric field intensity and magnetic field intensity (H-field).

6.4.5 Tensor Contractions: Tensor Vector on a Reference Plane

Tensor rank reduction technique can also be applied to second order tensors while preserving tensor character. It provides different perspectives of a tensor in terms of vectors and scalars with physical significance. Tensor contracted product is basically a form of quotient theorem for tensors. It transforms a tensor to a vector at a point by the fluxes it produces through an arbitrary infinitesimal surface defined by the surface unit normal vector $\hat{b}$. There is a reduction in tensor rank in these surface phenomena. From Cauchy's formula,

$$T = \tau_{ij} \cdot \hat{b} \qquad (6.18)$$

Above yields a tensor vector transmitted across the infinitesimal surface. It must be emphasized that the tensor vector direction is generally different from the surface unit normal vector. If there is a change in the direction of the surface unit normal vector, that is turning the plane about a point it is drawn through, we obtain a different vector at the same point. This allows us to examine the state of tensor in terms of a vector by varying the plane unit normal vector at a point.

6.4.6 Tensor Contractions: Tensor Vector at a Point

Alternatively, tensor vector is completely defined at a point by the fluxes it produces through three mutually perpendicular infinitessimal surfaces without reference to a plane. A tensor vector through Cauchy's formula and Gauss's theorem can conveniently represent the closed flux integral of a tensor. Mathematically, it expresses the fact that the net flux of a second order tensor field through a closed surface is equal to volume integral of divergence of tensor over the volume bounded by surfaces. This makes it possible to specify a tensor vector at a point on a surface of arbitrary orientation. Consider a stress tensor as an example,

$$F = \oint_A [\tau] \cdot \hat{b} dA = \int_v \nabla \cdot [\tau] dv \qquad (6.19)$$

$$\text{Or } Q = \frac{dF}{dv} = \nabla \cdot [\tau]$$

Above is a line vector representing the amount of stress flux transmitted through the bounding surface of a control volume in terms of a force vector at a point. In this contraction process, a surface phenomenon is transformed into a volumetric phenomenon. There is also a reduction in tensor rank.

6.5 Visualization of Symmetric Tensors

6.5.1 Tensor Invariants

Tensor rank reduction techniques can be extended to include the use of tensor invariants as visualization elements while tensor character is preserved. In tensor theory, a second order symmetric tensor can be contracted to yield three independent scalar parameters referred to as tensor invariants. In elasticity, strain tensor invariants are defined as:

$$I_1 = \sum_1^3 \gamma_{ii} = \varepsilon_x + \varepsilon_y + \varepsilon_z = \varepsilon_1 + \varepsilon_2 + \varepsilon_3 \qquad (6.20a)$$

$$I_2 = \frac{1}{2}(\gamma_{ii}\gamma_{kk} - \gamma_{ij}\gamma_{ji}) = \varepsilon_x\varepsilon_y + \varepsilon_y\varepsilon_z + \varepsilon_z\varepsilon_x - \gamma_{xy}^2 - \gamma_{yz}^2 - \gamma_{xz}^2$$

$$= \varepsilon_1\varepsilon_2 + \varepsilon_2\varepsilon_3 + \varepsilon_1\varepsilon_3 = \frac{(I_1^2 - \gamma_{ij}\gamma_{ij})}{2} \qquad (6.20b)$$

$$I_3 = \det|\gamma_{ij}| = \varepsilon_x\varepsilon_y\varepsilon_z + 2\gamma_{xy}\gamma_{yz}\gamma_{xz} - \varepsilon_x\gamma_{yz}^2 - \varepsilon_y\gamma_{xz}^2 - \varepsilon_z\gamma_{xy}^2 = \varepsilon_1\varepsilon_2\varepsilon_3 \quad (6.20c)$$

The single subscript associated with I_i denotes first, second and third strain tensor invariants respectively. Otherwise, it denotes principal strain ε_i or eigenvalues.

Tensor components depend on reference coordinate systems but each of these invariants being a scalar quantity remains unaltered regardless of coordinate orientation. Tensor invariants have an important role to play in tensor theory as they specify the state of a tensor completely through the characteristic equation. They could be employed as additional visual elements to compliment to principal states. For a strain tensor it is written as:

$$\varepsilon^3 - I_1\varepsilon^2 + I_2\varepsilon - I_3 = 0 \qquad (6.21)$$

The solution yields principal state with a set of three eigenvalues, each associated with an eigenvector. In this manner, all six degrees of freedoms are recovered in a system of six simultaneous equations. Information about strain at a point can be obtained when strain tensor invariants are given. In two-dimensional fields, the third invariant vanishes and above reduces to:

$$\varepsilon^2 - I_1\varepsilon + I_2 = 0 \qquad (6.22)$$

Tensor first invariant represents different aspects of tensor internal characteristics in terms of its trace. It is obtained by putting two indices equal and then carrying out summation in a contraction process. Sokolnikoff (1983) presented an interpretation of strain tensor first invariant by considering a rectangular parallelepiped whose edges are in the directions of principal strains. Such that,

$$I_1 = \frac{\Delta Vol}{Vol_0} = \sum \gamma_{ii} = \nabla \cdot U \qquad (6.23)$$

Tensor first invariant assumes the role of scalar source. Its physical interpretation is a measure of mean normal strain in elasticity and its distribution contours are called strain isopachs. When first invariant is positive or negative, it indicates expansion/tension (source) or contraction/compression (sink) respectively. When strain tensor first invariants vanish, it implies deformation is isochronic with no change in volume, such that strain in-flux and out-flux across a control volume is the same everywhere. From strain first invariant distributions, we are able to obtain information about volumetric distortion from the isotropic part of deformation.

First and second invariants of strain tensors are related to strain energy given by:

$$\Pi_\gamma = \frac{1}{2}[\tau] : [\gamma] = \Pi'_\gamma + \Pi''_\gamma \tag{6.24}$$

$$\text{Isotropic part: } \Pi'_\gamma = \frac{1}{2}(\lambda + 2\mu)I_1^2 \tag{6.25}$$

$$\text{Deviatorial part: } \Pi''_\gamma = -2\mu I_2 = -2\mu\varepsilon_1\varepsilon_2 \tag{6.26}$$

with $\mu\gamma_{ij}\gamma_{ij} = \mu(I_1^2 - 2I_2) > 0$.

The isotropic part characterized by first invariant square represents energy change associated with dilation distortion. The deviatorial part represented by second invariant describes energy change associated with shear deformation. This part usually plays a significant role in material strength and failure. By zooming onto these effects, we could assess them individually or collectively in quantitative terms.

In fluids, the rate of strain tensor invariants $\dot{I}_i$ have similar expressions, that is,

$$\dot{I}_1 = \sum \dot{\gamma}_{ii} = \dot{\varepsilon}_1 + \dot{\varepsilon}_2 + \dot{\varepsilon}_3 = \nabla \cdot V \tag{6.27}$$

$$\dot{I}_2 = \frac{1}{2}(\dot{\gamma}_{ii}\dot{\gamma}_{kk} - \dot{\gamma}_{ij}\dot{\gamma}_{ji}) = \dot{\varepsilon}_1\dot{\varepsilon}_2 + \dot{\varepsilon}_2\dot{\varepsilon}_3 + \dot{\varepsilon}_1\dot{\varepsilon}_3 = \frac{(\dot{I}_1^2 - \dot{\gamma}_{ij}\dot{\gamma}_{ij})}{2}$$

$$\dot{I}_3 = \det|\dot{\gamma}_{ij}| = \dot{\varepsilon}_1\dot{\varepsilon}_2\dot{\varepsilon}_3$$

$\dot{\varepsilon}_i$ is the eigenvalues of the rate of strain tensor. The rate of viscous dissipation energy is given by the viscous dissipation function.

$$\Pi = \frac{1}{\rho}[\tau] : [\dot{\gamma}]$$

$$= \Pi' + \Pi''$$

$$\text{where} \quad \Pi' = -\frac{4}{3}v\dot{I}_1^2 \tag{6.28}$$

$$\Pi'' = 4v\dot{I}_2 = 4v\dot{\varepsilon}_1\dot{\varepsilon}_2 \tag{6.29}$$

with $\dot{\gamma}_{ij}\dot{\gamma}_{ij} = (\dot{I}_1^2 - 2\dot{I}_2) > 0; \dot{I}_1^2 > 0$.

Viscous dissipation also consists of two parts; an isotropic part and a deviatorial part are associated with first invariant square and second invariant of the rate of strain tensor respectively. By comparing the distributions of the deviatorial part such as Π''_γ (or Π'') with those of strain energy Π_γ (or viscous dissipation energy Π), we can assess the way energy is being changed in the system due to shear deformation or dilation distortions.

6.5.2 Tensor Transformation

Representation of a second rank tensor in matrix form requires specifications of a reference coordinate system. It is necessary that the same results are obtained regardless of the orientations of reference coordinate systems. In matrix theory, a complete specification of a symmetric matrix could be established from the knowledge of principal values and principal axes. This comes from full analogy of tensors with the theory of real symmetric matrix such that a real symmetric tensor of second rank can be "diagonalized." The diagonal forms have provided a convenient means for intrinsic representation of tensors with reference to principal axes trajectories. Principal states are the simplest forms of symmetric tensors independent of coordinate orientations. All measurable effects of a change in fields are exclusively due to changes in principal states. Principal state transformation of tensors is regarded as a version of tensor rank reduction as the product only involves vector fields and scalar fields. Principal state transformation is of great value as it retains multidimensional information by a set of orthogonal vectors with a corresponding set of eigenvalues. Studies focus at principal axes transformation of a quadratic surface in Euclidean space of arbitrary domains has laid the fundamental connecting link between widely different physical phenomena. For example, principal states of a stress tensor in solids describe the ways in which force is transmitted along principal axes trajectories. Principal states of a viscous stress tensor or a convective momentum flux tensor in fluids reveal the ways in which momentum transfers by diffusion or convection along principal directions. The same applies to Maxwell stress tensor, which describes electromagnetic momentum transfer in electromagnetic fields. All these phenomena share the same analogy of second order symmetric stress tensor fields but in different field media. They are interpreted as contiguous action of stress waves propagating from point to point under tension or compression along respective principal axes trajectories. By correlating tensor fields, vector fields and scalar fields in mixed problems, we can gain insights into the field structure and behavior. Furthermore, principal states are useful in decoupling transport processes between mutually orthogonal directions when principal axes of field medium anisotropy coincide with spatial coordinates. It also plays a key role in tensor topological skeleton classifications in the study of critical points.

6.5.3 Principal States and Eigenanalysis

Geometrically we regard a real matrix as a representation of coordinates of points in multi-dimensional space and interpret eigenvalues and eigenvectors as an arrangement of these points. Consider a three dimensional tensor field represented by a symmetric matrix with reference to a Cartesian coordinate system. It is defined with respect to three sets of base vectors, each on a mutually orthogonal plane passing through a given point. When one of the eigenvalues is zero, the quadric surface degenerates from an ellipsoid to a plane ellipse in a class of plane problems. On (x, y) plane, the tensor vectors on x-face and y-face are given by the first and second column of the matrix respectively. Each of them defines a point in Figure 6.2 by a position vector from coordinate origin as follows:

$$\text{Vector on } x - \text{face}: \quad T = [\tau] \cdot \begin{bmatrix} 1 \\ 0 \end{bmatrix} = \begin{bmatrix} \sigma_x \\ \tau_{yx} \end{bmatrix} \qquad (6.30a)$$

$$\text{Vector on } y - \text{face}: \quad T = [\tau] \cdot \begin{bmatrix} 0 \\ 1 \end{bmatrix} = \begin{bmatrix} \tau_{xy} \\ \sigma_y \end{bmatrix} \qquad (6.30b)$$

where $\tau_{ij} = \begin{bmatrix} \sigma_x & \tau_{xy} \\ \tau_{yx} & \sigma_y \end{bmatrix}$ with $\tau_{xy} = \tau_{yx}$.

These vectors terminate on the boundary of ellipse whose center is located at origin. The orientation of ellipse axes is determined by principal angles $\theta_{1,2}$. The lengths of ellipse axes are defined by the corresponding principal values $\sigma_{1,2}$. The shapes of an ellipse may change from point to point and evolve into different conic forms depending on principal states. An ellipse is said to be elongated when one of the axes (major) is greater than the other (minor). With equal numeric eigenvalues, ellipse evolves into a circle. If two points coincide on ellipse boundary, or two column (row) vectors are parallel to each other, then the ellipse collapses into a line segment with one vanishing eigenvalue. Or it evolves into a degenerate point with identical eigenvalues and infinite principal axes crossing at the point. An ellipse will collapse into a degenerate point when all eigenvalues vanish. Any curve passing through a degenerate point is a principal axis. These geometric representations are

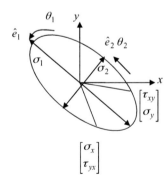

Figure 6.2 Geometric interpretation of principal states

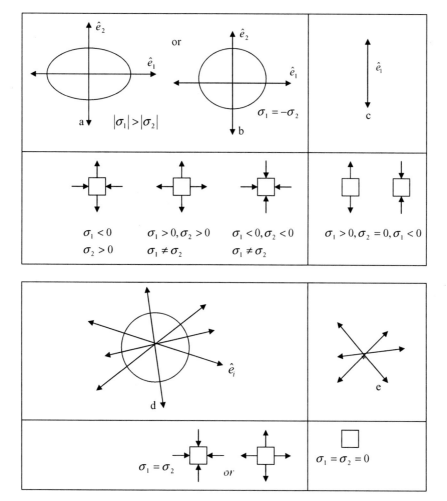

Figure 6.3 Evolving tensor ellipses at various principal states

depicted in Figure 6.3. Tensor ellipse breaks down at a singular point where field strength (eigenvalue) is infinite or undefined.

Principal states of a tensor are generally determined in Eigenanalysis by evaluating principal values and corresponding principal vectors or by solving the characteristic equation. For two-dimensional fields, the remaining two principal values can be determined explicitly below.

$$\sigma_{1,2} = \frac{(\sigma_x + \sigma_y)}{2} \pm \frac{[(\sigma_x - \sigma_y)^2 + 4\tau_{xy}^2]^{0.5}}{2} \tag{6.31}$$

$|\sigma_1| > |\sigma_2|$, or

$$\sigma_1 > 0 > \sigma_2 \text{ when } \sigma_1 = -\sigma_2$$

Subscripts 1 and 2 denote major and minor principal value (or principal axis) in accordance with the numbering system. Principal values may bear physical significance depending on the nature of tensor fields. For instance in a solid system, principal values of a stress tensor represent extreme normal stresses (principal stresses) acting on principal planes driven by strain fluxes. They describe force/load transmissions along respective principal axes trajectories. Figure 6.4 shows schematically the principal state of a stress tensor at which the off-diagonal components (shear stresses) vanish on principal planes and extreme principal stresses occur. They give rise to the action of contraction/stretching causing dilation distortion while maximum shear occurs along the slip line at 45 degrees from the principal plane unit normal. The sign of eigenvalues characterizes flux directions. The plus or minus sign represent out-going flux or in-coming flux on the outward surfaces of a control volume in Figure 6.3 respectively. It also classifies the type of normal stress as tension or compression by conventions. Principal values are usually presented in contours just like scalar quantities.

The inclinations of two principal planes on which principal stresses act are determined by principal angles within an integral multiple of $\pi/2$.

$$\tan 2\theta = \frac{2\tau_{xy}}{\sigma_x - \sigma_y} \tag{6.32}$$

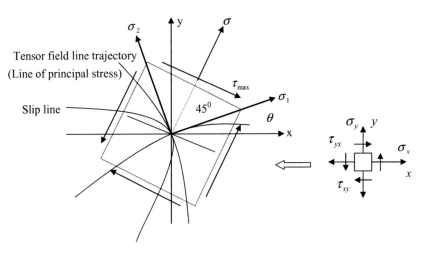

Figure 6.4 Principal state, tensor field line trajectory and slip line

Thus, $\theta_1 = 0.5 \tan^{-1} \left(\frac{2\tau_{xy}}{\sigma_x - \sigma_y} \right); \theta_2 = \theta_1 + \frac{\pi}{2}$.

Principal angle is defined by principal plane unit normal measured with respect to x-axis counterclockwise in right-handed coordinate systems. The above two directions are at right angles to each other and define the orientations of principal axes. By rotating a coordinate system through a principal angle about origin, the stress vectors are mapped from reference coordinate to principal axes coordinate. As a result, the tensor assumes diagonal form.

Thus, visualization of two-dimensional second order tensors requires simultaneously displays of two principal values and the orientation of a pair of orthogonal principal vectors continuously. Because of sign indeterminacy, principal vectors $\hat{e}_i$ are bi-directional and have a different structure from regular vectors. To determine the directions of load transmission, it is necessary to resolve the bi-direction nature of eigenvectors by means of conventions. Figure 6.5 shows one of the conventions that load/transfer process takes place in the direction of diverging neighboring flux trajectories in solenoidal tensor fields. The transport strength is diminishing as the space between two adjacent trajectories widens up in the same manner as streamlines. In this connection, the transport directions accompanied with decreasing eigenvalues could be visualized through the sizes of tensor conic icons.

To display the states of a stress tensor continuously, tensor field line (or lines of principal stress) trajectory was introduced (Woods, 1903, Dickinson, 1989, 1991) to describe the paths along which loading is being transmitted with a strength given by principal stresses. The principal axes trajectories are conveniently obtained from principal angles by tangent field method below:

$$\text{Major principal axis trajectory } \hat{e}_1 : \frac{dy}{dx} = \tan \theta_1$$

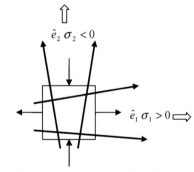

⟹ Principal axis trajectory direction

Figure 6.5 Principal axis direction

$$\text{Minor principal axis trajectory } \hat{e}_2 : \frac{dy}{dx} = \tan\theta_2$$

From the knowledge of principal states, other useful information is derived:

$$\text{Mean normal stress: } \sigma = \frac{\sigma_1 + \sigma_2}{2} = \frac{\Lambda_1}{2} \tag{6.33}$$

Mean normal stress defines the center of Mohr's circle along normal stress axis. It acts on the plane of maximum shear with a magnitude equal to half of the stress tensor first invariant.

$$\text{Maximum shear stress: } \tau_{max} = \frac{\sigma_1 - \sigma_2}{2} \tag{6.34}$$

Maximum shear stress defines the radius of Mohr's circle. It acts on a surface inclined at an angle of 45 degrees from principal axis. The plane of maximum shear bisects two principal planes. To visualize the paths of maximum shear stress, a slip line is constructed in Figure 6.4 whose tangents give the directions of maximum shear stress. Mohr's circles can be constructed from mean normal stress and maximum shear stress.

Manipulation and interpretation of tensors are simplified when they are transformed into diagonal forms corresponding to principal states. This immediately invokes the reduction of a matrix to conical form. From matrix theory, a symmetric two by two matrix can be written in quadratic form as follows:

$$\begin{aligned} Q &= x^T \cdot \tau_{ij} \cdot x \\ &= \sigma_x x_1^2 + \sigma_y x_2^2 + 2\tau_{xy} x_1 x_2 \end{aligned} \tag{6.35}$$

The loci of Q =constant are called stress quadric, also referred to as quadratic surfaces of second degree. It represents geometric shapes of second order conic sections. A simple method to display what quadratic surfaces look like is to obtain their intersections with planes parallel to coordinate planes. By rotating the coordinate system through a principal angle, we obtain principal state of matrix in diagonal form and the above is reduced to a standard quadratic equation:

$$Q = \lambda_1 x_1^2 + \lambda_2 x_2^2 \tag{6.36}$$

$$\text{Or } \frac{x_1^2}{\left(\sqrt{\frac{Q}{\lambda_1}}\right)^2} + \frac{x_2^2}{\left(\sqrt{\frac{Q}{\lambda_2}}\right)^2} = 1$$

Lines along which a plane intersects a right circular cone produce conic sections. There are four types of conic sections described by quadratic form depending on the coefficients:

$$\lambda_1 = \lambda_2 : \text{Circle}$$
$$\lambda_1\lambda_2 > 0, \lambda_1 \neq \lambda_2 : \text{Ellipse}$$
$$\lambda_1\lambda_2 < 0 : \text{Hyperbola}$$
$$\lambda_1\lambda_2 = 0 : \text{Parabola}$$

It is also possible to produce degenerate conic sections such as a line or a point. This occurs when intersecting plane assumes the role of a tangent plane on the napes of cone (surface) giving rise to a line of tangent. Or it cuts the cone at a single point of its vertex. The connection between conic sections of quadratic surfaces and principal states immediately leads to the ideas of representation of a two-dimensional tensor by an ellipse point icon. Ellipse provides a concise and an elegant way to visualize tensor fields with a simple shape but the same degrees of freedom as tensors. When tensor ellipse is constructed with its axes orientation coincide with principal axes and the lengths equal to principal values, it represents Lame's stress ellipse in Figure 6.2. Lame's stress ellipse has the property that stress vector of tensor on any plane is represented by a position vector from origin and ends on its boundary. Thus associated with every point in the field there is a stress quadric from which normal and tangential components of stress vector transmitted through an arbitrary surface with unit normal $\hat{b}$ may be determined in accordance with Cauchy's formula. To substantiate the above observations, we adopt principal axes as the reference coordinate system with major principal axis $\hat{e}_1$ parallel to x-axis and consider the stress vector on an arbitrary plane with a unit normal $\hat{b}$, that is,

$$T = \begin{bmatrix} T_x \\ T_y \end{bmatrix} = \begin{bmatrix} \sigma_1 & 0 \\ 0 & \sigma_2 \end{bmatrix} \cdot \hat{b} = \begin{bmatrix} \sigma_1 \cos(b, x) \\ \sigma_2 \cos(b, y) \end{bmatrix} \tag{6.37}$$

$$\text{where } \hat{b} = \begin{bmatrix} \cos(b, x) \\ \cos(b, y) \end{bmatrix}$$

$$\frac{T_x}{\sigma_1} = \cos(b, x), \frac{T_y}{\sigma_2} = \cos(b, y)$$

$$\text{Two-dimensional stress ellipse:} \frac{T_x^2}{\sigma_1^2} + \frac{T_y^2}{\sigma_2^2} = 1$$

$$\text{Three-dimensional stress ellipsoid:} \frac{T_x^2}{\sigma_1^2} + \frac{T_y^2}{\sigma_2^2} + \frac{T_z^2}{\sigma_3^2} = 1 \tag{6.38}$$

σ_i are eigenvalues. The stress vector describes loci of a stress quadric, which is specified uniquely by principal stresses. The stress quadric makes graphical determination of stress at a point possible. Stress vector on an arbitrary plane with a unit normal $\hat{b}$ is a position vector from origin and ends on quadric boundary. It has a magnitude given by its normal component σ and shear component τ acting parallel and perpendicular to $\hat{b}$ respectively:

$$|T| = (\sigma^2 + \tau^2)^{1/2}$$

For two-dimensional fields, the normal and shear components could be found in terms of the direction cosines of unit normal.

$$\sigma = \sigma_x \cos^2(b, x) + \sigma_y \cos^2(b, y) + 2\tau_{xy} \cos(b, x) \cos(b, y)$$
$$\tau = (\sigma_y - \sigma_x) \cos(b, y) \cos(b, x) + \tau_{xy}(\cos^2(b, x) - \cos^2(b, y))$$

Or

$$\sigma = \sigma_1 \cos^2(b, x) + \sigma_2 \cos^2(b, y)$$
$$\tau = (\sigma_2 - \sigma_1) \cos(b, x) \cos(b, y)$$

Above gives a complete description of stress variations at a point. By aligning unit normal $\hat{b}$ parallel to major principal axis $\hat{e}_1$ and hence x-axis, then

$$\cos(b, x) = \cos 0 = 1$$
$$\cos(b, y) = \sqrt{1 - \cos^2(b, x)} = 0$$
$$\rightarrow |T| = \sigma = \sigma_1; \ \tau = 0$$
$$\text{Or } T = \sigma_1 \hat{e}_1$$

Similarly,

$$\cos(b, y) = \cos 0 = 1$$
$$\cos(b, x) = \sqrt{1 - \cos^2(b, y)} = 0$$
$$\rightarrow |T| = \sigma = \sigma_2; \ \tau = 0$$
$$\text{Or } T = \sigma_2 \hat{e}_2$$

The stress vectors then correspond to principal stresses. It could be extended to three-dimensional fields, that is,

$$T = \sigma_3 \hat{e}_3$$

In general,

$$T = \sigma\hat{e} = [\tau] \cdot \hat{e}$$

Eigenanalysis is commonly applied to produce a summary of data structure from a given tensor when it is represented by a matrix with reference to a coordinate system. In general, principal values can be either real or complex depending on the nature of matrix. For a symmetric matrix, eigenvalues are real and orthogonal bases exist except at a singular/degenerate point. The above leads to exactly the same form in Eigenanalysis.

$$([\tau] - \sigma[I]) \cdot \hat{e} = 0$$

For two-dimensional fields, we have,

$$\hat{e}_2 \cdot [\tau] \cdot \hat{e}_1 = \hat{e}_2 \cdot \sigma_1\hat{e}_1$$
$$\hat{e}_1 \cdot [\tau] \cdot \hat{e}_2 = \hat{e}_1 \cdot \sigma_2\hat{e}_2$$
$$\text{Or } (\sigma_1 - \sigma_2)\hat{e}_1 \cdot \hat{e}_2 = 0$$

Thus, a general symmetric tensor is represented by its eigenvalues and eigenvectors through tensor transformation. The transformation is described by two consecutive contractions along two principal axes. The above result suggests the following possibilities, which determine the state of a given tensor. They are:

(1) $\sigma_1 - \sigma_2 \neq 0$, or
(2) $\sigma_1 - \sigma_2 = 0$

With finite distinct principal values in case (1) (Figure 6.3a), a pair of orthogonal eigenvectors is uniquely determined to within reversal of directions.

$$\rightarrow \hat{e}_1 \cdot \hat{e}_2 = 0$$

When $|\sigma_1| > |\sigma_2|$, the ellipse is elongated with major eigenvalue dominates the fields. The following cases are of special interest:

(1).a $\Lambda_1 = \sigma_1 + \sigma_2 = 0$

With equal principal values but opposite in signs, tensor ellipse becomes a circle (Figure 6.3b). The circle radius represents eigenvalue magnitude. This corresponds to equal contraction and expansion along respective principal axes with no dilation distortion. Furthermore, it implies that both tensor first invariant and metric coefficient vanish.

(1).b $\Lambda_2 = \sigma_1\sigma_2 = 0$

When one of the eigenvalues (or tensor second invariant) vanishes, tensor ellipse degenerates into a conic section of a line segment (Figure 6.3c). The tensor field is then regarded as a vector field since two row (or column) vectors in the given matrix are parallel with only one remaining active. The principal axis associated with the nonzero eigenvalue is parallel to vector field everywhere. Such condition corresponds to pure contraction or pure expansion along the remaining principal axis trajectory.

$$(1).c \quad \Lambda_2 = \sigma_1 \sigma_2 = \infty \quad \text{or} \quad \Lambda_1 = \sigma_1 + \sigma_2 = \infty$$

With one infinite eigenvalue, the ellipse stretches in opposite directions without bound. This corresponds to a singular point where fields/solutions break down.

$$(2). \quad \sigma_1 - \sigma_2 = 0$$

When equal finite principal values occur at a point, it is called a degenerate point, that is, a degenerate conic section. The tensor is then classified mathematically as having geometric multiplicity of unity less than the algebraic multiplicity of two. Orthogonal bases fail to exist since there is only one arbitrary eigenvector. There is infinite number of principal axes because any vector passing through this point is a principal vector (Figure 6.3d). Stresses at this point are principal stresses. The ellipse evolves into a circle implying that stresses are the same in all directions, that is, an isotropic tensor. The interpretation of a degenerate point in hydrodynamics is that velocity and shearing stress vanish there. They are points of hydrostatic pressure where flow is stagnant and pressure is the same in all directions. When both eigenvalues vanish ($\sigma_1 = \sigma_2 = 0$), we conceptually regard the tensor ellipse degenerating into just a point (Figure 6.3e).

In three-dimensional fields, the following possibilities arise:

(1) All three principal values are distinct. There is a unique set of three mutually orthogonal bases corresponding to principal directions. They define the orientation of ellipsoid axes.

(2) Two principal values are equal with opposite sign. The stress ellipsoid becomes an ellipsoid of revolution. The line of rotation symmetry coincides with principal axis of the remaining eigenvalue. If they have the same sign, the corresponding pair of eigenvectors may be chosen with any orientation in a plane normal to the remaining eigenvector.

(3) All three principal values are equal. Tensor ellipsoid becomes a sphere at this degenerate point. Any three orthogonal directions through degenerate point can be taken as principal axes. It will degenerate into a point if all equal principal values vanish.

(4) If one of the principal values vanishes, the third tensor invariant follows suit and tensor ellipsoid evolves into a plane ellipse. The stress vectors on all planes through the point lie on the plane ellipse and terminate on ellipse boundary in a class of plane stress.

(5) If two principal stresses vanish, the ellipsoid degenerates into a line of symmetry and the tensor field is regarded as a vector field. The principal axis associated with the non-zero eigenvalue is parallel to vector field everywhere and the system is in a state of simple tension or compression.

(6) With one infinite eigenvalue, the ellipsoid stretches in opposite directions without bound. This corresponds to a singular point.

To investigate the relations between eigenvalues, metric coefficient was introduced. Metric coefficient describes how different the transport rates can be along respective principal axes. This parameter gives a quantitative measure of degree of anisotropy in a given tensor field. There are several versions of metrics for anisotropy in tensor fields based on eigenvalues. A set of metric coefficients has been presented in literature for encoding eigenvalue distributions by normalizing the eigenvalue difference against major eigenvalue. For two-dimensional domains, the following definition is adopted:

$$c_l = \frac{|\sigma_1| - |\sigma_2|}{\sigma_1} \text{ with } -1 \leq c_l \leq 1$$

The degree of anisotropy also reflects the shape of tensor ellipse. High anisotropy with $|c_l|$ close to unity implies the transport rates vary greatly as a function of directions. This results in a highly elongated ellipse along its major axis, which dominates the state of tensors. It is said to be low with metric coefficient close to zero when the rates are about the same in all directions and ellipse becomes a circle.

6.5.4 Hybrid Method of Tensor Visualization

It is recognized that the most informative visualization does not necessarily come from striving to pack many dimensions of information into one image. Rather it may be desirable to create renderings of tensor data by displaying salient information using point icons at selected locations or regions. It is a common practice to visualize all degrees of freedom of a tensor locally at a point by glyph rendering. Because ellipse point icons are effective to show the local state of tensors, we could display them along selected principal axes trajectories systematically. Visualization of multidimensional data can be enhanced by independently moving visual point icons. On the other hand, principal axes trajectories are ideal line icons to describe tensor states from a given point continuously by combining graphs and point icons with accuracy of curves. As the principal state changes from point to point, load transmissions and/or transport processes could be visualized through evolving tensor ellipses displayed at selected points along principal axis trajectory. They reflect the change of principal values through elongated ellipse, circle, line or point in static displays or animations. By displaying evolving tensor ellipses in a sequence along a principal axis trajectory, we are able to take advantage of human visual systems. As the problem of cluster is removed, it is generally not difficult to interpret information at intermediate points. The method combines the arts of displays and information science through integration of local behaviors at discrete points into global phenomena.

6.6 Visualization of Antisymmetric Tensors

Since antisymmetric tensors possess axial vector characteristics, they are conveniently visualized by vector field techniques. Vorticity tensor is a typical example of antisymmetric tensors and because it plays an important part in hydrodynamics and has analogy in other fields, it becomes our study focus.

6.6.1 Vorticity Concepts and Dynamics

The concept of vorticity is of central importance in rotational fields. In the wake of a universal definition, we follow the classical definition of Cauchy and Stokes called the angular velocity of a fluid at a point in space "vorticity." A fluid particle is said to possess vorticity when it spins about its own axis as it travels along a streamline with change in position and orientation continuously. Vorticity is an infinitesimal microscopic vector quantity measuring fluid molecular rotations and defines the direction of rotational axis. It is a local phenomenon and, at its limit, approaches a point through which the axis of rotation (vortex line) passes. Vorticity is a measure of vector source strength equal to twice the angular speed and could be visualized physically by a vorticity meter. It characterizes the rotation component of tensor structure in addition to deformation component and translation component. Vortices are distributed throughout flow fields. This idea introduces the models of continuous distributions of vorticity and of vortex lines. Conceptually, one can think of vortex lines in the context of streamlines as stringing together a set of material particles about which fluid particles rotate during the course of movement. A vortex line is a line drawn in fluids such that the tangent to it at each point is in the direction of vorticity vector. It traces a trajectory of the field of vorticity vectors. A vortex line never begins or ends. It extends to infinity in opposite directions. Like streamlines, vortex lines satisfy trajectory differential equation. Vortex lines and vortex tubes must obey the same rules that streamlines and stream tubes do. Note however, that vorticity has the origin of antisymmetric tensors.

The mathematical concept of vorticity is developed from two approaches.

$$\text{Vorticity:} \; \omega = \nabla \times V \tag{6.39}$$

$$\text{Circulation:} \; \Gamma = \oint_C V \cdot d\vec{r} = \int_A \omega \cdot \hat{b} \, dA$$

In differential approach, vorticity fields derived from the curl of flow fields are regarded as the kinematics property of the fields and assume the role of vector sources. This leads to the geometric interpretation of rotation as the extent to which field vector 'curls' around an axis passing through an arbitrary point. Vorticity is solenoidal in accordance with vector identity. In integral approach, vorticity is linked to the concept of circulation around a closed path through Stokes theorem. Circulation of flow fields round a closed path c measures the net tangential velocity component and is invariant only for inviscid fluids in accordance with Kelvin's theorem. There is then no tendency to rotate or to form vortices. Circulation density per unit area defines the strength of an infinitesimal vorticity tube or a vortex filament in terms of amount of curl enclosed. The region that produces circulation is a vortex region where vorticity source is present. The sense of circulation direction is associated with its sign and convention. Following right hand rule, a two-dimensional counterclockwise circulation about z-axis defines the sense of positive direction of vorticity vector pointing in positive z-axis. When vorticity is positive in some parts of fluids and negative in other, it implies a state of steady motion. Circulation is independent of enclosed paths only for conservative fields. Vorticity affects flow behavior drastically since it is intimately related to the flow parameters in one way or other. It has far-reaching implications in energy conversion processes in flow systems.

In an incompressible, isotropic and homogenous medium, vorticity is related to viscous stress vector (frictional force) through vector identity. The existence of vorticity generally indicates that viscous effects are active in real fluids against fluid motions. Fluid viscosity provides the venue for vorticity activities in the processes of generation, diffusion and dissipation wherever the driving mechanisms of velocity gradients exist. Therefore, vorticity activities bound to occur at boundary and are closely associated with boundary layer flows. Vorticity cannot originate in the interior of a fluid system but must diffuse inwards from boundary. Thus, viscous effects are in part responsible for establishing vorticity distributions in the fields and are also connected to the distributions of viscous momentum, that is, the viscous stress tensor fields.

Vorticity behavior could be interpreted based on its structure in velocity gradients in the context of Lamb vector. Centripetal acceleration being one of components of Lamb vector will occur whenever trajectory is curved and is responsible for the change of velocity direction but not magnitude towards the center of curvature always. Centripetal acceleration requires other force(s) to maintain force balance in the trajectory normal direction in order to keep particles on curve tracks. Its magnitude is proportional to tangential speed square but inversely to radius of curvature. For counter-clockwise circular motions, curvature is positive and centripetal acceleration gives rise to the time rate of change of streamline tangent veering to the left of trajectory. From Frenet's formula,

$$\frac{\hat{n}}{R_s} = \frac{d\hat{s}}{dt}\frac{dt}{ds} \tag{6.40a}$$

$$\frac{v_s}{R_s}\hat{n} = \frac{d\hat{s}}{dt} = \frac{d\alpha}{dt}\hat{n} = \frac{d\theta}{dt}\hat{n}$$

$$\frac{v_s^2}{R_s} = v_s\frac{d\alpha}{dt} = v_s\frac{d\theta}{dt}$$

$$\frac{d\theta}{dt} = \frac{d\alpha}{dt}; \frac{d\theta}{ds} = \frac{1}{R_s} = \frac{d\alpha}{ds}$$

α and θ denote streamline tangent angle and angular rotation from positive x-axis counterclockwise. $\frac{v_s}{R_s}$ is interpreted as the rate of rotation of the particle equal to the rate of change of streamline tangent angle. Convective acceleration is the other component of Lamb vector and also contributes to the rate of change of streamline tangent in the same or opposite fashion depending on the sign of velocity gradient in normal direction. It may compete with centripetal acceleration or work together such as in free vortex or forced vortex motions. This term will be negative when velocity gradient is positive implying velocity increases when moving towards the center of curvature. Convective acceleration will cause a negative rate of change of streamline tangent as velocity direction veers to the right of trajectory in clockwise. This is attributed to the flow field structure that flows closer to center of curvature are moving faster on the left side than those on the right. Under this condition of free vortex motions, a pair of orthogonal lines rotates towards each other at same angular speed resulting in zero net rotation and zero vorticity.

$$-v_s\left(\frac{\partial v_s}{\partial n}\right) = v_s\left(-\frac{d\alpha}{dt}\right) \quad \text{when} \quad \frac{\partial v_s}{\partial n} > 0 \tag{6.40b}$$

$$\omega_z = \frac{v_s}{R_s} - \frac{\partial v_s}{\partial n} = \frac{d\alpha}{dt} - \frac{d\alpha}{dt} = 0$$

or

$$\frac{\partial v_s}{\partial n} = \frac{v_s}{R_s}$$

The flow is then classified as potential flow at the point. Otherwise, velocity decreases towards the center of curvature like forced vortex motions. Flows are then moving faster on the right than those on the left, such that the rate of change of streamline tangent becomes positive (counterclockwise). As a result, there is a net non-zero rotation in counterclockwise direction.

$$-v_s\left(\frac{\partial v_s}{\partial n}\right) = v_s\left(\frac{d\alpha}{dt}\right) \quad \text{when} \quad \frac{\partial v_s}{\partial n} < 0 \tag{6.40c}$$

$$\omega_z = \frac{v_s}{R_s} - \frac{\partial v_s}{\partial n} = \frac{d\alpha}{dt} + \frac{d\alpha}{dt} = 2\frac{d\alpha}{dt} = 2\dot{\Omega}_z$$

$$= \frac{\partial v_\theta}{\partial r} + \frac{v_\theta}{r} = \dot{\Omega}_z + \dot{\Omega}_z = 2\dot{\Omega}_z$$

$$\rightarrow \dot{\Omega}_z = \frac{d\alpha}{dt} = \frac{d\theta}{dt} \tag{6.40d}$$

If flow is uniform normal to streamlines or changes only in the flow direction, then

$$\frac{\partial v_s}{\partial n} = 0$$

Centripetal acceleration being a component of Lamb vector remains operative and determines vorticity strength.

$$\omega_z = \frac{v_s}{R_s} - \frac{\partial v_s}{\partial n} = \frac{v_s}{R_s}$$

The role of vorticity in interpreting fluid flows is that vorticity tracks not only frictional effect in streamline direction but also centripetal and convective effects in streamline normal direction as reflected by its structure in terms of velocity gradients. All these effects could exist simultaneously. By assessing each of them, we are able to identify regions of high shear/friction or high vorticity and to gain a better understanding of field behaviors.
    If streamline is straight, then

$$\omega_z = \frac{v_s}{R_s} - \frac{\partial v_s}{\partial n} = -\frac{\partial v_s}{\partial n}$$

Centripetal acceleration vanishes also at a point of inflection. Vorticity strength depends on velocity gradient in streamline normal direction. Nonetheless vorticity kinematic conditions (Equation 6.39) hold for all time.

To explore the nature and properties of vorticity, two representative types of circular vortex motions are presented: Forced vortex and free vortex.

6.6.2 Forced Vortex

A forced two-dimensional vortex on (x,y)-plane rotates like a solid body at constant angular speed of $\dot{\Omega}_z$ about z-axis. The mutual distance of any two points does not change but orientation. It possesses vorticity and circulation everywhere except the critical point at vortex center. Fluid in a steady rotating tank is one of examples of a forced vortex in a state of rigid body rotation maintained by the motions of boundaries. By placing the vortex center to coincide with polar coordinate origin, we consider an ideal fluid particle at a location defined by position vector $\vec{r}$ in the body of a counter-clockwise forced vortex. The motion of fluid particle is represented by its velocity relative to vortex center. For rigid body motion $r = |\vec{r}|$ is a constant quantity and the motion trajectory describes a circle of radius r.

$$\text{Velocity}: V = \frac{d\vec{r}}{dt} = \frac{d(r\hat{r})}{dt} = \frac{dr}{dt}\hat{r} + r\frac{d\theta}{dt}\frac{d\hat{r}}{d\theta} = \frac{dr}{dt}\hat{r} + r\dot{\Omega}_z\hat{\theta}$$

$$\text{where } v_\theta = r\dot{\Omega}_z; \quad v_r = \frac{dr}{dt} = 0; \quad \dot{\Omega}_z = \frac{d\theta}{dt}$$

$$\text{Angular velocity}: \dot{\Omega} = \begin{bmatrix} 0 & 0 & \dot{\Omega}_z \end{bmatrix}$$

$$\text{Acceleration}: a = \frac{dV}{dt} = \begin{bmatrix} -r\dot{\Omega}_z^2 & 0 \end{bmatrix}$$

$$a \cdot V = 0$$

$$\text{Continuity equation}: \frac{\partial v_\theta}{\partial \theta} = 0, \rightarrow v_\theta = v_\theta(r)$$

Euler equation:

$$\hat{\theta}\text{-direction}: \frac{\partial p}{\partial \theta} = 0 \rightarrow p = p(r)$$

$$\hat{r}\text{-direction}: \frac{\partial p}{\partial r} = \rho\frac{v_\theta^2}{r} = \rho\dot{\Omega}_z^2 r, \rightarrow p = \frac{\rho}{2}\dot{\Omega}_z^2 r^2 + C = \frac{\rho}{2}v_\theta^2 + C$$

The nonviscous flow model shows the flow fields are axisymmetric with respect to vortex center. In Table 6.1, circulation measures strength of vortex, which depends on angular speed and the distance from vortex center. Circulation density per unit area possessed by a forced vortex is equal to twice of angular speed. For circular motions at constant angular speed, acceleration and velocity are perpendicular to each other. The acceleration vector represents centripetal acceleration and points towards the vortex center. The

Table 6.1 Properties of forced vortex and free vortex

Vortex center	Forced vortex Critical point	Free vortex Singular point
Stream function:	$\psi = -\int v_\theta \partial r = -\dfrac{\dot{\Omega}_z r^2}{2}$	$\psi = -\int v_\theta \partial r = -\dfrac{\Gamma}{2\pi}\ln r$
Equipotential:	Circular motion Not exist	Circular motion $\phi = \int v_\theta r\partial\theta = \dfrac{\Gamma}{2\pi}\theta$
Complex potential		$F = -\dfrac{i\Gamma}{2\pi}(\ln r + i\theta) = -\dfrac{i\Gamma}{2\pi}\ln z$
Circulation: Circulation density:	$\Gamma = \oint v_\theta r d\theta = 2\pi r^2 \dot{\Omega}_z$ $\dfrac{d\Gamma}{dA} = 2\Omega_z = \omega_z$	$\Gamma = \oint v_\theta r d\theta = $ constant when singular point (origin) is enclosed by the path. Otherwise, $\Gamma = \oint v_\theta r d\theta = 0$ $\dfrac{d\Gamma}{dt} = 0$ $\omega_z = \dfrac{\partial v_\theta}{\partial r} + \dfrac{v_\theta}{r} - \dfrac{\partial v_r}{r\partial\theta} = -\dfrac{\Gamma}{2\pi r^2} + \dfrac{\Gamma}{2\pi r^2} = 0$ $\omega = \nabla \times V = 0$ Irrotational
Vorticity in two-dimensional fields:	$\omega = \nabla \times V = [0\ 0\ \omega_z]$ $|\omega| = \omega_z = \lim_{dA\to 0}\dfrac{d\Gamma}{dA} = \dfrac{\Gamma}{\pi r^2} = 2\dot{\Omega}_z$ $\omega_z = \dfrac{\partial(rv_\theta)}{r\partial r} - \dfrac{\partial v_r}{r\partial\theta} = \dfrac{\partial v_\theta}{\partial r} + \dfrac{v_\theta}{r} = 2\dot{\Omega}_z$ $\dfrac{d}{dt}\oint_S \omega \cdot dS = \dfrac{d}{dt}\int_v \nabla \cdot \omega dv = 0$ Rotational	

(Continued)

Table 6.1 (Continued)

Vortex center	Forced vortex Critical point	Free vortex Singular point
Angular momentum per unit mass in two-dimensional fields:	$Y \equiv \vec{r} \times Vm = \vec{r} \times \vec{\Omega} \times \vec{r}m = [\vec{\Omega}(\vec{r}\cdot\vec{r}) - \vec{r}(\vec{r}\cdot\vec{\Omega})]m$ $\dfrac{Y_z}{m} = r^2\dot\Omega_z$ $\dfrac{Y_x}{m} = \dfrac{Y_y}{m} = 0$ Or $\dfrac{Y_z}{m} = \dfrac{I\dot\Omega_z}{m} = r^2\dot\Omega_z$ $\dfrac{dY_z}{mdt} = 0$ where $I \equiv \int_m r^2 dm$ I denotes moment of inertia of a particle about rotational z-axis.	$Y = \vec{r} \times Vm$ $\dfrac{Y_z}{m} = r^2\dot\Omega_z = v_\theta r = \dfrac{\Gamma}{2\pi}$ $\dfrac{Y_x}{m} = \dfrac{Y_y}{m} = 0$ $\dfrac{dY_z}{mdt} = 0$ Angular momentum is conserved
Lamb vector	$L_n = \omega_z v_\theta$	$L_n = 0$
Kinetic energy per unit mass:	$\dfrac{I\dot\Omega_z^2}{2m} = \dfrac{1}{2}r^2\dot\Omega_z^2 = \dfrac{v_\theta^2}{2}$	$\dfrac{v_\theta^2}{2} = \dfrac{1}{8}\left(\dfrac{\Gamma}{\pi r}\right)^2$
Interaction energy:	$\dfrac{dI}{dA} = \dfrac{\psi\omega_z}{2} = -\dfrac{r^2\dot\Omega_z^2}{2}$	0
Enstrophy:	$\dfrac{d\Phi_\omega}{dv} \equiv \dfrac{\omega_z^2}{2} = 2\dot\Omega_z^2$	0
Helicity:	$\dfrac{dH}{dv} \equiv \omega \cdot V = [0\ 0\ \omega_z]\cdot[v_r\ v_\theta\ 0] = 0$	0
Rate of strain tensor components: $\dot\gamma_{ij} = \dfrac{1}{2}[\nabla V + \nabla V^T]$	$\dot\gamma_{11} = \dfrac{\partial v_r}{\partial r} = 0,$ $\dot\gamma_{22} = \dfrac{\partial v_\theta}{r\partial\theta} + \dfrac{v_r}{r} = 0$ $\dot\gamma_{12} = \dot\gamma_{21} = \dfrac{1}{2}\left(\dfrac{\partial v_\theta}{\partial r} + \dfrac{\partial v_r}{r\partial\theta} - \dfrac{v_\theta}{r}\right) = \dfrac{1}{2}(\dot\Omega_z - \dot\Omega_z) = 0$ Rigid body rotation	$\dot\gamma_{12} = \dfrac{1}{2}\left(\dfrac{\partial v_\theta}{\partial r} + \dfrac{\partial v_r}{r\partial\theta} - \dfrac{v_\theta}{r}\right) = -\dfrac{\Gamma}{2\pi r^2}\dot\gamma_{11} = \dfrac{\partial v_r}{\partial r} = 0,$ $\dot\gamma_{22} = \dfrac{\partial v_\theta}{r\partial\theta} + \dfrac{v_r}{r} = 0$ Pure shear and isochoric motion

stream function of vortex indicates streamlines resemble a system of concentric circles about vortex center. The looping nature of streamlines implies the flow fields are generated by vorticity sources, which could be determined by taking the curl of flow fields. Euler equation indicates that a positive pressure gradient exists in radial direction and maintains force balance with centripetal acceleration to keep particles moving on curved paths. The pressure contours exhibit a family of concentric circles similar to streamline patterns within a constant. It assumes minimum pressure at vortex center. The pressure profile in *p-r* diagram is a parabola with axis of symmetry coincide vertical *p*-axis. The parabola opens upward and its vertex is located at $(0, C)$ where pressure attains minimum with vanishing velocity. The flow field structure responses with a negative velocity gradient in normal direction towards vortex center such that convective acceleration and centripetal acceleration work together giving rise to a net nonzero rotation in counter-clockwise direction. As flow is rotational, equipotential does not exist.

The gradient of velocity fields is written in polar coordinate:

$$[\nabla V] = \sum_i \sum_j \left(\frac{\hat{e}_i \partial(v_j \hat{e}_j)}{h_i \partial x_i} \right)$$

$$= \begin{bmatrix} \dfrac{\partial v_r}{\partial r} & \dfrac{\partial v_\theta}{\partial r} \\ \dfrac{\partial v_r}{r \partial \theta} - \dfrac{v_\theta}{r} & \dfrac{\partial v_\theta}{r \partial \theta} + \dfrac{v_r}{r} \end{bmatrix}$$

$$\dot{\gamma}_{ij} = \frac{1}{2}[[\nabla V] + [\nabla V]^T]$$

$$\dot{\gamma}_{11} = \dot{\gamma}_{22} = 0$$

$$\dot{\gamma}_{12} = \dot{\gamma}_{21} = \frac{1}{2}\left(\frac{\partial v_\theta}{\partial r} - \frac{v_\theta}{r} \right) = 0$$

$$\rightarrow v_\theta = rC = r\dot{\Omega}_z$$

$$\Omega_{ij} = \frac{1}{2}[[\nabla V] - [\nabla V]^T] = \begin{bmatrix} 0 & \dot{\Omega}_z \\ -\dot{\Omega}_z & 0 \end{bmatrix}$$

$$\omega = \nabla \times V = \left(\frac{\partial(r v_\theta)}{r \partial r} - \frac{\partial v_r}{r \partial \theta} \right) \hat{k} = 2\dot{\Omega}_z \hat{k}$$

The field is driven by a vector source, which leads to the Lamb vector in Table 6.1. The rate of shear strain is produced due to combined centripetal effects and velocity gradient in streamline normal direction. The vanishing rate of strain tensor implies rigid body rotation with no change in shape or volume and that velocity is proportional to the distance from vortex center.

Angular momentum per unit mass: $r v_\theta = r^2 \dot{\Omega}_z$

$$= \frac{Y_z}{m} = |\vec{r} \times V|$$

Angular momentum is proportional to angular speed and radius squared. Its direction is perpendicular to the plane of motion. Angular momentum, kinetic energy, enstrophy and interaction energy are conserved respectively in forced vortex motion. A local streamline coordinate attached to a particle will rotate with it while undergoing changes in position and orientation continuously. The local coordinate remains orthogonal and spins one revolution for every round about vortex center.

The vortex center is a critical point and a point of symmetry. To depict the critical point topology (Chapter 7), the principal states of gradient of velocity tensor are required. The antisymmetric matrix yields pure complex eigenvalues. The critical point is classified as center (Figure 7.2). The imaginary part implies there is a vector source equal to angular speed of the forced vortex or half of vorticity.

$$[\nabla V]\,|_{r=0} = \begin{bmatrix} 0 & \dfrac{\partial v_\theta}{\partial r} \\ -\dfrac{v_\theta}{r} & 0 \end{bmatrix}_{r=0} = \begin{bmatrix} 0 & \dot{\Omega}_z \\ -\dot{\Omega}_z & 0 \end{bmatrix}_{r=0}$$

Eigenvalues : $\lambda_{1,2} = \pm \dot{\Omega}_z i$

$$\rightarrow |\mathrm{Im}| = \dot{\Omega}_z = \frac{\omega_z}{2}$$

Equation of streamline trajectory:

$$\psi = -\int v_\theta\, \partial r = -\frac{\dot{\Omega}_z r^2}{2}$$

or

$$r^2 = x^2 + y^2 = -\frac{2\psi}{\dot{\Omega}_z}$$

The position hodograph of trajectory is a system of circles about origin in implicit form as the stream function varies like a parameter. For a given angular speed, each circle corresponds to a streamline. The equation of hodograph representation of trajectory is obtained by mapping the stream function from physical plane onto hodograph plane.

$$v_\theta^2 = r^2 \dot{\Omega}_z^2 = -2\psi \dot{\Omega}_z$$

or

$$v_x = (-\sin\theta)v_\theta; \quad v_y = (\cos\theta)v_\theta$$
$$v_\theta^2 = v_x^2 + v_y^2 = -2\psi\dot{\Omega}_z$$

The hodograph representation of trajectory in forced vortex motion is also circles about origin. Each circle radius represents constant kinetic energy or speed possessed by a particle moving along a streamline. The kinetic energy level or speed is different from one streamline to another. The hodograph representation of isoclines is given by the equation of a straight line through origin at an inclination angle $\tilde{\theta}$ with respect to v_x-axis:

$$v_y = v_x \tan\tilde{\theta}$$

or

$$v_\theta \cos\theta = -v_\theta \sin\theta \tan\tilde{\theta}$$
$$\cot\theta = -\tan\hat{\theta}$$
$$\theta = \tilde{\theta} - \pi/2$$

Thus, the hodograph representation an isocline is mapped to a radial line of polar angle $\theta = \tilde{\theta} - \pi/2$ through origin on physical plane. The flow direction at each point on the radial line is constant equal to the inclination angle $\tilde{\theta}$. Table 6.1 summarizes the basic properties of forced vortex.

6.6.3 Free Vortex

A two-dimensional free vortex describes irrotational vortex motions. In irrotational flow, a particle has no rotation about its own axis but can have non-zero angular velocity in global rotations about an axis other than its own. A free vortex represents a classical example of circulating motions with a net circulation induced by a line vortex. Flow draining out of a pool is an example of a free vortex. The complex potential for free vortex is given in Table 6.1, which summarizes the basic properties of free vortex. A free vortex with its center at origin of polar coordinate rotating about z-axis is described by the same set of Euler equation and continuity equation, that is,

$$\frac{\partial p}{\partial r} = \rho\frac{v_\theta^2}{r} = \frac{\rho}{r^3}\left(\frac{\Gamma}{2\pi}\right)^2$$

$$\rightarrow p = -\frac{\rho}{2r^2}\left(\frac{\Gamma}{2\pi}\right)^2 + C = -\rho\frac{v_\theta^2}{2} + C$$

$$\rightarrow B = \frac{p}{\rho} + \frac{v_\theta^2}{2} = C$$

where

$$V = \begin{bmatrix} v_r & v_\theta \end{bmatrix} = \begin{bmatrix} \dfrac{\partial \phi}{\partial r} & \dfrac{\partial \phi}{r\,\partial \theta} \end{bmatrix} = \begin{bmatrix} 0 & \dfrac{\Gamma}{2\pi r} \end{bmatrix}$$

Acceleration is obtained from complex potential or by definition. It represents centripetal acceleration and points towards the vortex center:

$$a = \begin{bmatrix} -\dfrac{v_\theta^2}{r} & 0 \end{bmatrix}$$
$$a \cdot V = 0$$

The result leads back to Bernoulli's theorem, not only applicable along a streamline, but everywhere except the singular point at vortex center. Streamline patterns described by stream function resemble concentric circles whereas equipotential contours form a system of rays from the center. The two systems of contours cross each other orthogonally. With velocity perpendicular to acceleration, the circular motion is at constant speed. Euler equation shows that there is pressure reduction when moving across curved streamlines towards the center of curvature of streamlines. The pressure gradient in radial direction sets up force balance with centripetal force to keep particles moving on curved paths. The pressure contours resemble a system of concentric circles similar to streamlines. The hodograph representation of isobars is also concentric circles about origin in hodograph plane. The pressure profile is symmetric with respect to vertical p-axis in the p-r diagram. It has one vertical asymptote p-axis and one horizontal asymptote $p = C$. The vortex center is a singular point and a point of symmetry. A pressure core (rotational core) is developed about vortex center where pressure approaches vacuum as the limit while velocity approaches infinite. Normal stresses vanish and the motion is in a state of pure shear. For the purpose of illustration, consider a free vortex in anticlockwise motion induced by a positive circulation. A free vortex possesses constant circulation for any enclosed paths containing the singular point at vortex center. Otherwise, circulation vanishes. The tangential speed is inversely proportional to the distance from vortex center. With no vortices present, free vortex behaves in a manner opposite to that of forced vortex.

$$-\frac{\partial v_\theta}{\partial r} = \frac{v_\theta}{r}$$
$$\rightarrow v_\theta r = C = \frac{\Gamma}{2\pi} \tag{6.41}$$
$$= \frac{Y_z}{m} = |\vec{r} \times V|$$

Angular momentum per unit mass is constant for a given circulation and the direction is perpendicular to the plane of motion. Kinetic energy is also conserved. The pressure gradient force is directed towards the fixed point (vortex center) and depends only on distance from the vortex center. The streamline tangent rotates at a rate equal to $\frac{d\alpha}{dt} = \frac{v_s}{R_s}$ anticlockwise about binormal axis in response to velocity direction changes towards left of circular path, (that is, vortex center) while subject to centripetal acceleration. Because fluids close to the vortex center are moving faster on the left side of trajectory than those on the right, convective acceleration in normal direction $\frac{d\alpha}{dt} = \frac{\partial v_s}{\partial n}$ causes streamline normal to rotate to the right in clockwise direction. As a result, centripetal acceleration and convective acceleration compete with each other. Although a pair of mutually orthogonal lines rotates towards each another at same angular speed, the net rotation is zero and Lamb vector vanishes in irrotational flow.

$$L_n = v_s \omega_z = \frac{v_s^2}{R_s} - v_s \frac{\partial v_s}{\partial n} = v_s \frac{d\alpha}{dt} + \left(-v_s \frac{d\alpha}{dt}\right) = 0$$

$$\rightarrow \omega_z = 0$$

A local streamline coordinate attached to a particle will not rotate while undergoing changes in position continuously. It maintains orientation in spite of change in shape by shear deformation. The rate of shear strain is produced due to combined centripetal effects and velocity gradient in radial direction. It describes the rate at which a pair of mutually orthogonal lines rotates towards each another.

The equation of trajectory is:

$$\psi = -\int v_\theta \, \partial r = -\frac{\Gamma}{2\pi} \ln r$$

or

$$r = \exp\left(-\frac{2\pi \psi}{\Gamma}\right)$$

The position hodograph of trajectory is a system of circles about origin in implicit form as the stream function varies. Each circle corresponds to a streamline. The equation of hodograph representation of trajectory is obtained by mapping the stream function from physical plane onto hodograph plane.

$$v_\theta = \frac{\Gamma}{2\pi \exp\left(-\dfrac{2\pi \psi}{\Gamma}\right)} = C$$

or

$$v_x = (-\sin\theta)v_\theta; \quad v_y = (\cos\theta)v_\theta$$

$$v_\theta = \sqrt{v_x^2 + v_y^2} = C$$

$$r = \frac{\Gamma}{2\pi C}$$

Hodograph representation of trajectory is also a circle about origin. The circle of radius C on hodograph plane representing constant speed is mapped to a circle of radius $\Gamma/(2\pi C)$ on physical plane and conversely. The locus of constant flow direction represented by isocline at an inclination angle $\tilde{\theta}$ with respect to v_x-axis is mapped from hodograph plane to a radial line of polar angle $\theta = \tilde{\theta} - \pi/2$ through origin on physical plane similar to forced vortex. For both types of vortex motions, flow direction changes rapidly in the region around vortex center as the radial lines converge towards there. One can observe from Table 6.1 the difference in flow characteristics and the invariance of properties shared by two types of vortex motions.

6.6.4 Vortices Transport and Vorticity Function

Vortices transport is important processes. The governing equation that describes vortices transport is derived from momentum equation. By taking the curl of momentum equation, the pressure and body force terms are eliminated leading to the convection diffusion transport of vortices.

$$\frac{D\omega}{Dt} - (\omega \cdot \nabla)V - \nu\nabla^2\omega = 0 \tag{6.42}$$

This is in line with the fact that hydrostatic pressure forces and body forces (line vectors) act through the discrete fluid mass center and therefore do not affect vorticity. The first term in the above represents the effects of convective transport. Second term describes the nonlinear effects of vorticity motions in terms of spatial rate of change of velocity in vorticity line direction. It represents the combined effects of stretching and tilting of vortex lines. Material lines connecting two moving fluid particles along different streamlines both rotate and stretch. In two-dimensional (x, y) flows, vorticity stretching/tilting mechanisms fail to operate, as vorticity is perpendicular to the flows with only one nonzero z-component vorticity. In Cartesian coordinate systems, we have then,

$$(\omega \cdot \nabla)V = \omega_z \frac{\partial}{\partial z}\begin{bmatrix} v_x & v_y & 0 \end{bmatrix} = \begin{bmatrix} 0 & 0 & 0 \end{bmatrix} \tag{6.43}$$

Under these conditions, convection-diffusion transport of vorticity has the same form as heat transport, that is,

$$\frac{\partial \omega_z}{\partial t} + (V \cdot \nabla)\omega_z - \nu \nabla^2 \omega_z = 0 \tag{6.44}$$

The third term represents diffusion transport of vorticity due to molecular activity. Vorticity gradients drive the diffusion processes from high to low vorticity concentration. In this process, viscosity provides the venue for vorticity diffusion by spreading out existing vorticity in the same way that momentum diffuses. In inviscid flows, the diffusion mode of vorticity transport vanishes and vorticity is conserved in the motion.

$$\frac{D\omega_z}{Dt} = 0 \tag{6.45}$$

The connection between stream function and vorticity (Equation 3.13c) suggests vorticity may be tracked through other flow parameters if a relation of some kind exists between them. Indeed vorticity is functionally related to parameters like Lamb vector, velocity, frictional force and circulation. Circulation is also conserved according to Kelvin circulation theorem.

$$\frac{D\Gamma}{Dt} \equiv \frac{D}{Dt}\oint_c V \cdot d\vec{r} = \oint_c \left[-\nabla\left(\frac{p}{\rho} + \Omega\right) + \nu \nabla^2 V \right] \cdot d\vec{r} = \nu \oint_c \nabla^2 V \cdot d\vec{r}$$

$$= -\nu \oint_c (\nabla \times \omega) \cdot d\vec{r} = 0; \quad \text{if } \nu = 0$$

It implies flux of vorticity lines through any surface enclosed in a vortex tube is conserved for the motion of fluids as stated by Helmholtz vortex laws. Field properties such as vortex tubes/circulations, vorticity lines, streamlines and Lamb surfaces could be thought of material lines/surfaces compose of same material particles "frozen" with fluids in steady motions. The topological structure of fields remains invariant.

To visualize steady vorticity transports in two-dimensional fields, vorticity function ψ_ω is introduced in a manner similar to mechanical energy function. Vorticity function is defined in terms of vorticity flux vector in Cartesian coordinates:

$$\frac{\partial}{\partial x}\left[v_x \omega_z - \nu \frac{\partial \omega_z}{\partial x} \right] + \frac{\partial}{\partial y}\left[v_y \omega_z - \nu \frac{\partial \omega_z}{\partial y} \right] = 0$$

$$x - \text{direction} : q_x = \left[v_x \omega_z - \nu \frac{\partial \omega_z}{\partial x} \right] = \frac{\partial \psi_\omega}{\partial y}$$

$$y - \text{direction} : q_y = \left[v_y \omega_z - v \frac{\partial \omega_z}{\partial y} \right] = -\frac{\partial \psi_\omega}{\partial x}$$

$$\psi_\omega = \int q_x \partial y - \int q_y \partial x + C \tag{6.46}$$

The above could be integrated to yield the vorticity function. The difference in values of vorticity function determines the amount of vorticity or curl in a volume bounded by respective trajectories per unit depth per unit time.

$$\frac{Q}{\Delta z} = \int_1^2 q \cdot \hat{b} dl = \int_1^2 \nabla \psi_\omega \times \hat{k} \cdot \hat{b} dl = \psi_{\omega 1} - \psi_{\omega 2}$$

$\hat{b}$ and dl are the unit normal vector and arc length of an arbitrary surface bounded between two trajectories.

6.7 Example: 4.1c Convective Momentum Flux Tensor Visualization

This third part presents visualization of convective momentum flux tensor by a force vector q_2 through contraction then followed by principal state transformation and mapping. In this example, the force vector is balanced by pressure gradient force expressed in the momentum equation:

$$q_2 = \begin{bmatrix} q_{2x} & q_{2y} \end{bmatrix} = \nabla \cdot \rho[VV] = -\nabla p \tag{6.47}$$

$$q_{2x} = \rho \left(v_x \frac{\partial v_x}{\partial x} + v_y \frac{\partial v_x}{\partial y} \right) = 4\rho x; \quad q_{2y} = \rho \left(v_x \frac{\partial v_y}{\partial x} + v_y \frac{\partial v_y}{\partial y} \right) = 4\rho y$$

$$\nabla p = \begin{bmatrix} \frac{\partial p}{\partial x} & \frac{\partial p}{\partial y} \end{bmatrix} = -4\rho \begin{bmatrix} x & y \end{bmatrix}$$

$$|q_2| = |\nabla p| = 4\rho \sqrt{x^2 + y^2}$$

$$\frac{dy}{dx} = \frac{q_{2y}}{q_{2x}} = \frac{y}{x} \tag{6.48}$$

$$\rightarrow y = mx$$

The trajectory equation solution describes a system of straight lines passing through origin with slope equal to parametric constant m. The position hodograph representation of force trajectories are shown in Figure 6.6. The force vector is parallel to the pressure gradient vector everywhere and points radically outwards along which pressure decreases at maximum rate.

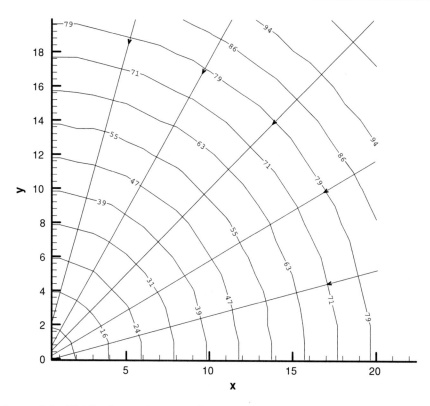

Figure 6.6 Distributions of convective momentum force vector trajectories (arrow) and magnitude contours in physical plane

Distributions of convective momentum force magnitude are also presented in Figure 6.6. They form a system of circular contours about origin similar to those of pressure contours. The force hodograph representation of constant force direction at an inclination angle β from q_{2x}-axis is mapped to a straight line through origin and coincides with the force vector at the same inclination angle on physical plane. The equation of force hodograph representation of isoclines is:

$$q_{2y} = q_{2x} \tan \beta$$

By substitution,

$$y = x \tan \beta$$
$$\tan \beta = m$$

The above position hodograph representation of force isoclines is then mapped from physical plane onto the velocity hodograph plane by substitution.

$$v_y = v_x \tan(-\beta)$$

From Example 4.1b, the hodograph representation of velocity isoclines is:

$$v_y = v_x \tan \tilde{\theta}$$
$$\rightarrow \beta = -\tilde{\theta}$$

Each force isocline is mapped from force hodograph plane/physical plane onto a radial line (velocity isocline) through origin and reflected about the horizontal axis on velocity hodograph plane.

Mapping of equipotential contours and streamline trajectories from physical plane to convective momentum force hodograph plane is presented in Figure 6.7. The state of force in the system along an arbitrary streamline could be determined by the position vectors as pencil of rays from origin. The force hodograph representation of streamlines is a system of rectangular hyperbola with center at origin and transverse axis coincides with line $q_{2y} = q_{2x}$, which is also a line of symmetry. Similarly, the force hodograph representation of equipotential is obtained by composite mapping from physical plane onto a system of hyperbola with center at origin and transverse axes coinciding with coordinate axes. The two systems of contours cross at right angle.

$$\phi(x, y) = \phi^*[x(q_{2x}, q_{2y}), \ y(q_{2x}, q_{2y})] = \frac{q_{2x}^2 - q_{2y}^2}{16\rho^2}$$

$$\psi(x, y) = \psi^*[x(q_{2x}, q_{2y}), \ y(q_{2x}, q_{2y})] = \frac{q_{2x}q_{2y}}{8\rho^2}$$

Because tensor vectors in convective momentum tensor are linearly dependent, the tensor can be described by a single vector parallel to velocity or streamline tangent.

$$m_{ij} = \rho[VV] = \rho \begin{bmatrix} v_x^2 & v_x v_y \\ v_y v_x & v_y^2 \end{bmatrix} = m_{ji}$$

$$\det |m_{ij}| = 0$$

$$\frac{v_x v_y}{v_x^2} = \frac{v_y}{v_x} = -\frac{y}{x} = \tan \tilde{\theta} = \tan \alpha = \frac{dy}{dx} \tag{6.49}$$

The tensor field is then regarded as a vector field. Convective momentum tensor is symmetric and yields the following real eigenvalues with a pair of

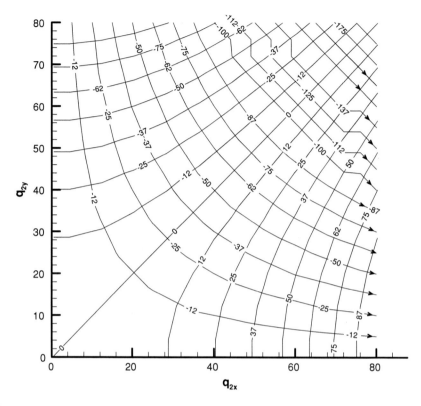

Figure 6.7 Mapping of streamlines (arrow) and equipotentials on convective momentum force hodograph

orthogonal eigenvectors:

$$\sigma_1 = \rho V^2; \quad \hat{e}_1 = \frac{1}{|V|}\begin{bmatrix} v_x & v_y \end{bmatrix} = \frac{V}{|V|} \tag{6.50}$$

$$\sigma_2 = 0; \hat{e}_2 = \frac{1}{|V|}\begin{bmatrix} v_y & -v_x \end{bmatrix}$$

Major principal angle: $\theta_1 = \tan^{-1}\left(\frac{v_y}{v_x}\right) = \alpha$

Major principal axes are parallel to fluid velocity vectors everywhere. To visualize convective momentum transport, principal axis trajectories are invoked by integrating the trajectory equation to result in:

$$xy = C$$

Figure 6.8 shows the distribution of major principal axis trajectories similar to those of streamlines within a constant in Figure 4.5. Symmetry test shows it has a line of symmetry about line $y = x$. By aligning the major principal axis parallel to x-axis to assume axis of reference, the tensor reduces to diagonal form. With a vanishing eigenvalue, the tensor ellipse degenerates to a line that coincides with the major principal axis trajectory. The positive nonzero eigenvalue indicates only out-going fluxes passing through the major principal plane carried by bulk fluid flows at a magnitude equal to kinetic energy density and that the system is under pure compression. It is observed that the critical point of velocity also corresponds to the degenerate point of convective momentum flux tensor, which yields identical vanishing eigenvalues. At this multiple point, tensor ellipse degenerates into a point and pressure is the same in all directions.

It is useful to compare major principal axis trajectories of convective momentum flux tensor with those of force vector in Figure 6.6, which was obtained from the same tensor by contraction. The former describes convective momentum transfer paths as a phenomenon of force transmission in terms of principal axes and principal values. It depicts a principal state of stress tensor fields through tensor transformation. The latter describes force balance

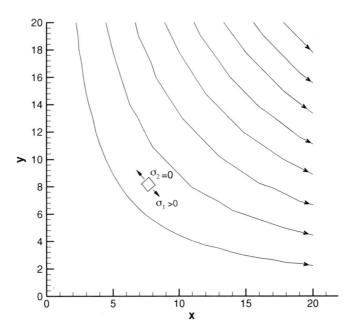

Figure 6.8 Major principal axes trajectories of the convective momentum tensor

at a point between pressure gradient force and convective momentum force through tensor contraction. It yields a stress tensor vector representing the net convective momentum flux across a control volume. There is a relation between the orientations of major principal axis (or streamline tangent), velocity isocline and those of force vector. They are mirror reflection images of force vector with respect to x-axis, that is,

$$\tan^{-1} m = -\tilde{\theta} = -\alpha = \beta$$

It is observed that the force vector is equal exactly to the product of density and acceleration vector in accordance with the Newton's law of motion. Given the direction of any one vector at a point (force vector, velocity vector, major principal axis or acceleration vector) the other directions could be determined. Furthermore, a radial line on physical plane corresponds to a force isocline at a direction equal to the inclination angle. The radial line also defines velocity isocline along which each point has the same velocity direction equal to the inclination angle reflected about x-axis. Correlation among streamlines, mechanical energy trajectories, heatlines and major principal axes trajectories of convective momentum flux tensor as well as its contraction force vectors reveals the connections and interactions between various transport processes of different nature. They are ready to be integrated to provide a better understanding of the physical phenomena.

7

Critical Point Topology, Classification and Visualization

7.1 Introduction

Field topology refers to the study of geometric figures that are invariant by transformation. Topological analysis and classification of critical points are concerned with local phenomena of field behaviors and aim to provide a compact global view of fields. We are interested in the topology of vector fields and its transformation in the neighborhood of a point where fields vanish. The interests are connected with those associated with critical points and system stability study. Critical points possess unique characteristics and topological structure and are referred to as stagnation (neutral) points in hydrodynamics. A significant amount of qualitative information could be obtained from a critical point analysis. Many of the interesting features and physical peculiarities of fields are indeed related closely to the distributions of critical points. Together with dividing trajectories crossing at critical points, they play an important part in the applications of field topology and visual sciences. Tensor topology is classified based on geometric characteristics and behavior of fields in the neighborhood of critical points. It is therefore essential to understand their properties. As simplified depictions of topology become possible by focusing near a critical point, this motivates the development of visualization techniques to capture their structure and features.

Visualization of Fields and Applications in Engineering, First Edition. Stephen Tou.
© 2011 John Wiley & Sons, Ltd. Published 2011 by John Wiley & Sons, Ltd.

By definition, a critical point is where vector field or gradient of fields vanishes and as a result, field line tangent becomes locally undefined causing failure of the trajectory equation. At a critical point, the directions of vector fields are indeterminate and conformal mapping breaks down. The geometric interpretation of critical point is the horizontal tangent plane where the graphs attain extreme value or have a flat spot (saddle point). When a streamline or equipotential ends on a critical point, it splits/rejoins abruptly into various directions giving rise to the so-called dividing streamlines or equipotentials. This implies in approaching a critical point, flow speeds will continue to diminish while flow directions and streamline curvatures are likely to change rapidly in the region close to a critical point. Dividing trajectories are a class of asymptotes. They guide vector fields to change direction and separate the fields into regions of distinct behaviors. A trajectory starting in one of the regions must remain in it. By integrating these regions, global field structure can be captured to reflect flow behaviors and boundary geometry/condition. The knowledge of dividing streamlines allows us to trace the nearby field patterns by successive lines starting from upstream, coinciding with the dividing streamline, and then diverting away from its course as vector field strengths continue to diminish making their turns to move away. In the neighborhood of a critical point, nearby streamlines widen and remain asymptotic to the dividing streamlines. The nature and occurrence of critical points vary. But, each of them exhibits a unique field topology useful for critical point analysis and visualization. The number and nature of critical points remain invariant under transformation. Mathematically, critical points of first order type in scalar field topology are points on the graph of a function at which the gradient of function vanishes and the function assumes extreme values. These points could be local maximum or local minimum or neither. To test for extreme values of functions and classify critical points, one could invoke the theory of calculus and Hessian matrix. With vanishing second and/or higher derivatives, it implies a higher order critical point. More complicated topological skeletons can occur. When there are infinite trajectories crossing at the critical point, it corresponds to the case of identical eigenvalues in the gradient of vector field tensor and is classified as a degenerate critical point. The gradient of vector field tensor carries information about the spatial rate of change of vector fields along mutually orthogonal directions at a point while its principal state describes field topology and structure independent of coordinates. Visualization/classification of critical points are based on analysis and transformation of gradient of vector field tensors in the neighborhood of critical point. It sets the focus on complex representation of fields by asymptotic analysis and principal state analysis of Hessian matrix. The study of critical point topology calls for critical point theory and principal state analysis of gradient of vector field tensors.

7.2 Complex Analysis of Critical Point

Asymptotic theory is useful to study the characteristics of a phenomenon when certain parameter approaches a limiting value. To investigate how a function behaves in the neighborhood of a point, we consider a complex analytic function expressed in Taylor expansion. The elegance of complex potentials is that the local behavior of an analytical function uniquely determines the local topology of vector fields and structure. The Taylor series gives the best linear approximation of a differentiable function near an arbitrary point z_0 about the spatial rate of change of field property.

$$F(z) = F(z_0) + (z - z_0)F^{(1)}(z_0) + \frac{1}{2!}(z - z_0)^2 F^{(2)}(z_0) + \cdots \qquad (7.1)$$

$$= \sum_{n=0} \frac{(z - z_0)^n F^{(n)}(z_0)}{n!}$$

where $F(z) = \phi + i\psi; z = x + iy$

The complex potential describes a vector field in terms of a pair of conjugate harmonic functions. At critical point z_0, the first derivative $(n = 1)$ of complex function vanishes,

$$\overline{V} = \frac{dF(z)}{dz} = F^{(1)}(z_0) = \frac{\partial \phi}{\partial x}\Big|_{\substack{x=x_0 \\ y=y_0}} + i\frac{\partial \psi}{\partial x}\Big|_{\substack{x=x_0 \\ y=y_0}} = 0$$

$$\rightarrow \frac{\partial \phi}{\partial y}\Big|_{\substack{x=x_0 \\ y=y_0}} = \frac{\partial \phi}{\partial x}\Big|_{\substack{x=x_0 \\ y=y_0}} = 0; \quad \frac{\partial \psi}{\partial y}\Big|_{\substack{x=x_0 \\ y=y_0}} = \frac{\partial \psi}{\partial x}\Big|_{\substack{x=x_0 \\ y=y_0}} = 0 \qquad (7.2)$$

Solutions from the above pairs of equations yield the location of critical point. The asymptotic behavior in the neighborhood of a critical point is obtained by neglecting terms higher than second order, that is,

$$F(z) - F(z_0) \approx \frac{(z - z_0)^2}{2} F^{(2)}(z_0) \qquad (7.3)$$

$$\rightarrow \arg[F(z) - F(z_0)] \approx 2\arg[(z - z_0)] + \arg[F^{(2)}(z_0)/2] \qquad (7.4)$$

where $F^{(2)}(z_0) \neq 0$

The nonvanishing second order derivatives indicate a critical point of first order type $(n = 1)$ with $(2n + 2)$ branches of trajectories crossing through at equal angles, that is, right angle. The number of branches implies the same number of regions separated by the crossing trajectories independent of each other. Consider z and z_0 to be neighboring points in z-plane with images $F(z)$ and $F(z_0)$ in F-plane under conformal mapping. A pair of curves C_1 and C_2 intersecting at an angle $\Delta\beta$ at the critical point z_0 in z-plane are mapped onto their corresponding images C_1', C_2' and angle $\Delta\gamma$ at $F(z_0)$ on F-plane. As $z \to z_0$ along curve C_1, the tangent defines the angle $\arg[(z - z_0)] = \beta_1$ measured with respect to x-axis. Similarly, along curve C_2, the tangent defines angle $\arg[(z - z_0)] = \beta_2$. On F-plane, as $F(z) \to F(z_0)$, then $\arg[(F(z) - F(z_0)] = \gamma_1$ and $\arg[(F(z) - F(z_0)] = \gamma_2$ measured with respect to real ϕ-axis when approaching along curves C_1' and C_2'. By substitution and subtraction, we obtain:

$$\gamma_2 - \gamma_1 = 2(\beta_2 - \beta_1)$$

$$\text{Or } \Delta\gamma = 2\Delta\beta$$

The results show that mapping is not conformal at the critical point as the angle between a pair of curves in z-plane is increased by a factor of two in the same sense when mapped onto F-plane. Since $F(z_0)$ and $F^{(2)}(z_0)/2$ assume the role of arbitrary constants in the neighborhood of z_0, the behavior of the complex potential near critical point is represented by quadratic approximation of the function, that is,

$$F(z) \approx (z - z_0)^2$$

Consider point z_0 is located at origin. A critical point will occur where vector field vanishes.

$$\overline{V} = \frac{dF}{dz}\bigg|_{z=z_0=0} = 2(z - z_0) = 0 \tag{7.5}$$

$$F^{(2)}(z_0) = 2 \neq 0$$

The above complex potential has only one first order critical point. Its real and imaginary parts correspond to potential function and stream function respectively. Both functions represent hyperbola conic sections of quadratic surfaces.

$$\phi = \mathrm{Re}F(z) = x^2 - y^2 \tag{7.6a}$$

$$\psi = \mathrm{Im}F(z) = 2xy \tag{7.6b}$$

$$V = \nabla\phi = \nabla\psi \times \hat{k} \tag{7.7}$$

To determine the behavior of functions of two variables near a critical point (a, b), we invoke Taylor series representation of the function, that is,

$$f(x_1, x_2) = \sum_{n=0}^{\infty} \sum_{k=0}^{n} \frac{\partial^n f(x_1, x_2)}{\partial x_k \partial x_{n-k}}\bigg|_{\substack{x_1=a \\ x_2=b}} \frac{(x_1 - a)^n (x_2 - b)^{n-k}}{k!(n-k)!}$$

For second order approximation,

$$f(x_1, x_2) \approx f(a, b) + [(x_1 - a)(x_2 - b)] \cdot \left. \frac{\partial f}{\partial x_i} \right|_{\substack{x_1 = a \\ x_2 = b}}$$

$$+ 0.5[(x_1 - a)(x_2 - b)] \frac{\partial^2 f}{\partial x_i \partial x_j} \bigg|_{\substack{x_1 = a \\ x_2 = b}} \begin{bmatrix} (x_1 - a) \\ (x_2 - b) \end{bmatrix}$$

where $\left. \dfrac{\partial f}{\partial x_i} \right|_{\substack{x_1 = a \\ x_2 = b}} = 0$

$$H_{ij} = \frac{\partial^2 f}{\partial x_i \partial x_j} \bigg|_{\substack{x_1 = a \\ x_2 = b}} = \begin{bmatrix} \dfrac{\partial^2 f}{\partial x_1^2} & \dfrac{\partial^2 f}{\partial x_1 \partial x_2} \\ \dfrac{\partial^2 f}{\partial x_2 \partial x_1} & \dfrac{\partial^2 f}{\partial x_2^2} \end{bmatrix}$$

Determinant of Hessian matrix: $D \equiv \det |H_{ij}|$

With vanishing gradient at critical point, it is natural to study the next available information about the field structure given by Hessian matrix H_{ij}. Hessian matrix describes field structure in terms of second order partial derivatives of the property function. It is interpreted as the spatial rate of change of field property flux acting on the i-face of a control volume in j-direction. Hessian matrix is useful for the test of extreme values of functions in general. The principal states of Hessian matrix provide the platform for critical point classifications. Since the mixed derivatives are continuous, Hessian matrix is symmetric and possesses real eigenvalues λ_i and a pair of orthogonal eigenvectors $\hat{e}_i$ crossing at critical point independent of coordinate systems. Eigenvectors define the orientations of two mutually orthogonal principal planes on which extreme spatial rate of change of field property fluxes occur. The following rules apply to characterize a first order critical point based on principal state of Hessian matrix:

Case 1: $D > 0$, this is an isolated critical point with two branches of trajectory crossing at it and separating the fields into four local regions. An isolated critical point refers to an infinitesimal circle that can be constructed in such a way that it contains no further critical points. It separates from the trajectory and exerts less influence on its neighborhood. The horizontal tangent plane does not intersect surface $f(x_1, x_2)$. Such critical point has a local minimum, $\lambda_1 > 0$; $\lambda_2 > 0$ or a local maximum, $\lambda_1 < 0$; $\lambda_2 < 0$.

Case 2: $D < 0$, there are two branches of trajectory passing through the critical point and separating the fields into four local regions. It possesses the characteristics of a saddle point. The horizontal tangent plane intersects surface $f(x_1, x_2)$. Such critical point has no local minimum or local maximum, $\lambda_1 < 0; \lambda_2 > 0$ or $\lambda_1 > 0; \lambda_2 < 0$.

Case 3: $D = 0$, this is a degenerate critical point with only one trajectory passing through, $\lambda_1 = 0$ or $\lambda_2 = 0$. It could be a terminal point or a cuspidal point or a point of contact with only one tangent. At a terminal point the curve terminates.

Case 4: A degenerate critical point, $\lambda_1 = \lambda_2$ with infinite trajectories crossing at the point.

Suppose field property function represents equipotential in a flow field, $f(x_1, x_2) = \phi$. Hessian matrix of equipotential is then equivalent to the gradient of velocity tensor in potential fields and shares the same interpretation. Hessian matrix is symmetric and yields the following eigenvalues and eigenvectors at critical point:

$$H_{ij} = \frac{\partial^2 \phi}{\partial x_i \partial x_j}\Big|_{\substack{x_1=a \\ x_2=b}} = \nabla V = \begin{bmatrix} \dfrac{\partial v_x}{\partial x_1} & \dfrac{\partial v_y}{\partial x_1} \\ \dfrac{\partial v_x}{\partial x_2} & \dfrac{\partial v_y}{\partial x_2} \end{bmatrix} = \begin{bmatrix} 2 & 0 \\ 0 & -2 \end{bmatrix}$$

$$\lambda_1 = 2; \hat{e}_1 = \begin{bmatrix} 1 & 0 \end{bmatrix}$$
$$\lambda_2 = -2; \hat{e}_2 = \begin{bmatrix} 0 & -1 \end{bmatrix}$$
$$D = -4 < 0$$

The critical point is classified as a saddle point. Field behaviors could be described by a point moving from critical point up or down the eigenvectors depending on the associated eigenvalues. The plus or minus sign of eigenvalues represent out-going flux or in-coming flux of a control volume. A pair of eigenvectors define the orientations of two mutually orthogonal principal planes coinciding with coordinate axes along each direction maximum spatial rate of change of potential occurs. Eigenvector trajectories crossing at critical point assume the role of dividing streamlines and serve as asymptotes for nearby streamlines. Figure 7.1 illustrates the critical point topology constructed from velocity complex potential consisting of a system of orthogonal curves. The real part represents equipotential and resembles families of hyperbola with center at origin. The foci are located at $(2\sqrt{C_1}, 0)$ and $(-2\sqrt{C_1}, 0)$ and the transverse axis coincides with x-axis when $C_1 > 0$. It has a conjugate hyperbola when $C_1 < 0$. Asymptotes for the real part are found to coincide with lines $y = \pm x$, which is also the line of antisymmetry. The trajectories are

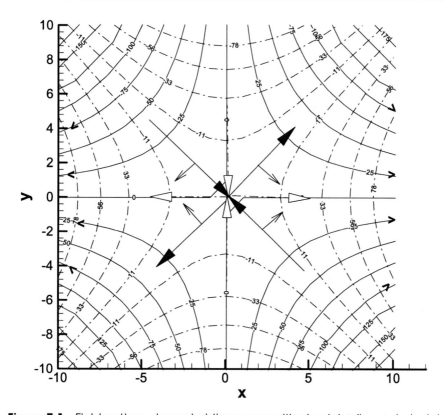

Figure 7.1 Field pattern characteristics near a critical point ψ-line ϕ-dash-dot

symmetric with respect to coordinate axes as well as origin. Along coordinate axes, we have

$$y\text{-axis: } V\,|_{x=0} = \nabla\phi\,|_{x=0} = \begin{bmatrix} 0 & -2y \end{bmatrix} \|\hat{e}_2$$
$$x\text{-axis: } V\,|_{y=0} = \nabla\phi\,|_{y=0} = \begin{bmatrix} 2x & 0 \end{bmatrix} \|\hat{e}_1$$

The landscape of equipotentials is that of a saddle point in space with valleys deepening away from origin along y-axis ($\hat{e}_2$) and mountains rising away from origin along x-axis ($\hat{e}_1$). Each of these directions marked with a hollow arrow corresponds to the maximum spatial rate of change of equipotential, that is, gradient of equipotential and points in the direction of increasing potential. On the upper x-y plane, flows approach critical point along y-axis downwards and depart from it along $\pm x$-axis in opposite directions. In this process, field potential goes through a maximum and simultaneously through a minimum following change of sign in eigenvalue from negative (in-coming flux) to positive (out-going flux). This is the basic characteristics of a saddle point with no local extreme.

Similarly, Hessian matrix of vector potential (stream function), $f(x_1, x_2) = \psi$ is symmetric and yields,

$$H_{ij} = \frac{\partial^2 \psi}{\partial x_i \partial x_j}\Big|_{\substack{x_1=a \\ x_2=b}} = \begin{bmatrix} \dfrac{\partial^2 \psi}{\partial x_1^2} & \dfrac{\partial^2 \psi}{\partial x_1 \partial x_2} \\ \dfrac{\partial^2 \psi}{\partial x_2 \partial x_1} & \dfrac{\partial^2 \psi}{\partial x_2^2} \end{bmatrix} = \begin{bmatrix} 0 & 2 \\ 2 & 0 \end{bmatrix}$$

$$\lambda_1 = 2; \; \hat{e}_1 = \begin{bmatrix} 1 & 1 \end{bmatrix}$$

$$\lambda_2 = -2; \; \hat{e}_2 = \begin{bmatrix} 1 & -1 \end{bmatrix}$$

$$D = -4 < 0$$

This is also classified as a saddle point. Eigenvalues represent extreme spatial rate of change of vector potential giving rise to the in-coming or out-going fluxes acting on respective principal planes. A pair of eigenvectors define the orientation of two mutually orthogonal principal planes coincide with lines $y = \pm x$ along each direction maximum spatial rate of change of vector potential occurs. Eigenvector trajectories crossing at critical point assume the role of dividing equipotentials and serve as asymptotes for nearby equipotentials. In Figure 7.1, the imaginary part represents vector potential and depicts streamline distributions. Streamlines resemble families of rectangular hyperbola with center at origin. The foci are at $(\sqrt{2C_2}, \sqrt{2C_2})$ and $(-\sqrt{2C_2}, -\sqrt{2C_2})$ and the transverse axis coincides with lines $y = x$ when $C_2 > 0$. It has a conjugate hyperbola and resembles flows in corner. The streamline patterns are symmetric with respect to lines $y = \pm x$ as well as origin. They are also antisymmetric about coordinate axes. The direction of flow fields is perpendicular to the gradient of stream function vector and z-axis as shown in plain arrows for right hand systems.

$$V = \nabla \phi = \nabla \psi \times \hat{k}$$

The landscape of streamlines is that of a saddle point with valleys deepening away from origin along line $y = -x$, that is, $\hat{e}_2$ and mountains rising away from origin along line $y = x$, that is, $\hat{e}_1$. Each of these directions marked with a solid arrow corresponds to the maximum spatial rate of increase of vector potential, that is, gradient of stream function. Upon approaching critical point along $\hat{e}_2$ trajectory, vector potential goes through a maximum and simultaneously a minimum following change of sign in eigenvalue and departing from it along $\hat{e}_1$ trajectory in opposite directions. Dividing streamlines are found by those passing through critical point, that is, $\psi(0, 0) = 2xy = 0$. They coincide with coordinate axes and the results are identical to those found by gradient of velocity tensor or Hessian matrix of equipotential. In the neighborhood of critical point, nearby streamlines asymptotic to the dividing streamline turn sharply and progress radically different from those with which they approach. Dividing streamlines separate the fields into each quadrant.

To visualize a saddle point from geometric viewpoint in terms of a graph of a function, that is, $z = f(x, y)$, the explicit function renders two-dimensional contour curves back to three-dimensional surface. A saddle point generated by a quadric surface is mapped onto the graph as a flat spot in space surrounded by a surface going up in opposite directions with positive eigenvalue like a U shape (mountain rising). The surface goes down from the flat spot in opposite directions dictated by the other eigenvector with negative eigenvalue like an inverted U shape (valley deepening). Each of these mutually orthogonal directions corresponds to the maximum spatial rate of change of function. Last but not the least, gradient mapping of equipotentials and streamlines is effective in displaying the graphs of saddle point. In Figure 7.1, the gradient systems in the region around critical point will give a qualitative picture of "runoff" of water flowing downhill towards local minima or away from local maxima by reversing the arrow directions.

7.3 Critical Point Theory and Classification

In critical point analysis, gradient of vector field tensors represents a general approach in nonharmonic fields where functions may not be analytic. Critical point theory was developed to study trajectories of dynamic systems in relation to field behavior and structure. Investigations on topology of simulated tensor fields have led to the method of representing field structure based on gradient of vector fields in the neighborhood of critical points. The gradient of vector field tensors carries information about the field characteristics and behaviors depending on the state of the tensors such as symmetry, antisymmetry or asymmetry. In this respect, the principal state of the captioned tensors plays an important role in the study of tensor field structure and critical point topology as well as classification. The studies showed that critical points delineated fields into regions of qualitative behavior independent of each other. The behavior of fields about a critical point can be analyzed by examining neighboring trajectories adjacent to the critical point. A particular set of trajectories crossing through critical points are of special interest because they define the topological skeleton, which reveals local phenomena in relation to boundary geometry or conditions in one way or another. The way these dividing trajectories split at a critical point and trace across field domain depicts global asymptotic behaviors of trajectories near to them. Field skeletons in neighborhood of critical points exist in various forms depending on the local gradient of vector field tensors. Several possible topological skeleton patterns and classifications have been presented in literature (Figure 7.2) based on principal states of gradient of vector field tensors at critical points.

The topology and structure of a manifold is reflected specifically by the differentiable function on the manifold. It sets the focus of critical point theory in vector field topology on the behavior of vector fields near a critical

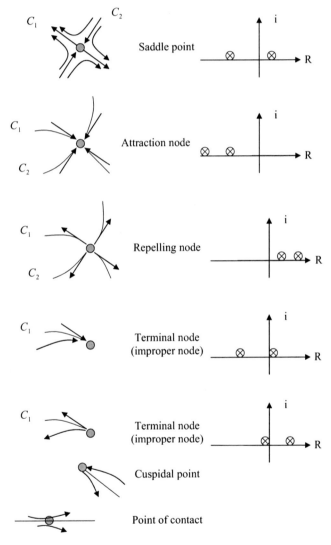

Figure 7.2   Classification of possible field skeleton patterns near a critical point

Degenerate node

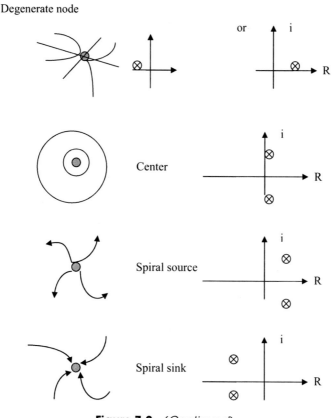

Figure 7.2 *(Continued)*

point, which is obtained from Taylor expansion of a vector function with reference to Cartesian coordinate:

$$V = V\big|_{(x_0.y_0)} + dr \cdot [\nabla V]\big|_{(x_0.y_0)} + \cdots$$

$$= \sum_{n=0} \frac{(dr)^n \cdot [\nabla^{(n)} V]_{(x_0.y_0)}}{n!} \tag{7.8}$$

(x_0, y_0) is a critical point with $V\big|_{(x_0.y_0)} = 0$. For $n = 1$,

$$[\nabla V] = \sum_{i=1} \sum_{j=1} \frac{\hat{e}_i \partial(v_j \hat{e}_j)}{h_i \partial x_i} = \sum_{i=1} \sum_{j=1} \frac{\partial v_j}{h_i \partial x_i} \hat{e}_i \hat{e}_j = \begin{bmatrix} \dfrac{\partial v_x}{\partial x} & \dfrac{\partial v_y}{\partial x} \\[2mm] \dfrac{\partial v_x}{\partial y} & \dfrac{\partial v_y}{\partial y} \end{bmatrix} \tag{7.9}$$

The gradient operation transforms the original vector fields into second order tensor fields. It expresses mathematically an approximation of a differentiable vector function at a given point. As the first term vanishes in Taylor series, the second term reveals information about field behavior and structure in the neighborhood of critical point. Depending on the nature of fields, the gradient vector tensor is interpreted as the spatial rate of change of vector fields on the i-face of a control volume in j-direction. If the gradient vector tensor vanishes, the critical point is second order or higher type. More complicated skeletons may arise.

Consider the same velocity fields in Example.4.1a, that is,

$$\nabla V \bigg|_{\substack{x=0 \\ y=0}} = \begin{bmatrix} 2 & 0 \\ 0 & -2 \end{bmatrix} \tag{7.10}$$

where $v_x = 2x; v_y = -2y$

Above symmetric matrix yields the following eigensystem:

$$\lambda_1 = 2; \hat{e}_1 = \begin{bmatrix} 1 & 0 \end{bmatrix} \tag{7.11a}$$

$$\lambda_2 = -2; \hat{e}_2 = \begin{bmatrix} 0 & -1 \end{bmatrix} \tag{7.11b}$$

$$\lambda_1 \lambda_2 < 0$$

The critical point is classified as a saddle point. Eigenvalues represent extreme strain rate fluxes acting on respective principal planes. A pair of eigenvectors defines the orientations of two mutually orthogonal principal planes coincide with coordinate axes. Such that,

$$\text{Along } y\text{-axis: } V\big|_{x=0} = \begin{bmatrix} 0 & -2y \end{bmatrix} \| \hat{e}_2$$

$$\text{Along } x\text{-axis: } V\big|_{y=0} = \begin{bmatrix} 2x & 0 \end{bmatrix} \| \hat{e}_1$$

The principal axes trajectories crossing at critical point assume the role of dividing streamlines and serve as asymptotes for nearby streamlines. On the upper x-y plane, flows approach critical point along y-axis ($\hat{e}_2, \lambda_2 < 0$) downwards and depart from it thereafter along $\pm x$-axis ($\hat{e}_1, \lambda_1 > 0$) in opposite directions. In this process, strain rate flux goes through a maximum and simultaneously through a minimum following the change of sign in eigenvalue from negative (in-coming flux) to positive (out-going flux). The results are consistent with those of complex potential analysis shown in Figure 7.1.

In general, gradient of vector field tensors are not necessary symmetric. For asymmetric tensors, eigenvalues are complex, that is,

$$\lambda_i = R_i + i\text{Im}_i \quad i = 1, 2$$

R_i and Im_i represent real and imaginary parts of an eigenvalue respectively. The following types of tensors in two-dimensional fields are considered: symmetric tensors, antisymmetric tensors and asymmetric tensors.

7.3.1 Symmetric Tensor: $(\nabla V) = (\nabla V)^T$; $Im_1 = Im_2 = 0$

Symmetric matrix yields real eigenvalues associated with a system of orthogonal eigenvectors. Spirals and centers will not occur. The sum of real eigenvalues of a symmetric tensor represents the tensor trace and is interpreted as a scalar source. Critical point classification is based on its eigenvalues as follows:

7.3.1.1 Saddle Point: $D < 0$ and $R_1 < 0$; $R_2 > 0$ or $R_1 > 0$; $R_2 < 0$

A first order saddle point is a nondegenerate critical point. There are two dividing trajectories crossing at critical point and nearby trajectories approach it asymptotically and turn away from it once they reach the minimum distance with respect to critical point. The dividing trajectories separate the fields into four local regions. In approaching a saddle point along eigenvector trajectory C_2 with negative eigenvalue $R_2 < 0$ in Figure 7.2, the in-coming flux points towards $\hat{e}_2$ principal plane at the critical point. It goes through a maximum and simultaneously through a minimum and points away from it on $\hat{e}_1$ principal plane along eigenvector trajectory C_1, $R_1 > 0$ in opposite directions as eigenvalue changes its sign from negative to positive. This is the basic characteristic of a saddle point with no local extremes. In this process, there is an abrupt change of vector field directions from $\hat{e}_2$ (incoming flux) to $\hat{e}_1$ (out-going flux) at critical point.

7.3.1.2 Attracting node: $R_1 < 0$, $R_2 < 0$; $D > 0$

This is an isolated critical point with the characteristics of a sink. It corresponds to a local maximum with a pair of principal axes trajectories (C_1, C_2) crossing at critical point. With both eigenvalues being negative, field line trajectories merge towards critical point following closely to the principal axes trajectories asymptotically. It is called a nodal sink at which fields vanish.

7.3.1.3 Repelling Node: $R_1 > 0$, $R_2 > 0$; $D > 0$

This is an isolated critical point with the characteristics of a source. It corresponds to a local minimum with a pair of principal axes trajectories (C_1, C_2) crossing at critical point. With both eigenvalues being positive, field line trajectories radiate outwards from the critical point following closely to the principal axes trajectories asymptotically. It is called a nodal source.

7.3.1.4 Terminal Nodes: $R_1 = 0$, or $R_2 = 0$; $D = 0$

When a symmetric gradient of vector field tensor yields one vanishing principal value, only the non-zero principal axis trajectory remains. It is classified as a degenerate critical point. Flows approach or depart from this terminal node along the remaining trajectory. Or it could be a cuspidal point with only one tangent and field vector direction reverses/converges/diverges. The orientation of trajectories on each side of the tangent may or may not change. Or it could be a point of contact when there is no change of vector field direction. There is a reverse of orientation of trajectories on each side of the tangent. Such critical point is also called improper node.

7.3.1.5 Degenerate Point: $R_1 = R_2$

This is a degenerate point with equal eigenvalues. Such tensor is classified mathematically as having geometric multiplicity of unity less than the algebraic multiplicity of two. At a degenerate point, trajectories cross in all directions. There is an infinite number of eigenvectors meaning that any curve passes through a degenerate point is a principal axis trajectory. If eigenvalues are negative, field directions point towards the degenerate point, otherwise the opposite applies. It is also called a proper node.

7.3.2 Antisymmetric Tensor: $\tau_{ii} = 0$, $i = j$; $\tau_{ij} = -\tau_{ji}$, $i \neq j$

Antisymmetric tensors have the properties that all diagonal components vanish and the off-diagonal components are equal but in opposite sign. Antisymmetric tensors yield pure imaginary eigenvalues and orthogonal base does not exist. They characterize the rotational aspect of the fields. The imaginary eigenvalues of an antisymmetric tensor is interpreted as vector source.

7.3.2.1 Center: $R_1 = R_2 = 0$, $Im_1 = -Im_2$

The imaginary part of eigenvalues of velocity gradient tensor represents rotational speeds equal to half of vector source (vorticity). In the neighborhood of critical point, flux lines form a system of concentric circles surrounding the critical point. This field topology is classified as a center.

7.3.3 Asymmetric Tensor

Asymmetric tensors yield complex eigenvalues with a real part and an imaginary part. Orthogonal base does not exist. Asymmetric tensors give rise to the following asymptotic points about which trajectories twist infinitely.

7.3.3.1 Spiral Source: $R_1 > 0$, $R_2 > 0$, $Im_1 \neq 0$, $Im_2 \neq 0$

Complex principal values with positive real parts indicate spiral nature of the fields swirling outwards from critical point. The field patterns resemble combination of an irrotational vortex and a source with which flow moves outwards in spiral paths. Neighboring flux lines spiral around the critical point and depart from it asymptotically.

7.3.3.2 Spiral Sink: $R_1 < 0$, $R_2 < 0$, $Im_1 \neq 0$, $Im_2 \neq 0$

Complex principal values with negative real parts indicate spiral nature of the fields swirling towards critical point. The field topology resembles combination of an irrotational vortex and a sink with which flow moves inwards in spiral paths. Neighboring flux lines spiral around critical point and approach it asymptotically.

The basic principle in critical point theory and analysis is to extract field topology by locating and classifying these points. This is then followed by drawing the dividing streamlines passing through the critical points whereby separating fields into regions of qualitative behavior. A small number of streamlines in the neighborhood are also drawn in to display field patterns and to envisage the main features of vector fields. These simple displays are useful to delineate the whole fields and to infer field structure. The same can be done for drawing equipotentials when field is harmonic.

7.4 Example 4.1d Critical Point Topology

This fourth part of Example 4.1 aims to study the topology of critical point, which occurs at origin (corner). At this point, conformal mapping breaks down and curvilinear square assumes the greatest size. The velocity critical point in this example is a first order type classified as a saddle point in Sections 7.2 and 7.3. The origin is multiple points with coincidence of velocity critical point and heat flux critical point. The gradient of heat flux tensor is interpreted as the spatial rate of change of heat flux acting on the i-face of a control volume in j-direction, that is,

$$\nabla q^* \bigg|_{\substack{x=0 \\ y=0}} = \begin{bmatrix} \dfrac{\partial q_x^*}{\partial x} & \dfrac{\partial q_y^*}{\partial x} \\[2mm] \dfrac{\partial q_x^*}{\partial y} & \dfrac{\partial q_y^*}{\partial y} \end{bmatrix}_{x=y=0} = kc_2 \begin{bmatrix} 0 & -1 \\ -1 & 0 \end{bmatrix}_{x=y=0}$$

q^* and ψ_t are given in part 4.1a. The symmetric matrix yields the following eigenvalues and eigenvectors:

$$\lambda_1 = kc_2, \; \hat{e}_1 = \begin{bmatrix} 1 & -1 \end{bmatrix}$$
$$\lambda_2 = -kc_2, \; \hat{e}_2 = \begin{bmatrix} -1 & -1 \end{bmatrix}$$
$$\lambda_1\lambda_2 < 0$$

It is classified as a saddle point. Eigenvalues represent extreme spatial rate of change of heat flux acting on respective principal planes. The plus or minus sign of eigenvalues represent out-going flux or in-coming heat flux of a control volume. A pair of eigenvectors define the orientations of two mutually orthogonal principal planes coincide with lines $y = \pm x$. Heat flux points towards critical point along $y = x$ (or $\hat{e}_2$) with $\lambda_2 < 0$ and points away from it along $y = -x$ (or $\hat{e}_1$) in opposite directions with $\lambda_1 > 0$. In this process, heat flux gradient goes through maximum and simultaneously minimum as the sign of eigenvalue changes from negative (in-coming flux) to positive (out-going flux). This is the basic characteristic of a saddle point with no local extreme. The principal axes trajectories crossing at critical point serve as dividing heatlines, which exist only in the infinitesimal neighborhood of critical point. This is because velocity critical point is also the critical point of convective heat flux. The dividing heatline is attributed to conduction heat transfer prevailing in the neighborhood of critical point.

Alternatively, we could study critical point topology by invoking the Hessian matrix of heatfunction and its principal state.

$$H_{ij} = \begin{bmatrix} \dfrac{\partial^2 \psi_t}{\partial x^2} & \dfrac{\partial^2 \psi_t}{\partial x \partial y} \\[2ex] \dfrac{\partial^2 \psi_t}{\partial y \partial x} & \dfrac{\partial^2 \psi_t}{\partial^2 y} \end{bmatrix}_{x=y=0} = kc_2 \begin{bmatrix} 1 & 0 \\ 0 & -1 \end{bmatrix}$$

$$\lambda_1 = kc_2, \; \hat{e}_1 = \begin{bmatrix} 1 & 0 \end{bmatrix}$$
$$\lambda_1 = -kc_2, \; \hat{e}_2 = \begin{bmatrix} 0 & -1 \end{bmatrix}$$
$$\lambda_1\lambda_2 = D = -k^2 c_2^2 < 0$$

This is also classified as a saddle point. Figure 4.9 depicts heatline trajectory also referred to as heat flux trajectory. The landscape of heatlines is that of a saddle point with cliff edges deepening away from critical point along y-axis and up-rising away from it along x-axis. Upon approaching critical point along $\hat{e}_2$ trajectory, heatfunction goes through a maximum and simultaneously a minimum following change of sign in eigenvalue from negative (in-coming flux) to positive (out-going flux) and departing from it along $\hat{e}_1$ trajectory in opposite directions. The direction of heat flux is perpendicular to the gradient of heatfunction vector and unit vector $\hat{k}$.

$$\text{Along vertical wall } y\text{-axis: } q^* = \nabla\psi_t \times \hat{k} = \hat{e}_2 \times \hat{k} = -\hat{i}$$
$$\text{Along horizontal wall } x\text{-axis: } q^* = \nabla\psi_t \times \hat{k} = \hat{e}_1 \times \hat{k} = -\hat{j}$$

Heat fluxes cross each wall perpendicularly as shown by arrows in Figure 4.9. The result tallies with those of gradient heat flux tensor analysis.

7.5 Singular Point Visualization and Mapping

A singular point in tensor fields is regarded as a material point where the field becomes infinite or undefined. In the neighborhood of singular points, field properties change rapidly and eventually break down. When approaching a singular point, flow speeds will continue to increase without bound. At this point, complex potential ceases to be analytic and its derivative is undefined. The gradient of vector field tensor also yields infinite eigenvalue. Singular points can appear in various forms with different nature. An isolated singularity is a singular point about which an infinitesimal circle can be constructed in such a way that it contains no further singular points. The function is bounded on an infinitesimal domain or it is said to have an essential singularity at a point where it becomes infinite. In physical plane, equipotentials (or streamlines) converge/diverge at a singular point while streamlines (or equipotentials) crowd together as the point is approached. Alternatively, they start and end instantly there. For example, flux lines emanating out from a singular point such as a source diverge in various directions or they converge into a sink whereas the circular equipotentials are densely packed around the singular point. Such singular point is classified as logarithmic singularity. When a source and a sink are combined, it gives rise to a doublet in hydrodynamics where flux lines start and end instantly. Singular points may behave like a free vortex with equipotentials radiating inwards or outwards from the vortex center while streamlines form a system of concentric circles crowd together. Such singular point is classified as vortex point. Singular points are important as they contribute significantly to the nature of fields. They are basic constituents of tensor topology and are useful to infer field structure and behavior in a manner similar to critical points.

Singular point is an improper point in F-plane or $\bar{q}$-plane. It is regarded as a point at infinity attached to F (or $\bar{q}$)-plane referred to as the extended complex planes. The idea implies such fields do not vanish at infinity and can only arise from sources located at infinity. Field maps break down at singular points since they require either the spacing of a curvilinear square to become infinite in complex F-plane or zero or vanish in physical plane. Analysis is simplified by considering fields being parallel at infinity with singularity removed. In this respect, the idea of a singular point at infinity is introduced to visualize singular points. It implies that parallel equipotentials (or parallel flux lines) will intersect each other at singular point located at infinity on complex F-plane. The same idea could be applied to the use of conjugate/hodograph plane on which singular points are also mapped to infinity as the vector field strength approaches infinite. This geometric interpretation suggests a way by which the concept of infinity can be visualized using singular images on certain planes.

Singular points could be produced due to discontinuity in boundary conditions or due to physical presence of source/sink in the domain/boundary. There are similarities and differences between singular points and critical points. Field directions are indeterminate and conformal mapping breaks down at both critical and singular points. Finite pairs of trajectories cross at a critical point but become infinite at a singular point. Field strength vanishes at critical point, but becomes infinite at singular point. This is reflected by the sizes of curvilinear cells in physical plane. Curvilinear cell assumes the biggest size in the neighborhood of critical point whereas it becomes infinitesimally small at singular point. Unlike critical points, singular points do not yield a valid solution. To circumvent the problems of field breakdown, it can be excluded by an extremely small circle around it and then interpolate between two adjacent points exterior to the circle. Alternatively, they could be mapped from one plane to another following the idea of singularity at infinity making graphical solution of shape factor possible.

7.6 Example 7.1 Mapping of a Point Source

Consider a singular point source at polar coordinate origin with a source strength C. The complex potential is given by a logarithm function:

$$F = f(z) = \frac{C}{2\pi} \ln z = \frac{C}{2\pi} \ln(r \exp(i\theta))$$
$$= \phi + i\psi$$

where $\psi = \frac{C}{2\pi}\theta$ and $\phi = \frac{C}{2\pi}\ln r$

$$\theta = \tan^{-1}\left(\frac{y}{x}\right); \quad r = \sqrt{x^2 + y^2}$$
$$0 \le \theta \le 2\pi; \quad r > 0$$

The stream function defines the position hodograph of streamlines implicitly.

$$y = x \tan \left(\frac{2\pi \psi}{C} \right) = x \tan \theta$$

Above is the equation of position hodograph for a point source trajectory. Each streamline is represented by a straight line through origin on x-y plane with a constant slope equal to $\tan(\frac{2\pi \psi}{C})$. The flow fields depict a system of straight streamlines issuing from origin like a pencil of rays as ψ varies. The streamline tangent is by definition equal to $\tan(\frac{2\pi \psi}{C})$.

$$\frac{dy}{dx} = \tan \left(\frac{2\pi \psi}{C} \right) = \tan \theta = \tan \alpha$$

α is streamline tangent angle measured from x-axis anticlockwise. Similarly, we obtain the position hodograph of equipotential from the equipotential function implicitly. For a given ϕ, the graph of equipotential is a circle of radius equal to $\exp(\frac{2\pi \phi}{C})$.

$$r = \exp \left(\frac{2\pi \phi}{C} \right)$$

Equipotentials resemble concentric circles about origin and cross streamlines at right angle to form a system of orthogonal curvilinear field map. The origin is a singular point and a point of symmetry. With crowding equipotentials around singular point, curvilinear squares become increasingly small and vanish there while velocity complex potential becomes infinite.

$$F \mid_{r=0} = \phi + i\psi = \infty$$
$$\bar{q} = \frac{dF}{dz} = \frac{\partial \phi}{\partial r} + i \frac{\partial \psi}{\partial r}$$
$$q_r = \frac{C}{2\pi r}; \ q_\theta = 0$$
$$q \mid_{r=0} = \infty$$

The gradient of velocity near singular point yields infinite eigenvalue.

$$[\nabla q]_{r=0} = \sum_{i=1}\sum_{j=1} \frac{\hat{e}_i \partial(q_j \hat{e}_j)}{h_i \partial x_i} = \begin{bmatrix} \dfrac{\partial q_r}{\partial r} & \dfrac{\partial q_\theta}{\partial r} \\[2ex] \dfrac{\partial q_r}{r \partial \theta} - \dfrac{q_\theta}{r} & \dfrac{\partial q_\theta}{r \partial \theta} + \dfrac{q_r}{r} \end{bmatrix}_{r=0}$$

$$= \begin{bmatrix} \dfrac{-C}{2\pi r^2} & 0 \\[2ex] 0 & \dfrac{q_r}{r} \end{bmatrix}_{r=0} = \begin{bmatrix} -\infty & 0 \\ 0 & \infty \end{bmatrix}_{r=0}$$

The singular point is mapped to a point at infinity on complex F-plane. On this plane, concentric equipotential contours are mapped to a family of vertical coordinate lines. Streamlines are mapped to a system of straight lines parallel to horizontal real axis on F-plane. Conceptually streamlines and equipotentials will intersect each other at singular point at infinity. The singular point is also mapped to infinity on velocity conjugate hodograph $\bar{q}$-plane as the vector assumes infinite magnitude extending to infinity.

By substitution, the position hodograph of streamlines is mapped to its hodograph representation as another system of straight lines through origin on hodograph plane with the same slope equal to $\tan(\frac{2\pi \psi}{C})$.

$$q_y = q_x \tan\left(\frac{2\pi \psi}{C}\right)$$

$$= q_x \tan \tilde{\theta}$$

Isocline inclination angle: $\tilde{\theta} = (2\pi \psi / C) = \alpha$

where

$$q_x = \frac{C \cos \theta}{2\pi r}; \quad q_y = \frac{C \sin \theta}{2\pi r}$$

$$x = r \cos \theta; \quad y = r \sin \theta$$

It is observed that hodograph representation of streamlines from a point source is equivalent to the hodograph representation of isoclines. The constant direction of motion along a straight streamline is specified by the isoclines inclination angle or streamline tangent measured from q_x-axis. The results are consistent with hodograph representation theory for straight impervious boundaries like straight streamlines. Hodograph representation of equipotential is also circular given by:

$$|V| = \sqrt{q_x^2 + q_y^2} = \frac{C}{2\pi r}$$

Or

$$|V| = \frac{C}{2\pi \exp(\frac{2\pi\phi}{C})}$$

The circle center is at the origin on hodograph plane, which corresponds to the critical point at infinity on physical plane. The circle radius represents the modulus equal to velocity magnitude and is inversely proportional to the distance from the point source. Each equipotential defines a circle and crosses streamlines also at right angle on hodograph plane. Furthermore,

$$r = \frac{C}{2\pi |V|}$$

The above is the position hodograph representation of constant speed giving rise to a system of circles about origin similar to those of equipotentials on physical plane. This is because each equipotential of a point source traces the locus of constant speed and conversely. The circle about origin with radius equal to $C/(2\pi r)$ on hodograph plane represents locus of constant speed $|V|$ is mapped a circle of radius $C/(2\pi|V|)$ about origin on physical plane.

The gradient of velocity tensor yields vanishing eigenvalues in the neighborhood of critical point classified as a degenerate point. From Equation 4.14a or 4.14b, the acceleration is determined.

$$a = \frac{dq}{dt} = \frac{dq_r}{dr}\frac{dr}{dt}\hat{r} = -\left(\frac{C}{2\pi}\right)^2\frac{1}{r^3}\hat{r}$$

$$a \cdot \frac{q}{|q|} = -\left(\frac{C}{2\pi}\right)^2\frac{1}{r^3} < 0$$

Tangents to the trajectory on physical plane and hodograph plane represent velocity and acceleration vector coinciding with each other and pointing in the opposite direction everywhere. The motion is on a straight line radiating outwards from the point source while speed decreases with distance from the source. The results are consistent with hodograph representation theory. Flow direction changes rapidly in the region close to the point source as all radial streamlines converge towards the origin.

The hodograph representation of constant kinetic energy K for a point source flow is a circle of radius $\sqrt{2K}$ about origin. It is mapped to a circle of radius $\frac{C}{\sqrt{8K\pi}}$ about origin on physical plane representing the position hodograph of kinetic energy contours (isotachs).

$$\frac{V^2}{2} = \frac{q_r^2}{2} = \frac{v_x^2 + v_y^2}{2} = \frac{1}{8}\left(\frac{C}{\pi r}\right)^2 = K$$

or

$$r^2 = \left(\frac{C}{\sqrt{8K\pi}} \right)^2$$

With vanishing curl of velocity fields, flow is potential. The pressure contours (isobars) are obtained from Bernoulli's theorem. Consider a point source on a horizontal plane with negligible potential energy. For a given Bernoulli's constant $B = \frac{p_0}{\rho}$, both hodograph representation of isobars and position hodograph representation of isobars are circles about origins in the hodograph plane and physical plane respectively. Pressure increases with distance from the point source and assumes Bernoulli's constant at infinity where velocity vanishes.

$$\frac{p_0}{\rho} - \frac{p}{\rho} = \frac{q_r^2}{2} = \frac{1}{8} \left(\frac{C}{\pi r} \right)^2$$

or

$$v_x^2 + v_y^2 = \left(\sqrt{2 \left(\frac{p_0}{\rho} - \frac{p}{\rho} \right)} \right)^2$$

$$r^2 = \left(\frac{C}{\pi \sqrt{8 \left(\frac{p_0}{\rho} - \frac{p}{\rho} \right)}} \right)^2$$

To investigate the nature of flow in ζ-plane when the point source is mapped from z-plane, we shift the singular point from origin to point z_0 and invoke Laurent series about the image point ζ_0.

$$F(\zeta) = F[\zeta(z)] = F(z) = \frac{C}{2\pi} \ln(z - z_0) = \frac{C}{2\pi} \ln(z(\zeta) - z(\zeta_0))$$

$$= \frac{C}{2\pi} \sum_{j}^{\infty} \frac{\ln \partial^j z(\zeta_0)}{j! \partial \zeta^j} (\zeta - \zeta_0)^j$$

$$\approx \frac{C}{2\pi} \frac{\ln \partial z(\zeta_0)}{\partial \zeta} (\zeta - \zeta_0)$$

$$\approx \frac{C}{2\pi} \ln(\zeta - \zeta_0) + c$$

where

$$c = \frac{C}{2\pi} \frac{\ln \partial z(\zeta_0)}{\partial \zeta}$$

$$\bar{q} = \frac{dF}{dz} = \frac{C}{2\pi} \frac{1}{z - z_0}$$

$$\bar{q}_\zeta = \frac{dF}{d\zeta} = \frac{C}{2\pi} \frac{1}{\zeta - \zeta_0}$$

$$\zeta = z$$

$$\bar{q} = \bar{q}_\zeta$$

The nature of point source and strength are invariant in the transformation. The complex potential could represent an electric field of an infinitely long charged wire passing through origin.

$$E = \frac{C\hat{r}}{2\pi r} = q_r \hat{r}$$

$\frac{C}{2\pi}$ is the charge per unit length. r is the distance from wire.

8

Engineering Application Examples

8.1 Example 8.1: Torsion of a Square Beam

The geometry of a square section beam with a side length equal to a and a longitudinal length $\Delta z = 3a$ is shown in Figure 8.1. One end of the beam is fixed at $z = 0$. At the other end, an external anticlockwise torque is applied about the longitudinal axis (z-axis). Body force F_b is negligible.

8.1.1 Displacement Analysis

Consider the subject beam at rest under the action of steady loads. The beam center axis is chosen to coincide with longitudinal z-axis of Cartesian coordinate system. The origin is located at the beam center at fixed end. The applied torque Γ_z twists the beam about center axis and causes a rotation measured in terms of angle of twist. Rotation is assumed to be linearly proportional to longitudinal distance, such that

$$\Omega = \begin{bmatrix} 0 & 0 & \theta_1 z \end{bmatrix} \tag{8.1}$$

$$\text{where } \theta_1 = \frac{\Gamma_z}{\mu J} \tag{8.2}$$

μ is the shear modulus of rigidity. J is the polar moment of the cross section about z-axis. θ_1 is the angle of twist per unit length. $\vec{r}$ is a position vector

Visualization of Fields and Applications in Engineering, First Edition. Stephen Tou.
© 2011 John Wiley & Sons, Ltd. Published 2011 by John Wiley & Sons, Ltd.

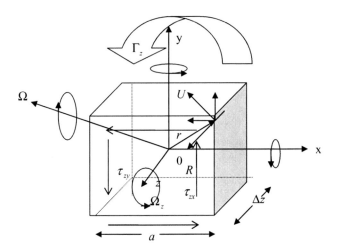

Figure 8.1   A square section beam subject to torsion

defining an arbitrary point in the beam. The applied torque gives rise to a displacement vector field in the beam in terms of angle of twist.

$$U = \Omega \times \vec{r}$$
$$= \begin{bmatrix} -\theta_1 zy & \theta_1 zx & 0 \end{bmatrix} \tag{8.3}$$
$$\text{where } \vec{r} = \begin{bmatrix} x & y & z \end{bmatrix} \tag{8.4}$$

The displacement vector is perpendicular to both position vector and rotation vector. Each displacement component on x-y plane appears to be a function of the other two spatial variables. With vanishing z-component displacement, each plane remains undistorted after a rotation. However for a non-circular section, the cross section warps in the absence of axial symmetry meaning that planes perpendicular to the axis of torsion do not remain planar after deformation. This has in fact attributed to the failure of earlier torsion theories before the introduction of the warping function method by Saint-Venant. Each point of a cross section may be displaced in the z-direction. Warping describes the patterns of a surface distortion in terms of vertical displacements measured with respect to the plane surface datum. In this method, a mode of deformation is assumed to be described by a warping function rendering the displacement fields three-dimensional. The assumption that each displacement component is a function of the other two mutually orthogonal directions implies warping of cross section is the same at all plane parallel sections. For a given angle of twist, the z-component displacement represents warping distortion is determined by warping function, $\eta(x, y)$.

$$u_z = \theta_1 \eta(x, y) \tag{8.5}$$

$$\text{Or } U^* = \frac{U}{\theta_1 a^2} = \begin{bmatrix} -z^* y^* & z^* x^* & \eta^* \end{bmatrix} \tag{8.6}$$

$$\nabla^* \cdot U^* = 0 \tag{8.7}$$

The non-dimension forms are defined as:

$$x^* = \frac{x}{a}, \ y^* = \frac{y}{a}, \ z^* = \frac{z}{a}, \ \Delta z^* = \frac{\Delta z}{a} = 3, \ u_x^* = \frac{u_x}{\theta_1 a^2},$$

$$u_y^* = \frac{u_y}{\theta_1 a^2}, \ u_z^* = \frac{u_z}{\theta_1 a^2} = \eta^* = \frac{\eta}{a^2}$$

$$\psi^* = \frac{\psi}{\mu\theta_1 a^2}, \ \Delta s^* = \frac{\Delta s}{a}, \ \Delta n^* = \frac{\Delta n}{a}, \ \tau_{ij}^* = \frac{\tau_{ij}}{\mu\theta_1 a}, \ \nabla^* = a\nabla, \ \nabla^{*2} = a^2\nabla.$$

$$-1/2 \le x^* \le 1/2; \ -1/2 \le y^* \le 1/2; \ 0 \le z^* \le 3$$

The displacement fields are solenoidal and the strain tensor first invariant vanishes accordingly. From displacement fields, the rotation vector is obtained from curl operation:

$$\Omega = \frac{1}{2}\varpi = \frac{1}{2}\nabla \times U \tag{8.8}$$

where

$$\Omega_x = \frac{\theta_1}{2}\left[\frac{\partial\eta}{\partial y} - x\right]; \quad \Omega_y = -\frac{\theta_1}{2}\left[y + \frac{\partial\eta}{\partial y}\right]; \quad \Omega_z = \theta_1 z$$

or

$$\Omega_x^* = \frac{\Omega_x}{\theta_1 a} = \frac{1}{2}\left[\frac{\partial\eta^*}{\partial y^*} - x^*\right]; \quad \Omega_y^* = \frac{\Omega_y}{\theta_1 a} = -\frac{1}{2}\left[y^* + \frac{\partial\eta^*}{\partial y^*}\right]; \quad \Omega_z^* = \frac{\Omega_z}{\theta_1 a} = z^*$$

It is observed that in addition to rotation about beam center axis, there are rotations about x and y-axes as well. The x- and y-component rotations are likely nonlinear due to the presence of warping displacement gradients. The axis of rotation defined by the resultant rotation vector varies from point to point. The problem solution hinges on the search for a suitable warping function which is required to satisfy equation of equilibrium as well as the boundary conditions presented next.

8.1.2 Governing Equation

For an isotropic homogeneous solid system, the equation of equilibrium is:

$$(\lambda + \mu)\nabla I_1 + \mu\nabla^2 U + F_b = 0 \tag{8.9}$$

$$\text{where } F_b = 0$$

λ is an elastic constant. I_1 denotes the strain tensor first invariant. As the strain tensor first invariant vanishes, the equation is uncoupled. Each displacement component is required to satisfy Laplace equation. It can be shown by substitutions that both x and y-displacement components satisfy the equation of equilibrium. The z-component displacement is required to do likewise, such that

$$\mu\nabla^2 u_z = \mu\theta_1\nabla^2\eta = \nabla \cdot q = 0 \tag{8.10}$$

$$\text{where } q = \mu\theta_1\nabla\eta$$

$$\text{Or } \nabla^{*2}u_z^* = \nabla^{*2}\eta^* = 0$$

The physical interpretation of the governing equation is force balance, which requires the net warping stress flux across a control volume to vanish. By introducing warping flux function ψ, the governing equation can be written in terms of warping stress flux of z-component in the respective directions:

$$x\text{-direction}: q_x = \frac{\mu\theta_1\partial\eta}{\partial x} = \frac{\partial\psi}{\partial y} \tag{8.11a}$$

$$y\text{-direction}: q_y = \frac{\mu\theta_1\partial\eta}{\partial y} = -\frac{\partial\psi}{\partial x} \tag{8.11b}$$

$$\text{Or } q^* = \frac{q}{\mu\theta_1 a} = \begin{bmatrix} \dfrac{q_x}{\mu\theta_1 a} \\ \dfrac{q_y}{\mu\theta_1 a} \end{bmatrix} = \begin{bmatrix} \dfrac{\partial\eta^*}{\partial x^*} \\ \dfrac{\partial\eta^*}{\partial y^*} \end{bmatrix} = \begin{bmatrix} \dfrac{\partial\psi^*}{\partial y^*} \\ -\dfrac{\partial\psi^*}{\partial x^*} \end{bmatrix} \tag{8.12}$$

The above Cauchy-Riemann relation shows that the warping function assumes the role of potential. Warping flux function is analogous to stream function. Warping stress flux points in the direction of increasing warping function. If warping function is known then warping flux function can be found. They lead to the construction of warping complex potential.

$$F = f(z^*) = \eta^*(x^*, y^*) + i\psi^*(x^*, y^*) \tag{8.13}$$

8.1.2.1 Boundary Conditions

In a physical sense, boundary conditions are set up by a traction force vector on the beam's lateral surfaces. They are required to vanish by equilibrium conditions. The traction force acting on the beam lateral surfaces with a surface unit normal vector $\hat{b}$ is obtained from Cauchy's formula, that is,

$$\tilde{T}^* = \tau_{ij}^* \cdot \hat{b} = \begin{bmatrix} 0 \\ 0 \\ 0 \end{bmatrix}$$

$$\text{where } \hat{b} = \begin{bmatrix} \cos(b, x) \\ \cos(b, y) \\ \cos(b, z) \end{bmatrix}$$

$$\hat{b} \cdot \hat{k} = 0; \rightarrow \cos(b, z) = 0$$

$$\tilde{T}_x^* = \tau_{xx}^* \cos(b, x) + \tau_{xy}^* \cos(b, y) + \tau_{xz}^* \cos(b, z) = 0 \tag{8.14a}$$

$$\tilde{T}_y^* = \tau_{yx}^* \cos(b, x) + \tau_{yy}^* \cos(b, y) + \tau_{yz}^* \cos(b, z) = 0 \tag{8.14b}$$

$$\tilde{T}_z = \cos(b, x)\tau_{zx} + \cos(b, y)\tau_{zy} = 0$$

Or

$$\frac{dy^*}{dx^*}\Big|_{Boundary} = \frac{\tau_{zy}^*}{\tau_{zx}^*} = -\frac{\cos(b, x)}{\cos(b, y)} \tag{8.14c}$$

8.1.3 Determination of Warping Function

Sokolnikoff (1983) has given an appropriate warping function in terms of a series based on harmonic analysis, that is,

$$\eta = xy - \frac{8a^2}{\pi^3} \sum_{n=0}^{\infty} \frac{(-1)^n \sinh(k_n y) \sin(k_n x)}{(2n+1)^3 \cosh(k_n a/2)} \tag{8.15}$$

where $k_n = \dfrac{(2n+1)\pi}{a}$

Or $u_z^* = \eta^* = x^* y^* - \dfrac{8}{\pi^3} \sum_{n=0}^{\infty} \dfrac{(-1)^n \sinh((2n+1)\pi y^*) \sin((2n+1)\pi x^*)}{(2n+1)^3 \cosh((2n+1)\pi/2)}$ \hfill (8.16)

Symmetry Test

$$\eta^*(x^*, y^*) = -\eta^*(-x^*, y^*) = -\eta^*(x^*, -y^*) = \eta^*(-x^*, -y^*)$$

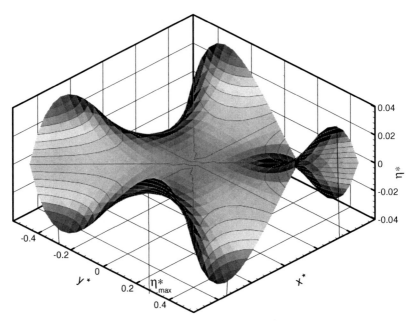

Figure 8.2 Three-dimensional view of warping distortion mesh

Figure 8.2 shows the graph of warping function $u_z^* = \eta^*(x^*, y^*)$, which describes surface distortion in terms of z-component displacement explicitly in a three-dimensional elevation mesh plot. The z^*-axis represents the dependent variable of surface distortion that is, $z^* = u_z^*$. Symmetry test shows that the graphs of warping function are symmetric with respect to origin. The graphs are also antisymmetric about each coordinate axis and behave in a completely opposite manner on either side of it. If one side goes up, the other goes down. η^* is positive indicating upwards displacements, otherwise it represents downwards displacements with η^* negative. This is reflected by a concave and a convex warping pattern in each quadrant due to the action of compression and tension. Maximum z-component displacements occur at eight locations along beam periphery, that is, $(x^* = \pm 0.3,\ y^* = \pm 0.5)$ and $(x^* = \pm 0.5,\ y^* = \pm 0.3)$. This agrees with the maximum principle for harmonic functions which ensures that warping function attains extreme values on the domain boundary. To zoom into the local phenomena of warping, we deploy the tangent plane through a point $(x_0^*,\ y_0^*)$ in terms of the linear function.

$$u_z^* = \eta^*(x_0^*, y_0^*) + \frac{\partial \eta^*}{\partial x^*}\Big|_{x_0^*, y_0^*} (x^* - x_0^*) + \frac{\partial \eta^*}{\partial y^*}\Big|_{x_0^*, y_0^*} (y^* - y_0^*)$$

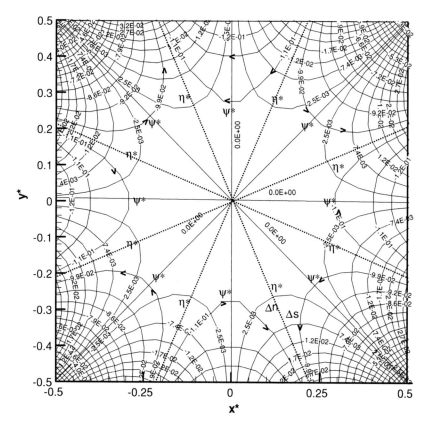

Figure 8.3 Warping equipotentials and warping flux trajectories field map

The coefficients of second and third term describe the slopes at a point along x^* and y^*-directions and approximate the behavior of warping in the neighborhood of the point. The orientation of tangent plane could be visualized by its normal vector at the point, that is,

$$\hat{n} = \nabla f(x^*, y^*, z^*) = \left[\frac{\partial \eta^*}{\partial x^*} \quad \frac{\partial \eta^*}{\partial y^*} \quad -1 \right]_{x_0^*, y_0^*}$$

where $f(x^*, y^*, z^*) = \eta^*(x^*, y^*) - u_z^* = 0$

The radius of curvature of warping could be determined from Equation 2.41 by substituting the function with η^*. Figure 8.3 depicts the warping equipotential contours in two-dimensions at equal intervals by changing the role of dependent variable u_z^* to that of a parametric constant. Warping patterns exhibit a family of hyperbolas. Warping strength intensifies as warping equipotentials squeeze towards corners. The difference between two warping equipotentials indicates the change of warping displacements per unit angle of twist.

$$\Delta \eta^* = \eta_2^* - \eta_1^*$$

The beam center (origin) is a critical point where warping displacement vanishes.

$$U^* \big|_{x^*=y^*=z^*=0} = 0$$

Since,

$$\frac{dF}{dz^*} \big|_{z^*=0} = \frac{\partial \eta^*}{\partial x^*} + i \frac{\partial \psi^*}{\partial x^*} = 0$$

$$\frac{d^2 F}{dz^{*2}} \big|_{z^*=0} = 0 + i(-1+1) = 0$$

$$\frac{d^3 F}{dz^{*3}} \big|_{z^*=0} = 0 + i(0) = 0$$

With vanishing third derivatives of complex potential at origin, it is classified as a third order critical point $n = 3$. There are $2n + 2 = 8$ branches of trajectories crossing through at equal angles, that is, $\pi/4$. It is observed that z-component displacement vanishes along coordinate axes and two diagonals crossing at critical point. They assume the role of dividing equipotentials and separate the displacement fields into eight distinct regions equal to the number of branches. The dividing equipotentials also serve as asymptotes for nearby equipotentials.

The beauty of warping complex potential is that its imaginary part describes the directions along which surface warps. It is obtained by integration of Equation 8.12:

$$\psi^* = 0.5 y^{*2} - 0.5 x^{*2} - \frac{8}{\pi^3} \sum_{n=0}^{\infty} \frac{(-1)^n \cosh((2n+1)\pi y^*) \cos((2n+1)\pi x^*)}{(2n+1)^3 \cosh((2n+1)\pi/2)} \tag{8.17}$$

$$\frac{\partial \psi^*}{\partial x^*} \big|_{x^*=0} = \frac{\partial \psi^*}{\partial y^*} \big|_{y^*=0} = 0$$

$$\psi^*(x^*, y^*) = \psi^*(-x^*, y^*) = \psi^*(x^*, -y^*) = \psi^*(-x^*, -y^*)$$

Figure 8.3 presents warping flux line trajectories, which are also displayed at equal intervals in the field map. Tangents along the trajectories define the directions of warping stress flux vectors along which a cross section of the beam is distorted. The trajectories depict a family of hyperbola and intersect coordinate axes at right angles. The trajectories are symmetric with respect to coordinate axes and origin. Warping flux trajectories and warping equipotential contours are orthogonal except at the critical point. Dividing warping flux trajectories $\psi^* = 0$ split at the critical point into two pairs of straight lines (dotted). They serve as asymptotes for nearby flux trajectories and separate the fields into eight local regions different from those of equipotentials (Figures 8.6 and 8.3). In the neighborhood of critical point, nearby warping flux lines asymptotic to the dividing warping flux lines turn sharply and progress radically different from those with which they approach. Warping strength intensifies as warping flux trajectories squeeze towards corners in a manner similar to squeezing equipotentials. It also implies rapid change of warping stress flux directions. The difference between values of two warping flux trajectories indicates the warping force transmission per unit longitudinal length per unit angle of twist bounded by the trajectories.

$$Q^* = \frac{Q}{\mu \theta_1 a^2 \Delta z} = \psi_1^* - \psi_2^*$$

It is observed that the square field domain geometry possesses point symmetry about origin and line-symmetry (coordinate axes and diagonals) and that the same kind of symmetry may exist in the warping patterns in one form or another. It appears that field patterns in one of the quadrants are sufficient to infer the global field patterns by reflection mapping with respect to coordinate axes or rotation mapping through multiple/right angle about origin.

8.1.4 Mapping of Warping Field Map on Hodograph Space

Hodograph representation of warping potential and warping flux trajectories is presented in three-dimensional displacement hodograph space in Figure 8.4. The three-dimensional contour surfaces are rendered from two-dimensional contour curves with η^* assuming the role of z^*-axis.

$$\eta^*(x^*, y^*) = \eta^*[x^*(u_x{}^*, u_y{}^*, u_z{}^*), y^*(u_x{}^*, u_y{}^*, u_z{}^*)] = \eta^*(u_x{}^*, u_y{}^*, u_z{}^*)$$
$$\psi^*(x^*, y^*) = \psi^*[x^*(u_x{}^*, u_y{}^*, u_z{}^*), y^*(u_x{}^*, u_y{}^*, u_z{}^*)] = \psi^*(u_x{}^*, u_y{}^*, u_z{}^*)$$

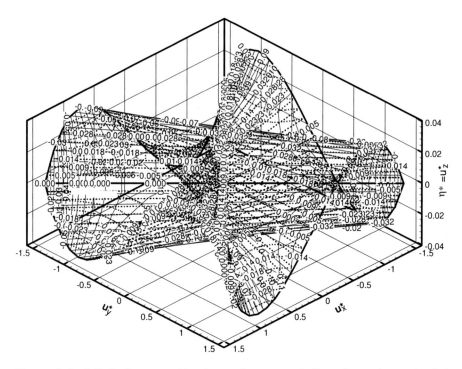

Figure 8.4 3-D displacement hodograph representation of warping potential (dotted) and warping flux trajectories (line)

where

$$u_x^* = -z^* y^*$$
$$u_y^* = z^* x^*$$
$$u_z^* = \eta^*(x^*, y^*)$$

8.1.5 Mapping of Warping Field Map on Complex F-Plane

To facilitate graphical evaluation of shape factor, warping equipotentials and warping trajectories are mapped from the physical plane onto complex F-plane. The result is shown in Figure 8.5 from which the shape factor is found equal to tube ratio in close agreement with transfer number.

$$\frac{S}{\Delta z} = D = \zeta = \frac{18}{14} = 1.3$$

$$D = \frac{\Delta \psi_{max}^*}{\Delta \eta_{max}^*} = \frac{-0.033 - (-0.124)}{0.036 - (-0.036)} = 1.26$$

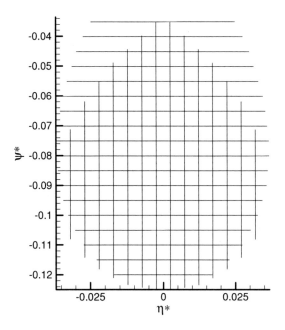

Figure 8.5 Mapping of warping field map on complex *F*-plane

8.1.6 Warping Phenomenon Analysis and Visualization
8.1.6.1 Warping Strength

As a measure of warping strengths, warping stress fluxes are obtained from warping complex potential via differentiation:

$$\overline{q^*} = \frac{dF}{dz^*} = \frac{\partial \eta^*}{\partial x^*} + i\frac{\partial \psi^*}{\partial x^*}$$

where

$$q_x^* = \frac{\partial \eta^*}{\partial x^*} = y^* - \frac{8}{\pi^2}\sum_{n=0}^{\infty}\frac{(-1)^n \sinh((2n+1)\pi y^*)\cos((2n+1)\pi x^*)}{(2n+1)^2 \cosh((2n+1)\pi/2)} \tag{8.18a}$$

$$q_y^* = \frac{\partial \eta^*}{\partial y^*} = x^* - \frac{8}{\pi^2}\sum_{n=0}^{\infty}\frac{(-1)^n \cosh((2n+1)\pi y^*)\sin((2n+1)\pi x^*)}{(2n+1)^2 \cosh((2n+1)\pi/2)} \tag{8.18b}$$

$$|q^*| = \sqrt{q_x^{*2} + q_y^{*2}}$$

$$q_x^*(0,0) = 0; \quad q_y^*(0,0) = 0$$

$$\frac{dy^*}{dx^*} = \frac{q_y^*}{q_x^*}$$

Figure 8.6 presents the distributions of warping stress field line trajectories obtained from the above trajectory equation. They are equivalent to those of warping flux function for solenoidal fields. Tangent at a point along a trajectory gives the direction of stress flux vector. The magnitudes of stress flux fields are presented in contour forms in the same figure. Its distribution resembles a system of concentric circles about origin on physical plane. The

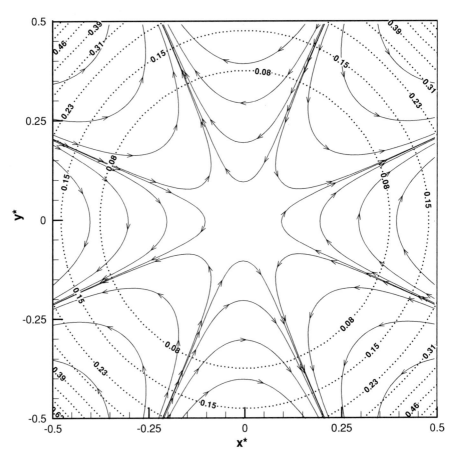

Figure 8.6 Distributions of warping stress flux field line trajectories (arrow) and magnitude contours (dotted)

hodograph representation of constant warping stress flux magnitude is also a family of circles about origin on warping stress flux hodograph plane with radii equal to the magnitudes. Furthermore the origins of two planes are mapped to coincide with each other and assume the role of warping stress flux critical point. Warping strength is negligible in the central part of the beam and builds up rapidly around beam peripheral. Warping strength increases with distance from origin and attains maximum at four corners. The results tally with those of field maps in Figure 8.3 where equipotential and warping flux trajectories squeeze towards corners.

8.1.6.2 Warping Energy

Warping energy is interpreted as a form of elastic potential energy due to warping distortion and is a component of total internal energy in the system. Warping energy storage per unit depth in a curvilinear cell is evaluated from warping stress flux fields. The integral forms of field energy storage are written with reference to streamline coordinate as:

$$\frac{\Delta W}{\Delta z} = \frac{1}{2\mu} \int_{\Delta n} \int_{\Delta s} |q|^2 \, dn \, ds$$

$$\text{where } |q| = |q_s| = \frac{\mu \theta_1 \partial \eta}{\partial s}$$

With the aid of the Cauchy-Riemann relation, it is written as:

$$\frac{\Delta W}{\Delta z} = \frac{\mu \theta_1 a^4}{2} \oint_S \eta^* \frac{\partial \psi^*}{\partial n^*} dn^* = \frac{\mu \theta_1 a^4}{2} \Delta \eta^* \Delta \psi^*$$

$$\Delta W^* = \frac{\Delta W}{\mu \theta_1^2 a^4 \Delta z} = \frac{\Delta \eta^* \Delta \psi^*}{2}.$$

Warping energy density per unit volume is given by:

$$K^* = \frac{\Delta W^*}{\Delta n^* \Delta s^*} = \frac{\Delta n^* \Delta \psi^*}{2 \Delta n^* \Delta s^*} = \frac{|q^*|^2}{2} \tag{8.19}$$

Warping energy density is proportional to its image size on complex F-plane but inversely on the physical plane. For cells of the same kind, warping energy density is greater on cells with smaller size on the physical plane. Distributions of warping energy density contours are presented in Figure 8.7 as a system of concentric circles. The patterns are similar to those of warping stress flux magnitudes in Figure 8.6. Warping energy density is also negligible

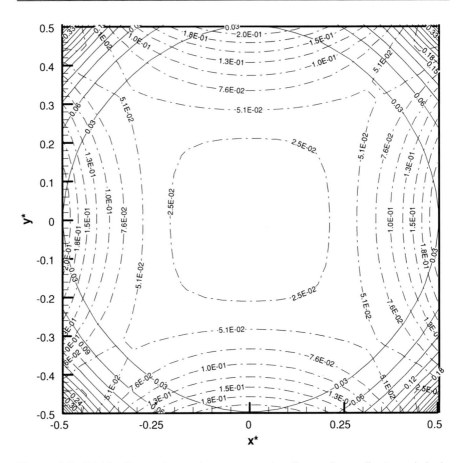

Figure 8.7 Distributions of warping energy density contours (line) and strain energy density contours (dash dot)

in the central part of beam. It increases with distance from origin and assumes maximum at four beam corners similar to warping strengths.

8.1.7 Strain/Stress Field Evaluation

The applied load generates a strain/ stress field in the beam which maintains the system in equilibrium. The strain tensor is obtained with the aid of Equation 6.4, namely:

$$\gamma_{ij} = \begin{bmatrix} 0 & 0 & \gamma_{xz} \\ 0 & 0 & \gamma_{yz} \\ \gamma_{zx} & \gamma_{zy} & 0 \end{bmatrix} \tag{8.20}$$

$$\text{where } \gamma_{zx} = \frac{1}{2}\left(\frac{\partial u_x}{\partial z} + \frac{\partial u_z}{\partial x}\right) = \frac{\theta_1}{2}\left(\frac{\partial \eta}{\partial x} - y\right)$$

$$\gamma_{zy} = \frac{1}{2}\left(\frac{\partial u_y}{\partial z} + \frac{\partial u_z}{\partial y}\right) = \frac{\theta_1}{2}\left(\frac{\partial \eta}{\partial y} + x\right)$$

Shear strain components are responsible for change in shapes. With vanishing diagonal components, it implies the system is in pure shear strain with no change in volume. There are only two unknown off-diagonal components on the z-face of a control volume. A rectangular element parallel to coordinate axes is not stretched/contracted at all but changes to parallelogram of the same size under the action of shear stress components. From strain tensor, the corresponding stress components are obtained from Hooke's law for an isotropic and homogenous material. The beam is subject to pure shear as shown in Figure 8.1.

$$\tau_{ij} = \lambda I_1 \delta_{ij} + 2\mu\gamma_{ij}$$

$$= 2\mu\gamma_{ij} \tag{8.21}$$

$$\text{where } \tau_{zx} = \mu\left(\frac{\partial u_x}{\partial z} + \frac{\partial u_z}{\partial x}\right) = \mu\left(\theta_1\frac{\partial \eta}{\partial x} - \theta_1 y\right)$$

$$\tau_{zx}^* = \frac{\tau_{zx}}{\theta_1 a \mu} = \frac{\partial \eta^*}{\partial x^*} - y^*$$

$$= -\frac{8}{\pi^2}\sum_{n=0}^{\infty}\frac{(-1)^n \sinh((2n+1)\pi y^*)\cos((2n+1)\pi x^*)}{(2n+1)^2 \cosh((2n+1)\pi/2)} \tag{8.22a}$$

$$\tau_{zy} = \mu\left(\frac{\partial u_y}{\partial z} + \frac{\partial u_z}{\partial y}\right) = \mu\left(\theta_1\frac{\partial \eta}{\partial y} + \theta_1 x\right)$$

$$\tau_{zy}^* = \frac{\tau_{zy}}{\theta_1 a \mu} = \frac{\partial \eta^*}{\partial y^*} + x^* \tag{8.22b}$$

$$= 2x^* - \frac{8}{\pi^2}\sum_{n=0}^{\infty}\frac{(-1)^n \cosh((2n+1)\pi y^*)\sin((2n+1)\pi x^*)}{(2n+1)^2 \cosh((2n+1)\pi/2)}$$

$$\frac{\partial \tau_{zx}^*}{\partial x^*}\Big|_{x^*=0} = 0, \quad \frac{\partial \tau_{zy}^*}{\partial y^*}\Big|_{y^*=0} = 0$$

It can be shown that the stress fields satisfy the equation of equilibrium as well as the boundary conditions Equation 8.14, that is,

$$\tilde{T}_x^* = 0\cos(b, x) + 0\cos(b, y) + \tau_{xz}^* 0 = 0$$

$$\tilde{T}_y^* = 0\cos(b, x) + 0\cos(b, y) + \tau_{yz}^* 0 = 0$$

$$\text{Along vertical boundaries}: \left.\frac{dy^*}{dx^*}\right|_{x^*=\pm1/2} = \frac{\tau^*_{zy}}{\tau^*_{zx}}\Big|_{x^*=\pm1/2} = \frac{\tau^*_{zy}}{0}\Big|_{x^*=\pm1/2} = \infty$$

$$\text{Along horizontal boundaries}: \left.\frac{dy^*}{dx^*}\right|_{y^*=\pm1/2} = \frac{\tau^*_{zy}}{\tau^*_{zx}}\Big|_{y^*=\pm1/2} = \frac{0}{\tau^*_{zx}}\Big|_{y^*=\pm1/2} = 0$$

$$\tau^*_{xz}(0,0) = 0; \quad \tau^*_{yz}(0,0) = 0$$

The distributions of shear stress components are depicted in Figure 8.8. Shear stress component τ^*_{zx} intersects y^*-axis orthogonally, which serves as a line of symmetry. It is antisymmetric about x^*-axis as well as origin. Consider the positive z^*-face (free end) of a control volume, the positive direction of shear stress τ^*_{zx} points, by convention, towards positive x^*-axis direction. With $\tau^*_{zx} > 0$ in the lower half section and $\tau^*_{zx} < 0$ in the upper half section, the directions of τ^*_{zx} are marked with arrow in Figure 8.8 in opposite directions in the two sections. The shearing stresses give rise to a couple in counterclockwise directions about z^*-axis. Shear stress component τ^*_{zy} intersects x^*-axis orthogonally serving as line of symmetry. It is antisymmetric about y^*-axis as well as origin. On the positive z^*-face, the positive direction of shear stress τ^*_{zy} points towards positive y^*-axis direction. With $\tau^*_{zy} < 0$ on the left half section and $\tau^*_{zy} > 0$ on the right half section, the directions of τ^*_{zy} are marked with arrow in Figure 8.8 in opposite directions. The shearing stresses give rise to a couple in counter-clockwise directions about z^*-axis. Both shear stress components vanish at origin (critical point) and appear to have simple relationships with the warping stress flux.

Recall two of the six Beltrami-Michell compatibility equations,

$$\nabla^2 \tau_{zx} + \frac{1}{1+\nu}\frac{\partial^2 \Lambda_1}{\partial z \partial x} + \frac{\partial F_x}{\partial z} + \frac{\partial F_z}{\partial x} = 0 \tag{8.23}$$

$$\nabla^2 \tau_{zy} + \frac{1}{1+\nu}\frac{\partial^2 \Lambda_1}{\partial z \partial y} + \frac{\partial F_z}{\partial y} + \frac{\partial F_y}{\partial z} = 0 \tag{8.24}$$

where $\Lambda_1 = (3\lambda + 2\mu)I_1 = 0$, $F_x = F_y = F_z = 0$.

$$\sum_{i=1}^{3} \tau_{ij} = 0$$

With negligible body forces, the above reduces to Laplace equation. It can be shown by substitution that the stress fields satisfy Beltrami-Michell compatibility requirements and that each shear stress component assumes the role of a harmonic function. Extreme shearing stresses are expected to occur on peripheral boundaries in accordance with the maximum principle for harmonic functions. The stress tensor fields are solenoidal.

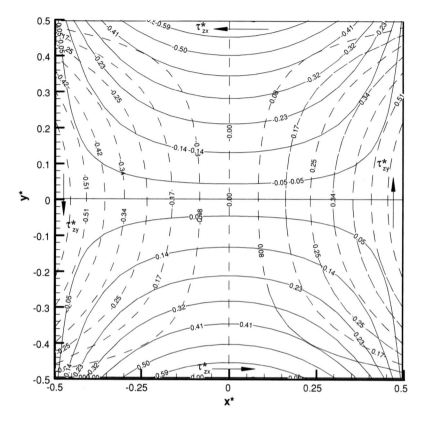

Figure 8.8 Distributions of z^*-face shear stress components (τ_{zx}^* ----, τ_{zy}^* ·········)

8.1.8 Stress Tensor Contraction and Stress Vector Potential (Prandtl Stress Function)

Although the displacement fields are three-dimensional, it is possible to visualize a pure shear stress tensor on one reference coordinate plane by stress trajectories similar to streamlines. Firstly, the stress vector on z^*-face can be obtained from Cauchy's formula through contraction. It reduces to a planar vector.

$$T^* = \tau_{ij}^* \cdot \hat{k} = \tau_{ij}^* \cdot \begin{bmatrix} 0 \\ 0 \\ 1 \end{bmatrix} = \begin{bmatrix} \tau_{zx}^* \\ \tau_{zy}^* \\ 0 \end{bmatrix} \tag{8.25}$$

$$\nabla^* \cdot T^* = 0$$

$$|T^*| = \sqrt{\tau_{zx}^{*2} + \tau_{zy}^{*2}} \tag{8.26}$$

Stress vector flux function exists in solenoidal stress tensor fields. It is to be found from the definition below.

$$T^* = \begin{bmatrix} \tau^*_{zx} \\ \tau^*_{zy} \end{bmatrix} = \begin{bmatrix} \dfrac{\partial \varphi^*}{\partial y^*} \\ -\dfrac{\partial \varphi^*}{\partial x^*} \end{bmatrix}$$

By integration,

$$\varphi^* = \frac{\varphi}{a^2 \mu \varphi_1} = -x^{*2} - \frac{8}{\pi^3} \sum_{n=0}^{\infty} \frac{(-1)^n \cosh((2n+1)\pi y^*) \cos((2n+1)\pi x^*)}{(2n+1)^3 \cosh((2n+1)\pi/2)}$$

$$\frac{\partial \varphi^*}{\partial x^*}\Big|_{x^*=0} = \frac{\partial \varphi^*}{\partial y^*}\Big|_{y^*=0} = 0$$

The boundary conditions lead to:

$$\left(\frac{\partial \varphi^*}{\partial y^*}\right)\Big/\left(-\frac{\partial \varphi^*}{\partial x^*}\right) = \frac{dx^*}{dy^*}\Big|_{Boundary} \tag{8.27}$$

$$\text{Or } d\varphi^*\big|_{Boundary} = 0, \rightarrow \varphi^*\big|_{Boundary} = C$$

The result is identical to the well-known Prandtl stress function. Figure 8.9 shows the distributions of Prandtl stress function trajectories, also referred to as stress trajectories. The stress trajectories are symmetric with respect to coordinate axes as well as origin (beam center). They assume a family of concentric loops about origin and intersect coordinate axes at right angles. Prandtl stress function is constant along the peripheral boundary, which is one of the stress trajectories. Stress vector at a point is tangential to the stress trajectory with directions marked by arrows. The results tally with the action of shear stresses presented in Figure 8.8. Furthermore stress vector magnitude increases with distance from origin. There is a vector source generating the looping stress trajectories, which could be derived from the curl of stress vector fields, that is,

$$\Theta^* = \nabla^* \times T^*$$

$$= [0 \quad 0 \quad 2] \tag{8.28}$$

$$= 2\hat{k}$$

The vector source direction is parallel to z-axis with only one nonzero z-component. The stress vector magnitude contours are presented in

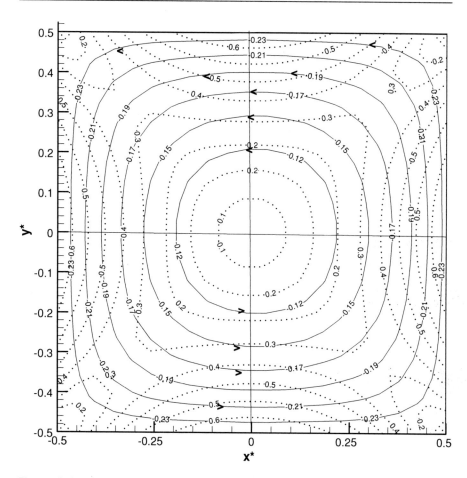

Figure 8.9 Distributions of stress trajectories (line) and stress vector magnitude contours (dotted)

Figure 8.9. The distribution of stress vector magnitude is symmetric with respect to coordinate axes. It is observed that maximum shearing stresses occur at the mid-points of four edges along beam peripheral, that is, $(x^* = \pm 0.5, \ y^* = 0)$ and $(x^* = 0, \ y^* = \pm 0.5)$. The results are in agreement with the maximum principle analysis. Its value is equal to:

$$|T^*|_{max} = 0.68 \tag{8.29}$$

Maximum shear stress (or maximum strain energy) :

$$(x^* = \pm 0.5, \ y^* = 0) \text{ and } (x^* = 0, \ y^* = \pm 0.5)$$

Maximum warping distortion : $(x^* = \pm 0.3,\ y^* = \pm 0.5)$ and

$(x^* = \pm 0.5,\ y^* = \pm 0.3)$

Maximum warping energy (or maximum warping strength) :

$(x^* = \pm 0.5,\ y^* = \pm 0.5)$

The results indicate that maximum shear stress corresponds to maximum strain energy density and that maximum shear stress, maximum warping distortion and maximum warping energy are at different locations. Such information is useful in the study of warping phenomena and the mode of fracture/failure.

Stress vector flux function is analogous to stream function. Stress force bounded between two trajectories is given by the difference in trajectory line values. It represents the magnitude of force transmitted between two trajectories per unit length.

$$Q^* = \frac{Q}{\mu \theta_1 a^2 \Delta z} = \int_1^2 T^* \cdot \hat{b}\, dl^* = \int_1^2 \nabla^* \varphi^* \times \hat{k} \cdot \hat{b}\, dl^* = -\int_1^2 \nabla^* \varphi^* \cdot \hat{l}\, dl^* = \varphi_1^* - \varphi_2^*$$

$\hat{b}$ and $\hat{l}$ represent unit normal vector and unit tangent vector of an arbitrary surface bounded by two stress trajectories. dl^* is the surface differential arc length.

8.1.9 Stress Vector Critical Point Topology

The origin is a multiple point with coincidence of stress vector critical point of first order and displacement vector critical point of third order. The gradient of stress vector tensor assumes an antisymmetric matrix,

$$[\nabla^* T^*]\Big|_{\substack{x^*=0 \\ y^*=0}} = \begin{bmatrix} \dfrac{\partial \tau_{zx}^*}{\partial x^*} & \dfrac{\partial \tau_{zy}^*}{\partial x^*} \\[2mm] \dfrac{\partial \tau_{zx}^*}{\partial y^*} & \dfrac{\partial \tau_{zy}^*}{\partial y^*} \end{bmatrix}\Bigg|_{\substack{x^*=0 \\ y^*=0}}$$

$$= \begin{bmatrix} 0 & 1 \\ -1 & 0 \end{bmatrix}$$

It yields two pure imaginary eigenvalues opposite in sign indicating a vector source in the stress vector fields. Orthogonal bases do not exist.

$$\sigma_{1,2} = \pm 1 i$$

The critical point is classified as a center with circular trajectories of rotational characteristics around it (Figure 7.2). The pure imaginary eigenvalue and looping nature of stress trajectories suggest there is a vector source. It has a magnitude equal to half of the curl of stress vector fields.

8.1.10 Prandtl Stress Function and Warping Flux Function Relation

There exists a relation between Prandtl stress function and warping flux function via Cauchy-Riemann equation. They are written in exactly the same way as those between shear stress and warping stress flux, that is,

$$\tau_{zx}^* = \frac{\partial \varphi^*}{\partial y^*} = \frac{\partial \eta^*}{\partial x^*} - y^* = \frac{\partial \psi^*}{\partial y^*} - y^*$$

$$\tau_{zy}^* = -\frac{\partial \varphi^*}{\partial x^*} = \frac{\partial \eta^*}{\partial y^*} + x^* = -\frac{\partial \psi^*}{\partial x^*} + x^*$$

By integration, above yields the relation between Prandtl stress function and warping flux function.

$$\varphi^* = \psi^* - \frac{x^{*2} + y^{*2}}{2} \tag{8.30}$$

Taking the Laplacian of the above results in:

$$\nabla^{*2}\varphi^* + 2 = \nabla^{*2}\psi^* = 0 \tag{8.31}$$

Thus, Prandtl stress function is not harmonic while warping flux function is. This means, by taking Prandtl stress function approach, the problem becomes Poisson's type. It can be seen that the vector source is the same as the curl of stress vector fields given by Equation 8.28.

8.1.11 Stress Vector Hodograph

In a manner similar to displacement hodograph, we can map Prandtl stress function from the physical plane to stress vector hodograph. They are presented in Figure 8.10 with components τ_{zx}^* and τ_{zy}^* assuming the role of x and y-axis respectively.

$$\varphi^*(x^*, y^*) = \varphi^*[x^*(\tau_{zx}^*, \tau_{zy}^*), y^*(\tau_{zx}^*, \tau_{zy}^*)] = \varphi^*(\tau_{zx}^*, \tau_{zy}^*)$$

The figure shows that maximum shear stresses at the mid point of four edges along beam peripheral in z^*-plane are mapped to the corresponding corner points of maximum shear stresses in stress vector hodograph. The stress

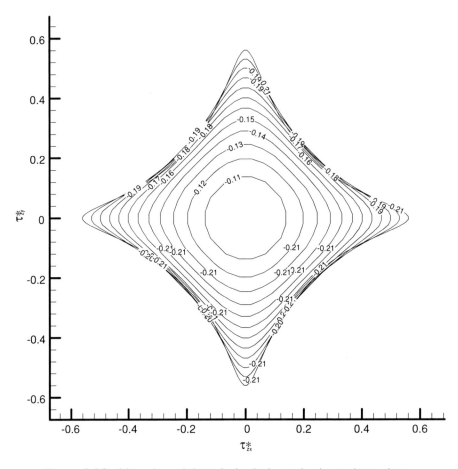

Figure 8.10 Mapping of stress trajectories onto shear stress plane

vector critical point is mapped from beam center to origin in this figure. In consideration of boundary conditions, two vertical peripheral boundaries symmetric with respect to y^*-axis are mapped to a vertical straight line passing through origin on stress vector hodograph plane. Similarly, two peripheral horizontal boundaries symmetric with respect to x^*-axis are mapped to a horizontal straight line passing through origin. The same results could be obtained from the hodograph representation of straight boundaries/streamlines. Maximum shear stresses are located at the end points of these lines in stress vector hodograph. We could determine shear stress variations along an arbitrary stress trajectory as pencil of rays from origin to the points of interest in respective figures. For instance, the state of stress at a point along an arbitrary stress trajectory on z^*-plane in Figure 8.9 could be mapped graphically to its image on the stress plane. This is accomplished by drawing a position vector in the stress plane parallel to the tangent at the selected point along a stress trajectory on z^*-plane. The intersection of the position vector and corresponding image stress trajectory defines the point in stress vector hodograph plane and hence the state of stress at the point of interest. Alternatively, by mapping circles of radius C with center at origin on stress vector hodograph plane, we could visualize lines of constant stress vector magnitude on the physical plane as shown in Figure 8.9. In the neighborhood of beam center, a circle is mapped approximately to the same circle about origin on physical plane. We could trace loci of stress vector isoclines on physical plane by mapping lines through hodograph origin at the inclination angles equal to the constant stress vector directions. By applying the same approximation in the region about origin, each stress vector isocline is mapped from hodograph plane to another line through origin perpendicularly on physical plane.

$$|T^*| = \sqrt{\tau_{zx}^{*2} + \tau_{zy}^{*2}} = C$$

$$x^* = x^*(\tau_{zx}^*, \tau_{zy}^*)$$

$$y^* = y^*(\tau_{zx}^*, \tau_{zy}^*)$$

With $x^*, y^* \to 0$

$$|T^*| = \sqrt{\tau_{zx}^{*2} + \tau_{zy}^{*2}}\,\big|_{x^*,y^*\to 0} \approx \sqrt{(-y^*)^2 + x^{*2}}\,\big|_{x^*,y^*\to 0} = C$$

$$\tau_{zy}^* = \tau_{zx}^* \tan\gamma; \quad \to y^* \approx -x^* \cot\gamma \approx x^* \tan(\gamma - \pi/2)$$

where

$$\tau_{zx}^* = \frac{\partial \eta^*}{\partial x^*}\,\big|_{x^*,y^*\to 0} - y^* \approx -y^*\,\big|_{y^*\to 0}$$

$$\tau_{zy}^* = \frac{\partial \eta^*}{\partial y^*}\,\big|_{x^*,y^*\to 0} + x^* \approx x^*\,\big|_{x^*\to 0}$$

8.1.12 Stress Tensor Visualization

8.1.12.1 Principal Axes Trajectories and Load Transmission

To visualize stress phenomena and load transmissions in the beam, the principal states of stress tensor are required. Eigenvalues of a stress tensor represent extreme normal stresses (principal stresses) acting on principal planes as forces transmit along respective principal axes trajectories. Eigenvalues, eigenvectors and stress tensor invariants are evaluated below:

$$\text{Eigenvalues}: \sigma_1 = \sqrt{\tau_{zx}^{*2} + \tau_{zy}^{*2}} = |T^*|, \sigma_2 = -\sqrt{\tau_{zx}^{*2} + \tau_{zy}^{*2}} = -|T^*|, \sigma_3 = 0$$

$$\text{Eigenvectors}: \hat{e}_1 = \left[\frac{\tau_{zx}^*}{\sqrt{2}|T^*|} \quad \frac{\tau_{zy}^*}{\sqrt{2}|T^*|} \quad \frac{1}{\sqrt{2}}\right], \hat{e}_2 = \left[\frac{\tau_{zx}^*}{\sqrt{2}|T^*|} \quad \frac{\tau_{zy}^*}{\sqrt{2}|T^*|} \quad \frac{-1}{\sqrt{2}}\right],$$

$$\hat{e}_3 = \left[\frac{-\tau_{zy}^*}{|T^*|} \quad \frac{\tau_{zx}^*}{|T^*|} \quad 0\right]$$

Tensor invariants:

$$\Lambda_3 = \det|\tau_{ij}^*| = 0, \Lambda_1 = \sum \tau_{ii}^* = \sigma_1 + \sigma_2 + \sigma_3 = 0$$

$$\Lambda_2 = \frac{1}{2}(\tau_{ii}^* \tau_{kk}^* - \tau_{ij}^* \tau_{ij}^*) = \sigma_1 \sigma_2 = -|T^*|^2 \tag{8.32}$$

Such that,

$$\hat{e}_3 = \hat{e}_1 \times \hat{e}_2, \hat{e}_i \cdot \hat{e}_j = \delta_{ij}$$

$$\hat{e}_3 \cdot T^* = 0 \tag{8.33}$$

With one of the eigenvalues vanishing and a pair of equal eigenvalues in opposite sign, the three-dimensional tensor ellipsoid degenerates into a two-dimensional circle with the radius given by the equal eigenvalues. In Figure 8.11, load transmission is reflected by the size of the evolving circles equal to stress vector magnitude. The figure depicts deformation and stress patterns symmetric with respect to coordinate axes, which separate the beam section into four identical quadrants. Each quadrant separated by a diagonal line warps into concave or convex shapes reflecting respectively the state of compressive or tensile stresses in the beam. Principal axes $\hat{e}_3$ trajectories associated with the vanishing eigenvalue are planar curves on x^*y^*-plane. No loads transmit along $\hat{e}_3$ trajectories, which end on peripheral boundaries orthogonally in accordance with boundary conditions. This is attributed to the fact that $\hat{e}_3$ is perpendicular to stress vector T^* everywhere. Load transmission takes place on all surfaces spanned by $\hat{e}_1$ and $\hat{e}_2$ principal axes including all four peripheral surfaces. Figure 8.12 displays the distributions and orientations of principal axis $\hat{e}_1$ and principal axis $\hat{e}_2$. They are shown on two cross sections at different locations to avoid cluster. The three-dimensional trajectories of $\hat{e}_1$ and $\hat{e}_2$ are presented in Figures 8.13 and 8.14 respectively. Since,

$$\tau_{zx}^* < |T^*|; \tau_{zy}^* < |T^*|$$

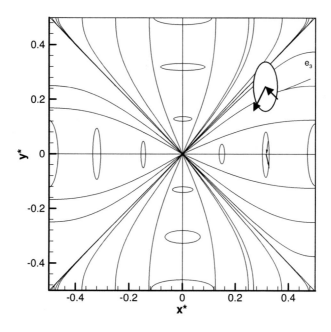

Figure 8.11 Principal axis $\hat{e}_3$ trajectories and evolving tensor circles

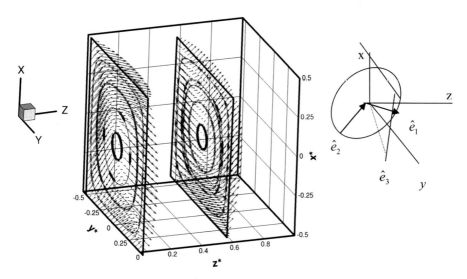

Figure 8.12 Orientations of principal axis $\hat{e}_1(z^* = 0)$ and principal axis $\hat{e}_2$ $(z^* = 0.6)$

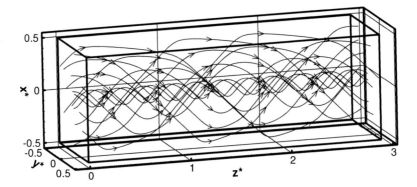

Figure 8.13 Principal axis $\hat{e}_1$ trajectories

The directions of principal axes $\hat{e}_1$ and $\hat{e}_2$ are dominated by the z^*-component giving rise to a system of spirals about z^*-axis. Tensile stresses (positive eigenvalue) and compressive stresses (negative eigenvalue) transmit along principal axes trajectories $\hat{e}_1$ and $\hat{e}_2$ and cause the beam section to warp into convex and concave shapes respectively in accordance with the warping function in Figure 8.2. Load transmission paths spiral about z-axis from peripheral towards the beam center like a conical helix in response to the twisting action by the applied torque. Load transmissions are predominant along beam peripherals and diminish to nil towards center axis. The transmission patterns tally with those of warping flux trajectories (Figures 8.3 and 8.6) and the looping patterns of stress trajectories (Figure 8.9). Furthermore, both metric coefficient and tensor trace (tensor first invariant) are required to vanish everywhere. The beam center is a degenerate point where all principal stresses vanish. The tensor ellipsoid degenerates into a point with infinite principal axes.

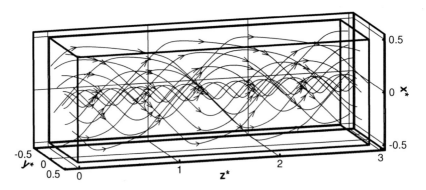

Figure 8.14 Principal axis $\hat{e}_2$ trajectories

8.1.13 Strain Energy Visualization

Strain energy is part of energy storage in the beam. Strain energy density in the twisting beam could be evaluated from definition as follows:

$$\Pi_\gamma = \frac{1}{2}[\sigma_x \varepsilon_x + \sigma_y \varepsilon_y + \sigma_z \varepsilon_z] + [\tau_{xy}\gamma_{xy} + \tau_{zx}\gamma_{zx} + \tau_{zy}\gamma_{zy}] \qquad (8.34)$$

$$= \frac{1}{4\mu}[\sigma_1^2 + \sigma_2^2 + \sigma_3^2]$$

$$= \frac{1}{2\mu}[\tau_{zx}^2 + \tau_{zy}^2] = \frac{|T|^2}{2\mu}$$

$$\text{Or } \Pi_\gamma^* = \frac{\Pi_\gamma}{\mu\theta_1^2 a^2} = \frac{1}{2}[(\tau_{zx}^*)^2 + (\tau_{zy}^*)^2] = \frac{|T^*|^2}{2}$$

The distribution of strain energy density in Figure 8.7 is independent of longitudinal direction. It is the same for every section. With vanishing first invariant, strain energy is contributed entirely by the deviatorial part in terms of second invariant, that is,

$$\Pi_\gamma' = 0$$
$$\Pi_\gamma'' = -2\mu I_2$$

$$\Pi_\gamma = \Pi_\gamma'' = -2\mu\varepsilon_1\varepsilon_2 = -\frac{\sigma_1\sigma_2}{2\mu} = -\frac{\Lambda_2}{2\mu} = \frac{|T|^2}{2\mu} \qquad (8.35)$$

Strain energy is proportional to shear stress magnitude square. Maximum strain energy density occurs at the mid points of four edges along beam peripheral. At these locations maximum shear stresses also occur there. One could observe and compare the energy storage patterns of strain energy and warping energy. Both appear to be concentrated along beam peripheral.

8.1.14 Energy Transfer Visualization

Strain energy transfer trajectories are presented to visualize energy transfer paths.

$$\frac{\partial}{\partial x^*}[\tau_{zx}^* u_z^*] + \frac{\partial}{\partial y^*}[\tau_{zy}^* u_z^*] + \frac{\partial}{\partial z^*}[\tau_{zx}^* u_x^* + \tau_{zy}^* u_y^*] - 2\Pi_\gamma^* = 0 \qquad (8.36)$$

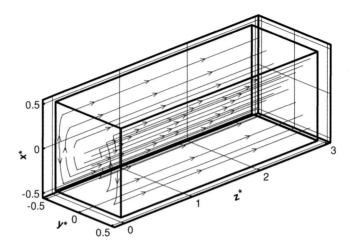

Figure 8.15 Distributions of strain energy transfer trajectories

The trajectory equation is written as:

$$\frac{dx^*}{\tau_{zx}^* u_z^*} = \frac{dy^*}{\tau_{zy}^* u_z^*} = \frac{dz^*}{\tau_{zx}^* u_x^* + \tau_{zy}^* u_y^*}$$

Results are presented in Figure 8.15. It is observed that energy transfers with warping distortions from the fix end ($z^* = 0$) towards free end (arrows). Transport paths are nearly parallel to z^*-axis. This implies warping energy represented by the third term in the energy transfer trajectory is a dominant part of the energy being transferred. They appear to correlate with displacements as well as the rotational axes presented next.

8.1.15 Visualization of Displacement and Rotation

Displacement trajectories and rotational axis field line trajectories are obtained from the trajectory equations, that is,

$$\frac{dx^*}{-z^* y^*} = \frac{dy^*}{z^* x^*} = \frac{dz^*}{\eta^*}$$

$$\frac{dx^*}{\Omega_x^*} = \frac{dy^*}{\Omega_y^*} = \frac{dz^*}{\Omega_z^*}$$

$$\text{where } \Omega^* = \frac{\Omega}{\theta_1 a} = \left[\frac{1}{2} \left[\frac{\partial \eta^*}{\partial y^*} - x^* \right] - \frac{1}{2} \left[y^* + \frac{\partial \eta^*}{\partial x^*} \right] z^* \right] \qquad (8.37)$$

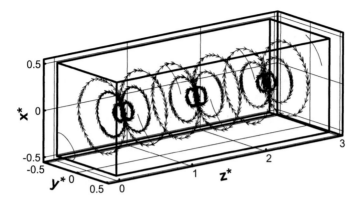

Figure 8.16 Displacement trajectories

Displacement trajectories displayed in Figure 8.16 resemble a spiral pattern. This is due to the combined effects of circular movements on x^*y^*-plane plus warping distortions along longitudinal z^*-axis. Figure 8.17 presents the orientations of the axis of rotations. There are rotations about all three coordinate axes driven by the applied torque. Rotation is highly complex and three-dimensional near the fixed end ($z^* = 0$) featuring a system of spirals. Away from this end, rotation axis trajectory changes rapidly into unidirectional. As the z^*-component of rotation becomes dominant, rotation is practically about longitudinal axis for the most part of the beam except the fixed end portion. One could observe if correlation exists between displacement, shear stress vector and principal axes trajectory from their distribution patterns.

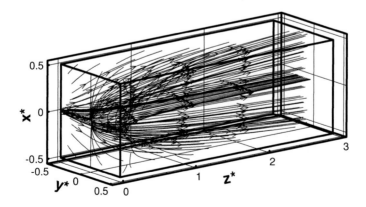

Figure 8.17 Orientations of axis of rotation

8.2 Example 8.2: Bending of a Cantilever Beam Subject to a Point Load

A narrow rectangular beam of uniform depth Δz is loaded at the free end by a point load, p as shown in Figure 8.18. The origin of a left-hand coordinate is placed at the point of intersection of beam's fixed end and neutral x-axis. The beam center axis coincides with x-axis. The y-axis and z-axis point downwards and out of the paper respectively.

The following model parameters are assumed:

$$L = 1.5m, \quad h = 0.5m, \quad p = 1000N, \quad I = 3e - 6m^4, \quad E = 2e11\,pa,$$
$$v = 0.12, \quad \Delta z << h, \quad 0 \le x \le L, \quad -h/2 \le y \le h/2$$

v is the Poisson's ratio, E is the modulus of elasticity. I is area moment of inertia about center axis.

8.2.1 Assumptions and Boundary Conditions

As a first approximation, bending of beams could be considered essentially two-dimensional with negligible body forces. That is, stresses will vary along the length of beam and through its width, but are assumed invariant through the depth (thickness in z-direction). The x-y plane remains parallel under loading. The displacement fields are written in terms of an arbitrary function as follows:

$$U = \begin{bmatrix} u_x & u_y & u_z \end{bmatrix}$$
$$U(0, 0, 0) = \begin{bmatrix} 0 & 0 & 0 \end{bmatrix} \tag{8.38}$$

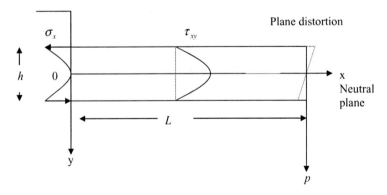

Figure 8.18 A cantilever beam loaded at one end

Displacement is required to vanish at origin. Furthermore,

$$\frac{\partial u_y}{\partial x}\Big|_{x=0, y=0} = 0 \tag{8.39}$$

Along the central axis (x-axis), the x-component normal stress vanishes and the central axis is in the state of maximum pure shear. The theory of elasticity then asserts that the directions of principal axes will assume 45 degrees to the center axis. This implies,

$$\sigma_x\Big|_{y=0} = 0, \quad \frac{\partial \tau_{xy}}{\partial y}\Big|_{y=0} = \frac{\partial \tau_{xy}}{\partial x}\Big|_{y=0} = 0$$

$$\frac{\partial^2 \tau_{xy}}{\partial y^2}\Big|_{y=0} < 0, \quad \rightarrow \tau_{xy}\Big|_{y=0} = \tau_{xy}\Big|_{max} \tag{8.40}$$

Shear stresses cross center axis perpendicularly and are symmetric with respect to center axis. Since shearing stress vanishes at the top and bottom free boundary surfaces, one of principal planes must coincide with the free boundaries and the other principal axes cross them at right angle. This means a free boundary is itself one of principal stress trajectories and principal stress crossing the free boundary satisfies the homogenous Neumann condition. As x-component normal stress becomes major principal stress at free boundary, y-component normal stress is required to vanish. Such that,

$$\sigma_x\Big|_{y=\pm h/2} + \sigma_y\Big|_{y=\pm h/2} = \sigma_1 + \sigma_2 = \sigma_1 \tag{8.41a}$$

$$\sigma_y\Big|_{y=\pm h/2} = \sigma_2 = 0 \tag{8.41b}$$

$$\frac{\partial \sigma_2}{\partial y}\Big|_{y=\pm h/2} = 0 \tag{8.41c}$$

$$\tau_{xy}\Big|_{y=\pm h/2} = 0 \tag{8.41d}$$

Consider beam's thickness is small and the stress fields are approximated as plane stresses with negligible stresses on z-face (cross section), that is, $\sigma_z = 0$ and $\varepsilon_z \neq 0$. Furthermore, condition expressed by Equation 8.41b is assumed to prevail everywhere in Saint-Venant's method. The stress tensor reduces to plane stress as follows:

$$\tau_{ij} = \begin{bmatrix} \sigma_x & \tau_{xy} & 0 \\ \tau_{yx} & 0 & 0 \\ 0 & 0 & 0 \end{bmatrix} \tag{8.42}$$

$$\Lambda_3 = \sigma_1 \sigma_2 \sigma_3 = \det\Big|\tau_{ij}\Big| = 0$$

$$\rightarrow \sigma_3 = 0$$

$$\Lambda_2 = \sigma_1\sigma_2 = \sigma_x\sigma_y + \sigma_x\sigma_z + \sigma_y\sigma_z - \tau_{xy}^2 - \tau_{yz}^2 - \tau_{xz}^2 = -\tau_{xy}^2 \qquad (8.43)$$

$$\Lambda_1 = \sigma_x + \sigma_y + \sigma_z$$

$$= \sigma_1 + \sigma_2 = \sigma_x$$

The x-component normal stress is then equal to stress first invariant. With non-vanishing first invariant, there is a change in volume and the stress fields are not solenoidal. The above assumptions and boundary conditions are useful in obtaining stress function solution.

8.2.2 Stress Function Solution

In the spirit of stress function approach, Airy stress function is used to describe stress fields in elastic deformation. It is common practice to select a suitable polynomial function with unknown coefficients to assume the role of Airy function. Because stress fields can be determined from beam bending theory, they are used to derive Airy stress function, φ directly.

$$\sigma_x = \frac{\partial^2\varphi}{\partial y^2} = -\frac{My}{I} = \frac{p(x-L)y}{I} \qquad (8.44)$$

$$\text{where } M = p(L-x) \qquad (8.45)$$

M is the bending moment about z-axis. Figure 8.19 depicts the distribution of x-component normal stress antisymmetric about center axis. x-component normal stress acts in the positive and negative x-axis directions in the upper and lower sections as the sign changes from positive to

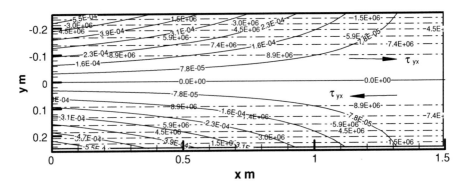

Figure 8.19 Distributions of shear stress (dash-dot) and x-component normal stress (line)

negative across center axis. It vanishes on the loading end surface as well as along center axis. By direct integration, we obtain,

$$\varphi = -\frac{pLy^3}{6I} + \frac{pxy^3}{6I} + f(x)y + g(x)$$

$f(x)$ and $g(x)$ are determined by the state of stress as follows:

$$\sigma_y = \frac{\partial^2 \varphi}{\partial x^2}$$

$$= \frac{\partial^2 [f(x)y]}{\partial x^2} + \frac{d^2 g(x)}{dx^2} = 0$$

$$\tau_{xy}\big|_{y=\pm h/2} = -\frac{\partial^2 \varphi}{\partial x \partial y}$$

$$= -\frac{py^2}{2I} - \frac{df(x)}{dx} = 0$$

$$\text{Or } \frac{df(x)}{dx} = -\frac{ph^2}{8I}$$

$$\rightarrow f(x) = -\frac{ph^2 x}{8I} + c_1$$

$$f(x)y = -\frac{ph^2 xy}{8I} + c_1 y$$

$$\frac{d^2 g(x)}{dx^2} = 0$$

$$\rightarrow g(x) = c_2 x + c_3$$

Since linear terms are not required in Airy stress function, they are set to vanish.

$$g(x) = c_2 x + c_3 = 0; \; c_1 y = 0$$

This leads to the following Airy stress function, which satisfies biharmonic equation:

$$\varphi = -\frac{p[4Ly^3 - 4xy^3 + 3h^2 xy]}{24I} \tag{8.46}$$

From Airy stress function, stress components are obtained accordingly:

$$\sigma_y = \frac{\partial^2 \varphi}{\partial x^2} = 0 \tag{8.47}$$

$$\tau_{yx} = -\frac{\partial \varphi}{\partial x \partial y}$$

$$= \frac{p}{2I} \left(\frac{h^2}{4} - y^2 \right) \tag{8.48}$$

$$\frac{\partial \tau_{yx}}{\partial y} \Big|_{y=0} = 0; \quad \frac{\partial^2 \tau_{yx}}{\partial y^2} = -\frac{p}{I} < 0$$

The stress fields satisfy the equation of equilibrium as well as the boundary conditions. Shear stress is an even function in y independent of x. It represents a quadratic curve and its distribution profiles resemble a parabolic shape symmetric with respect to center axis. Shear stresses cross center axis orthogonally at maximum magnitude. The boundary conditions are satisfied and shear stresses vanish on top and bottom as shown in Figures 8.18 and 8.19. In left-handed coordinates, positive shearing stresses acting on positive y-faces in the lower half section point towards negative x-axis. Positive shearing stresses acting on negative y-faces in the upper half section point towards positive x-axis. With shear stresses being positive everywhere, the directions of shear stresses are marked with arrows in Figure 8.19 giving rise to a clockwise couple about z-axis. The result tallies with the loading applied to the beam. Since $L \gg h$, deformations caused by shearing stresses are dominated by x-component displacement gradient.

$$\rightarrow \quad \left| \frac{\partial u_x}{\partial y} \right| > \left| \frac{\partial u_y}{\partial x} \right|$$

At any cross section, there is no axial loading in x-direction and the shear force is equal to the applied load as required by equilibrium condition. The results are consistent with those of beam bending theory.

$$F_x = \int_{-h/2}^{h/2} \sigma_x dy \Delta z = \frac{p}{2I}(x - L)y^2 \Delta z \Big|_{-h/2}^{h/2} = 0$$

$$F_y = \int_{-h/2}^{h/2} \tau_{xy} dy \Delta z = \int_{-h/2}^{h/2} \frac{p}{2I} \left(\frac{h^2}{4} - y^2 \right) dy \Delta z = p$$

$$M = -\int_{-h/2}^{h/2} \sigma_x y dy \Delta z = \frac{p(L-x)\Delta z h^3}{12I} = p(L-x)$$

where $I = \dfrac{\Delta z h^3}{12}$

From stress fields and Hooke's laws, the strain components could be found accordingly:

$$\frac{\partial u_x}{\partial x} = \varepsilon_x = \frac{\sigma_x}{E} - \frac{v(\sigma_y + \sigma_z)}{E} = \frac{p(x-L)y}{EI} \tag{8.49a}$$

$$\frac{\partial u_y}{\partial y} = \varepsilon_y = \frac{\sigma_y}{E} - \frac{v(\sigma_x + \sigma_z)}{E} = \frac{vp(L-x)y}{EI} \tag{8.49b}$$

$$\varepsilon_z = \frac{\sigma_z}{E} - v\left(\frac{\sigma_y}{E} + \frac{\sigma_x}{E}\right) = \frac{vp(L-x)y}{EI} \tag{8.49c}$$

$$\gamma_{yz} = \frac{\tau_{yz}}{2\mu} = 0 \tag{8.49d}$$

$$\gamma_{xz} = \frac{\tau_{xz}}{2\mu} = 0 \tag{8.49e}$$

$$\gamma_{xy} = \frac{\tau_{xy}}{2\mu} \tag{8.49f}$$

μ and λ are elastic constants. It is observed that z-component normal strain is independent of z and does not vanish in plane stress approximation. With the assumptions, $\sigma_y = \sigma_z = 0$, the strain first invariant is obtained and satisfies Laplace equation:

$$I_1 = \varepsilon_x + \varepsilon_y + \varepsilon_z = \frac{p(x-L)y(1-2v)}{EI} \tag{8.50a}$$

$$\Lambda_1 = (3\lambda + 2\mu)I_1$$

$$= \frac{EI_1}{1-2v} = \frac{p(x-L)y}{I} = \sigma_x \tag{8.50b}$$

The results and assumptions could be verified by substitution. Strain components are not independent and they are related by a set of Saint-Venant's equations of compatibility given below:

$$\frac{\partial^2 \varepsilon_x}{\partial y^2} + \frac{\partial^2 \varepsilon_y}{\partial x^2} = \frac{\partial^2 \gamma_{xy}}{\partial x \partial y}$$

$$\frac{\partial^2 \varepsilon_y}{\partial z^2} + \frac{\partial^2 \varepsilon_z}{\partial y^2} = \frac{\partial^2 \gamma_{yz}}{\partial y \partial z}$$

$$\frac{\partial^2 \varepsilon_x}{\partial z^2} + \frac{\partial^2 \varepsilon_z}{\partial x^2} = \frac{\partial^2 \gamma_{xz}}{\partial x \partial z}$$

$$2\frac{\partial^2 \varepsilon_x}{\partial y \partial z} = \frac{\partial}{\partial x}\left(-\frac{\partial \gamma_{yz}}{\partial x} + \frac{\partial \gamma_{zx}}{\partial y} + \frac{\partial \gamma_{xy}}{\partial z}\right)$$

$$2\frac{\partial^2 \varepsilon_y}{\partial x \partial z} = \frac{\partial}{\partial y}\left(-\frac{\partial \gamma_{zx}}{\partial y} + \frac{\partial \gamma_{xy}}{\partial z} + \frac{\partial \gamma_{yz}}{\partial x}\right)$$

$$2\frac{\partial^2 \varepsilon_z}{\partial y \partial x} = \frac{\partial}{\partial z}\left(-\frac{\partial \gamma_{xy}}{\partial z} + \frac{\partial \gamma_{yz}}{\partial x} + \frac{\partial \gamma_{zx}}{\partial y}\right)$$

It can be shown that the strain fields satisfy the above first five equations except the last one. In order to satisfy the severe compatibility requirement, it is necessary to introduce a small term being a function of z into Equation 8.49f, that is,

$$\gamma_{xy} = \frac{\tau_{xy}}{2\mu} + Az^2$$

$$\frac{\tau_{xy}}{2\mu} \gg Az^2 \tag{8.51}$$

This implies, in plane stress approximation, the beam is relatively thin compared to the other two-directions. The stress fields then remain approximately invariant in the z-direction in line with the assumption. Upon substitution into the last compatibility equation, the result gives,

$$A = \frac{vp}{EI} \tag{8.52}$$

The displacement components are then obtained by performing integrations on the above system of equations to result in:

$$u_x = \frac{p}{EI}\left[-\left(L - \frac{x}{2}\right)xy\right] + f(y) \tag{8.53a}$$

$$u_y = \frac{pv}{2EI}(L - x)y^2 + g(x) \tag{8.53b}$$

$$u_z = \frac{pv}{EI}(L - x)y \int \partial z \approx 0 \tag{8.53c}$$

$$\frac{df(y)}{dy} = \frac{vpy^2}{2EI} - \frac{(1+v)py^2}{EI} + \frac{(1+v)ph^2}{4EI}$$

$$\frac{dg(x)}{dx} = \frac{p}{EI}\left[Lx - \frac{x^2}{2}\right]$$

Above is integrated to yield,

$$f(y) = \frac{vpy^3}{6EI} - \frac{(1+v)py^3}{3EI} + \frac{(1+v)ph^2y}{4EI} + a$$

$$g(x) = \frac{p}{EI}\left[\frac{Lx^2}{2} - \frac{x^3}{6}\right] + b$$

The coefficients are determined by boundary conditions, that is,

$$a = b = 0$$

Hence,

$$u_x = \frac{p}{EI}\left[-\left(L - \frac{x}{2}\right)xy - \frac{(2+v)y^3}{6} + \frac{(1+v)h^2y}{4}\right] \qquad (8.54)$$

$$u_y = \frac{p}{EI}\left[\frac{v}{2}(L-x)y^2 + \frac{Lx^2}{2} - \frac{x^3}{6}\right] \qquad (8.55)$$

As z-component of displacement is considered negligible for a thin beam, the displacement fields have two non-zero components being a function of two spatial variables. There is a source in the displacement fields equal to the strain first invariant resulting in a volumetric change.

$$\nabla \cdot U = \frac{(1-v)p(x-L)y}{EI} = I_1$$

$$|U| = \sqrt{u_x^2 + u_y^2}$$

It can be shown by substitution that the above results satisfy the equation of equilibrium as well as the boundary conditions Equations 8.38 and 8.39. u_x produces surface distortions in opposite directions with respect to center axis. u_y measures beam deflections, which are in the same direction everywhere. The displacement trajectories are obtained from trajectory equation. Distributions of displacement trajectories and strength are presented in Figure 8.20. Maximum displacement occurs at loading end. The shape of beam deflections can be determined with $y = 0$ for the center axis.

$$u_y\big|_{y=0} = \frac{p}{EI}\left[\frac{Lx^2}{2} - \frac{x^3}{6}\right] \qquad (8.56)$$

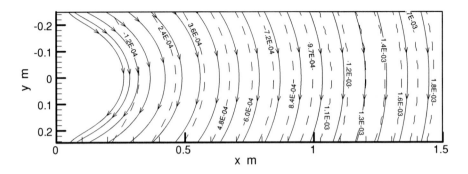

Figure 8.20 Distributions of displacement field line trajectories (arrows) and field strength (dash)

It is a third order curve in x and is referred to as beam deflection equation. The curvature of this line is given by:

$$\frac{\partial^2 u_y}{\partial x^2} = \frac{p}{EI}[L - x]$$

Curvature assumes maximum at the fixed end and diminishes linearly to nil at the loading end. To visualize plane surface distortions in x-direction due to bending, we arbitrarily choose planes at the middle section and at the loading end for comparisons.

$$u_x \big|_{x=L/2,L} \text{ and } u_y \big|_{x=L/2,L}$$

Both displacement components do not obey a simple linear relationship with spatial variables. If points are chosen from a simple shape like a plane, the ends of vectors can be connected to trace a deformed surface. Figure 8.21 presents surface distortions at the middle plane and at the loading end. The distortion patterns reflect the actions of tension and compression developed in the beam. As displacement fields are not solenoidal, there is a volumetric change except at center axis and vertical line $x = L$. Volumetric expansion caused by tension occurs in the upper section while compression is responsible for contraction in the lower section. For a given plane along longitudinal axis, the section surface distorts as a third order curve in y. As a result, plane sections do not remain plane on y-z plane.

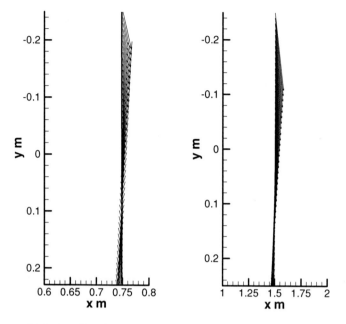

(a) Surface distortions at mid section   (b) Surface distortions at loading end

Figure 8.21 Surface (y-z plane) distortions at mid section and loading end. (a) Surface distortions at mid section (b) Surface distortions at loading end

A critical point occurs at the origin where displacements vanish. The gradient of displacement tensor is then used to visualize the critical point topology.

$$[\nabla U]\Big|_{\substack{x=0 \\ y=0}} = \begin{bmatrix} \dfrac{\partial u_x}{\partial x} & \dfrac{\partial u_y}{\partial x} \\[2mm] \dfrac{\partial u_x}{\partial y} & \dfrac{\partial u_y}{\partial y} \end{bmatrix}\Bigg|_{\substack{x=0 \\ y=0}} \qquad (8.57)$$

$$= \begin{bmatrix} 0 & 0 \\[2mm] \dfrac{p(1+v)h^2}{4EI} & 0 \end{bmatrix}$$

$$\lambda_1 = \lambda_2 = 0$$

Above matrix yields vanishing eigenvalues. The critical point is a degenerate point.

8.2.3 Strain/Stress Tensor Visualization

8.2.3.1 Strain Tensor Contraction and First Invariant

Strain tensor is useful to study the external characteristics of deformation under loading. Although the z-component normal strain does not vanish in plane stress approximation, it is independent of z. It is possible to visualize the deformation process by strain vector trajectories similar to streamlines. Firstly, the strain tensor is contracted to yield a strain vector at a point. The result is a planar vector on x-y plane.

$$\gamma = \begin{bmatrix} \gamma_x & \gamma_y & \gamma_z \end{bmatrix} = \nabla \cdot \gamma_{ij} \tag{8.58}$$

$$\text{where } \gamma_x = \frac{\partial \varepsilon_x}{\partial x} + \frac{\partial \gamma_{xy}}{\partial y} = -\frac{p}{EI}(1+2v)y$$

$$\gamma_y = \frac{\partial \gamma_{xy}}{\partial x} + \frac{\partial \varepsilon_y}{\partial y} = \frac{p}{EI}(L-x)v$$

$$\gamma_z = 0$$

Since divergence of strain vector vanishes, strain vector flux trajectory represented by an integral curve is obtained by integration, that is,

$$\psi_\gamma = \int (-\gamma_y \partial x + \gamma_x \partial y) + C \tag{8.59}$$

$$= \frac{p}{EI} \left(\frac{v}{2}(x^2 - 2xL) - \frac{(1+2v)}{2} y^2 \right) + C$$

$$\frac{\partial \psi_\gamma}{\partial y} \bigg|_{y=0} = 0$$

$$\frac{\partial \psi_\gamma}{\partial x} \bigg|_{x=L} = 0$$

Figure 8.22 depicts the distributions of strain vector trajectories of conic curves. The transverse axes of hyperbola coincide with center axis and vertical line $x = L$ respectively. A system of hyperbolic trajectories crosses center axis orthogonally, which serves as line of symmetry. The patterns correlate closely with those of displacements (Figure 8.20) and surface distortions (Figure 8.21). The conjugate system of hyperbolic trajectories cross the vertical line $x = L$ at right angle. It is also symmetric with respect to center axis. The strain vector magnitude contours are presented in Figure 8.30 resembling a system of ellipses (left half portions) with center at $(x = L, y = 0)$. This point corresponds to the maximum beam deflection displacement and is also a critical point where strain vector vanishes. The amount of strain flux bounded

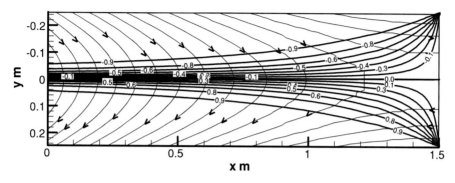

Figure 8.22 Distributions of strain vector trajectories (arrows) and metric coefficient (dotted)

between two trajectories could be conveniently evaluated from the difference of trajectory values.

Strain/stress first invariant is a measure of scalar source and they are within a proportional constant. Figure 8.23 presents distributions of strain first invariant (strain isopachs) to display the state of deformation. The distribution patterns suggest that volume changes are the greatest at fix end's top and bottom corners and propagate towards the center axis. Strain first invariant vanishes along center axis and assumes maximum at fixed end's top and bottom corners in accordance with maximum principle for harmonic function. Strain first invariant assumes positive in upper section indicating that there is a volumetric expansion caused by tension. It becomes zero and then changes to negative when crossing center axis. As a result, volumetric contraction occurs in lower section subject to compression. The state of beam deformation agrees with those presented in Figure 8.21. It is noted from Equation 8.49c

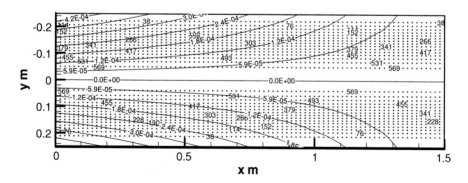

Figure 8.23 Distributions of strain first invariant I_1 (line) and deviatorial part of strain energy Π_γ'' (dotted)

that z-component normal strain exists giving rise to an expansion/contraction in z-direction. This is because stress/strain in one direction is accompanied by effects in perpendicular directions. Normal stresses cause a change in normal strain perpendicular to the plane depending on Poisson's ratio. In this example z-component normal strain varies in proportion to strain first invariant (or σ_x). It describes thickness variations as its sign and value may change from point to point. There is a contraction of thickness in the upper section and expansion in the lower half section in response to the loading.

$$\varepsilon_z = \frac{\sigma_z}{E} - v\left(\frac{\sigma_y}{E} + \frac{\sigma_x}{E}\right)$$

$$= -v\frac{\Lambda_1}{E} = -v\frac{\sigma_x}{E} \tag{8.60}$$

8.2.3.2 Principal States and Visualization

Principal states of stress tensors are useful in the study of the internal characteristic of deformation in terms of stress phenomena and load transmissions. Eigenvalues of a stress tensor represent extreme normal stresses (principal stresses) acting on principal planes as forces transmit along respective principal axes trajectories. Principal states of a plane stress tensor are determined as per Equations 6.31 and 6.32. Distributions of principal stresses are presented in Figure 8.24. Along top and bottom free boundaries, shear stresses are zero and normal stresses become principal stresses. The results tally with the analysis of Equation 8.41a.

$$\text{Upper half section}: \ \sigma_x \bigg|_{\substack{x \leq L \\ -h/2 < y < 0}} > 0 \ ; \ \sigma_1 \bigg|_{\substack{x \leq L \\ -h/2 < y < 0}} = \frac{\sigma_x}{2} + \sqrt{\frac{\sigma_x^2}{4} + \tau_{xy}^2} > 0$$

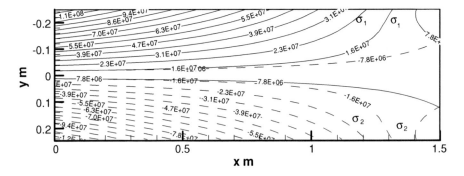

Figure 8.24 Distributions of principal stresses σ_2 (dashed) and σ_1 (solid)

By convention, x-component normal stress and principal stress are tensile (positive) in the upper section. Tensile stresses are dominant in the upper section and maximum tensile stresses occur at fixed end's top corner ($x = 0$, $y = -h/2$) where principal stress attains the greatest magnitude. Along center axis, principal tensile stress and principal compressive stress are equal to the maximum shear stress numerically.

$$\text{Center axis}: \sigma_1 \big|_{y=0} = -\sigma_2 \big|_{y=0} = \tau_{yx} \big|_{y=0} = \tau_{yx} \big|_{\max} = \frac{\sigma_1 - \sigma_2}{2}$$

Stress phenomena reverse when crossing center axis into lower half section where compressive (negative) stress becomes dominant. Maximum compressive stresses occur at fixed end's bottom corner ($x = 0$, $y = h/2$). The gradient mapping of principal stresses will depict the paths from two opposite free surfaces towards the center axis along which principal stresses change most rapidly.

$$\text{Lower half section}: \sigma_x \bigg|_{\substack{x \leq L \\ h/2 > y > 0}} < 0 \; ; \; \sigma_2 \bigg|_{\substack{x \leq L \\ -h/2 < y < 0}} = \frac{\sigma_x}{2} - \sqrt{\frac{\sigma_x^2}{4} + \tau_{xy}^2} < 0$$

Distributions of maximum shear stress and mean normal stress are presented in Figure 8.25. The distributions of maximum shear stress trace the paths of slip lines, which could be visualized by bisecting the principal axes trajectory in Figure 8.26. The distributions of mean normal stress are the same as those of strain first invariant in Figure 8.23 within a proportional constant.

Principal axes trajectories are presented in Figure 8.26. Trajectory tangents define the directions of load transmissions accompanied by decreasing principal stresses. Tensile stresses propagate along principal axis $\hat{e}_1$ trajectories

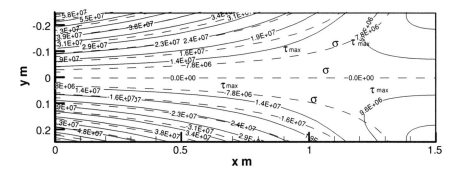

Figure 8.25 Distributions of the maximum shear stress (solid lines) and mean normal stress (dotted lines)

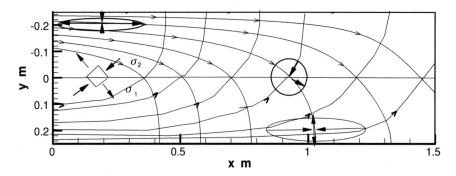

Figure 8.26 Principal axes trajectories with static display of evolving tensor ellipse along a tensile principal axis trajectory $\hat{e}_1$

(clockwise arrows) from its peak value at upper fix end across center axis and end perpendicularly on the bottom free surface where they vanish. The x-component normal stress then assumes principal compressive stress and acts parallel to bottom free surface, which is a principal axis trajectory $\hat{e}_2$. Compressive stresses propagate along principal axis $\hat{e}_2$ trajectories (counterclockwise arrows) from its peak value at lower fix end across center axis and terminate perpendicularly on the top free surface where they vanish. The x-component normal stress becomes tensile principal stress and acts parallel to top free surface, which is also a principal axis trajectory $\hat{e}_1$. This is attributed to the fact that off-diagonal components in the stress tensor vanish on both free surfaces in accordance with the boundary conditions. The right top and bottom corners are degenerate points where both principal stresses are equal and vanish. As a result, the tensor ellipse degenerates into a point with infinite principal axes. Principal axes trajectories converge/squeeze as the stress field intensifies towards fix end due to constraints of movement. The lines that bisect the right angles between principal directions are lines of maximum shear (slip lines). Slip lines can be constructed to visualize the paths of maximum shear stresses depicted in Figure 8.25 and to assess failure modes. For example, center axis is one of such lines.

Figure 8.26 displays the evolving ellipses at selected points along a tensile principal axis trajectory. The lengths of the ellipse axes are a measure of principal stresses. A sequence of stress tensor transformation is summarized in Table 8.1a, b. Similar evolving patterns are obtained along the compressive principal axes trajectories (not shown to avoid cluster). Metric coefficient is presented in Figure 8.22. Metric coefficient vanishes along the center axis. This corresponds to a tensor circle with compression equal to tension. Along the top surface, metric coefficient assumes positive unity with a highly elongated ellipse indicating that tension is dominant. Along the bottom surface, metric coefficient becomes negative unity with highly elongated ellipse

Table 8.1a Summary of tensor transformation along a tensile principal axes trajectories

Line →	Ellipse →	Circle →	Ellipse →	Line
Tension	Tension	Compression	Compression	Compression
$c_l = 1$	> compression	= tension	> tension	$c_l = -1$
	$1 > c_l > 0$	$c_l = 0$	$-1 < c_l < 0$	
$\sigma_1 = \sigma_x > 0,$				$\sigma_1 = \sigma_x < 0,$
$\sigma_2 = \sigma_y = 0$	$|\sigma_1| > |\sigma_2|$	$\sigma_1 = -\sigma_2$	$|\sigma_1| > |\sigma_2|$	$\sigma_2 = \sigma_y = 0$
$|\sigma_1| > |\sigma_2|$	$\sigma_2 < 0, \sigma_1 > 0$	$= \tau_{max}$	$\sigma_2 > 0, \sigma_1 < 0$	$|\sigma_1| > |\sigma_2|$
$y = -h/2$	$-h/2 < y < 0$	$y = 0$	$h/2 > y > 0$	$y = h/2$

implying compression is dominant. The evolving tensor ellipse patterns along a principal axis trajectory provide a comprehensive source of information about stress phenomena.

8.2.4 Energy Storage and Transfer Visualization

Work done by external forces is converted into strain energy storage in the system given by the strain function.

$$\Pi_\gamma = \frac{1}{2}[\sigma_x \varepsilon_x + \sigma_y \varepsilon_y + 2\tau_{xy}\gamma_{xy}]$$

$$= \frac{p^2}{2EI^2}\left[(L-x)^2 y^2 + (1+v)\left(\frac{h^2}{4} - y^2\right)^2\right]. \tag{8.61}$$

Strain energy density distributions are presented in Figure 8.27. The distribution patterns are symmetric with respect to center axis. It is seen that strain energy assumes maximum at two corners of fixed end section where stress is the greatest. Strain energy vanishes at two right corners of degenerate points.

Table 8.1b Summary of tensor transformation along compressive principal axes trajectories

Line →	Ellipse →	Circle →	Ellipse →	Line
Compression	Compression	Compression	Tension	Tension
$c_l = -1$	> Tension	= tension	> compression	$c_l = 1$
	$-1 < c_l < 0$	$c_l = 0$	$1 > c_l > 0$	
$\sigma_1 = \sigma_x < 0,$				$\sigma_1 = \sigma_x > 0,$
$\sigma_2 = \sigma_y = 0$	$|\sigma_1| > |\sigma_2|$	$\sigma_1 = -\sigma_2$	$|\sigma_1| > |\sigma_2|$	$\sigma_2 = \sigma_y = 0$
$|\sigma_1| > |\sigma_2|$	$\sigma_2 > 0, \sigma_1 < 0$	$= \tau_{max}$	$\sigma_2 < 0, \sigma_1 > 0$	$|\sigma_1| > |\sigma_2|$
$y = h/2$	$h/2 > y > 0$	$y = 0$	$-h/2 < y < 0$	$y = -h/2$

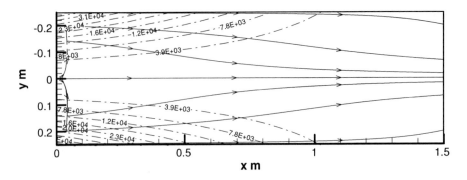

Figure 8.27 Distributions of strain energy (dash dot) and strain energy transfer trajectories (arrow)

This is attributed to the fact that the stress tensor vanishes at these two points. The first invariant square provides a measure of the isotropic part of strain energy. The deviatorial part of strain energy is given in terms of tensor second invariant.

$$\text{Isotropic part}: \Pi'_\gamma = \frac{1}{2}(\lambda + 2\mu)I_1^2$$

$$\text{Deviatorial part}: \Pi''_\gamma = -2\mu I_2 = -2\mu\varepsilon_1\varepsilon_2 \tag{8.62}$$

$$= -\frac{\sigma_1\sigma_2}{2\mu} = -\frac{\Lambda_2}{2\mu} = \frac{\tau_{xy}^2}{2\mu}$$

Distributions of the deviatorial part of strain energy are presented in Figure 8.23. This part of energy density is contributed by shearing stresses and attains maximum along the center axis. It decreases uniformly towards top and bottom surfaces and vanishes there. On the contrary, strain energy due to isotropic part vanishes along the center axis and assumes maximum on free surfaces. Such information is useful in the study of energy storage processes associated with dilation distortion or shear deformation.

Strain energy transfer trajectories are used to describe energy transfer paths in terms of energy flux vector. They are obtained from the trajectory equation below:

$$\frac{dy}{dx} = \frac{\tau_{xy}u_x}{\sigma_x u_x + \tau_{xy}u_y} \tag{8.63}$$

The results are presented in Figure 8.27. Energy transfers primarily from fix end to loading end. Energy is also transferred from free surfaces towards origin locally near the fix end portion. Energy transfer trajectories appear to have a close correlation with principal stress distributions in Figure 8.24 which depicts that principal stresses decrease in the direction of energy transfer. One could also display the arrow icons of displacement vector, strain vector, principal axes and strain energy transfer trajectory at a point simultaneously to study the deformation process.

8.2.5 Mapping of Field Patterns

Mappings of strain vector trajectories from the physical plane onto displacement hodograph plane and strain vector hodograph are presented in Figures 8.28 and 8.29 respectively. The critical point at z-plane origin is mapped to origins in these hodograph planes. The distribution patterns in Figure 8.28

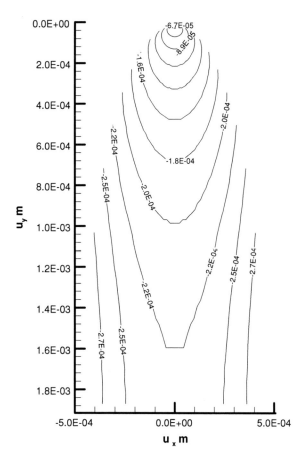

Figure 8.28 Mapping of strain vector trajectory on displacement hodograph

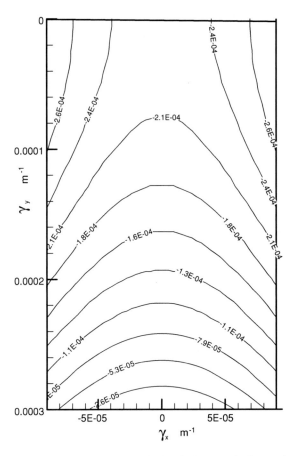

Figure 8.29 Mapping of strain vector trajectories on strain vector hodograph

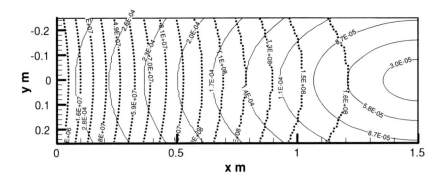

Figure 8.30 Distributions of rotation (dotted) and strain vector magnitude contours (line) on physical plane

are symmetric with respect to u_y-axis (axis of symmetry). Strain vector trajectories in the upper and lower section of the physical plane are mapped to the right and left of line of symmetry. The distribution patterns in Figure 8.29 are symmetric with respect to γ_y-axis (axis of symmetry). Strain vector trajectories in the upper and lower section are mapped to the right and left of line of symmetry respectively.

$$\psi_\gamma(x, y) = \psi_\gamma[x(u_x, u_y), y(u_x, u_y)] = \psi_\gamma(u_x, u_y)$$
$$\psi_\gamma(x, y) = \psi_\gamma[x(\gamma_x, \gamma_y), y(\gamma_x, \gamma_y)] = \psi_\gamma(\gamma_x, \gamma_y)$$
$$x = x(u_x, u_y)$$
$$y = y(u_x, u_y)$$
$$x = x(\gamma_x, \gamma_y)$$
$$y = y(\gamma_x, \gamma_y)$$

The hodograph representation and mapping techniques could be applied to visualize loci of constant strain vector magnitude by mapping a circle of radius C about origin on strain vector hodograph plane. With the aid of Equation 8.58, we obtain the equation of position hodograph representation of constant strain vector magnitude, which describes an ellipse on physical plane in terms of strain vector magnitude C. The ellipse center is located at $(L, 0)$ and has its major and minor axis coinciding with x-axis and line $x = L$. A system of ellipses is generated by varying parameter C and the results confirm those presented in Figure 8.30.

$$\gamma_x^2 + \gamma_y^2 = C^2$$
$$\rightarrow \frac{(L - x)^2}{a^2} + \frac{y^2}{b^2} = 1$$

where

$$a = C\sqrt{\frac{EI}{p\nu}}$$
$$b = C\sqrt{\frac{EI}{p(1 + 2\nu)}}$$

The mapping process may be carried over to strain vector isoclines. By mapping a straight line (strain vector isoclines) through origin at an inclination angle γ equal to the constant strain vector direction, we obtain the equation of position hodograph representation of isoclines by another straight line through point $(L, 0)$ on physical plane (Figure 8.22). The inclination angle is defined with respect to positive γ_x-axis clockwise in left hand coordinate and assumes the role of a parametric constant. For example, the center axis is one of isoclines that define the constant strain vector direction at an inclination angle $\gamma = \pi/2$.

$$\gamma_y = \gamma_x \tan \gamma$$

$$\rightarrow y = \frac{v}{(1 + 2v)\tan \gamma} x - \frac{Lv}{(1 + 2v)\tan \gamma}$$

Likewise, the techniques could be applied to trace loci of constant displacement magnitude or displacement isoclines on physical plane by mapping circles or lines about origin on displacement hodograph in Figure 8.28. The circle radius and isoclines inclination angle correspond to constant displacement magnitude and constant displacement direction respectively.

8.2.6 Rotation

Rotation vector is obtained by taking the curl of displacement fields. In two-dimensional fields, it has only one nonzero z-component and the axis of rotation is parallel to z-axis.

$$\Omega = \frac{1}{2} \nabla \times U = \begin{bmatrix} 0 & 0 & \Omega_z \end{bmatrix} \tag{8.64}$$

$$\text{where } \Omega_z = \frac{1}{2}\left(\frac{\partial u_y}{\partial x} - \frac{\partial u_x}{\partial y}\right)$$

$$= \frac{p}{2EI}\left(2xL - x^2 + y^2 - \frac{(1+v)h^2}{4}\right) \tag{8.65}$$

$$\frac{\partial \Omega_z}{\partial y}\bigg|_{y=0} = 0; \quad \frac{\partial \Omega_z}{\partial x}\bigg|_{x=L} = 0$$

Distribution of rotation contours is presented in Figure 8.30. Rotation contours resemble a system of hyperbolas with center $(L, 0)$ and transverse axis coinciding with center axis. The rotation contours cross center axis at right angle serving as the axis of symmetry. Rotation is positive in clockwise direction and assumes maximum value at boundary right corner points.

8.3 Example 8.3: Squeezing Flow and Vorticity Transport

Figure 8.31 illustrates the two-dimensional slow viscous motions of an incompressible fluid between two horizontal plates, which travel at a constant velocity V_0 towards each other. A channel is formed by two plates of length $2L$ and a width of $2h$. Squeezing flows propagate within the channel and change directions due to obstacles. The flow phenomenon is called Stefan's squeezing flow. For slow motions, the system is considered quasi steady and viscous creeping flow approximation is assumed. Potential energy and kinetic energy are assumed negligible in comparison with other mechanical energy components. Cartesian coordinate origin and x-axis are placed to coincide with channel center and center axis respectively to take advantage of symmetry. The coordinate axes are lines of symmetry and origin is a point of symmetry. The model non-dimensional parameters and physical dimensions are defined as follows:

$$v_x^* = \frac{v_x}{V_0}, \ v_y^* = \frac{v_y}{V_0}, \ p^* = \frac{(p - p_0)}{\rho V_0^2}, \ x^* = \frac{x}{h}, \ y^* = \frac{y}{h}, \ z^*$$

$$= \frac{z}{h}, \ \sigma_x^* = \frac{\sigma_x}{\rho V_0^2}, \ \sigma_y^* = \frac{\sigma_y}{\rho V_0^2}$$

$$\tau_{xy}^* = \frac{\tau_{xy}}{\rho V_0^2}, \ \nabla^{*2} = h^2 \left(\frac{\partial^2}{\partial x^2} + \frac{\partial^2}{\partial y^2} \right), \ \nabla^* = h \left[\frac{\partial}{\partial x} \ \frac{\partial}{\partial y} \right]$$

$$= \left[\frac{\partial}{\partial x^*} \ \frac{\partial}{\partial y^*} \right], \ \mathrm{Re} = \frac{\rho V_0 h}{\mu}, \ \omega^* = \frac{\omega h}{V_0}$$

$$\psi^* = \frac{\psi}{V_0 h}, \ \Pi^* = \frac{\Pi \rho h^2}{\mu V_0^2}, \ \Pi''^* = \frac{\Pi'' \rho h^2}{\mu V_0^2}, \ [\dot{\gamma}^*] = [\dot{\gamma}] \frac{h}{V_0} = -\frac{\mathrm{Re}[\tau^*]}{2}.$$

$$\mathrm{Re} = 1, h = 0.25m, L = 0.75m \cdot \frac{L}{h} = 3, \ -\frac{L}{h} \le x^* \le \frac{L}{h}, \ -1 \le y^* \le 1$$

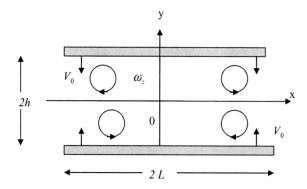

Figure 8.31   A fluid flow channel formed by two opposite moving plates

8.3.1 Governing Equation

Consider bidirectional flows of incompressible fluids between two horizontal planes. The equation of continuity is:

$$\frac{\partial v_x^*}{\partial x^*} + \frac{\partial v_y^*}{\partial y^*} = 0$$

v_x^* and v_y^* are velocity components. The flow fields are isochronic everywhere and could be modeled by Stokes flows, which describe force balance between pressure and friction:

$$\nabla^* p^* = \frac{\nabla^{*2} V^*}{Re}$$

$$= -\frac{\nabla^* \times \omega^*}{Re} \tag{8.66}$$

$$x\text{-direction} : \quad -\frac{\partial p^*}{\partial x^*} + \frac{1}{Re}\left[\frac{\partial^2 v_x^*}{\partial x^{*2}} + \frac{\partial^2 v_x^*}{\partial y^{*2}}\right] = 0 \tag{8.67}$$

$$y\text{-direction} : \quad -\frac{\partial p^*}{\partial y^*} + \frac{1}{Re}\left[\frac{\partial^2 v_y^*}{\partial x^{*2}} + \frac{\partial^2 v_y^*}{\partial y^{*2}}\right] = 0 \tag{8.68}$$

$$z\text{-direction} : \quad \frac{\partial p^*}{\partial z^*} = 0 \tag{8.69}$$

$$\rightarrow p^* = p^*(x^*, y^*)$$

8.3.2 Boundary Conditions

In addition to domain geometric symmetry, the boundary conditions also show the same type of symmetry about coordinate axes.

$$v_x^*(x^*, \pm 1) = 0$$
$$v_y^*(x^*, \pm 1) = \mp 1$$
$$\frac{\partial v_y^*}{\partial y^*}\Big|_{y^*=\pm 1} = 0$$
$$p^*(\pm L/h, \pm 1) = 0$$

8.3.3 Solution and Analysis

To decouple flow fields and pressure fields, we take divergence of Stokes equation to result in,

$$\frac{\partial^2 p^*}{\partial x^{*2}} + \frac{\partial^2 p^*}{\partial y^{*2}} = 0 \tag{8.70}$$

Pressure field is harmonic in Stokes flow. Solutions to Laplace equation hinge on the selection of an appropriate function which satisfies the given boundary conditions. The pressure field developed in the channel could be described by the equipotential obtained from corner flow within a constant to reflect the kind of symmetry in the system, that is,

$$p^* = C_1(x^{*2} - y^{*2}) + C \tag{8.71}$$

$$p^*(x^*, y^*) = p^*(-x^*, y^*) = p^*(x^*, -y^*) = p^*(-x^*, -y^*) = -p^*(y^*, x^*)$$

Jaeger (1969) presented a technique to solve the squeezing flow problem and a solution for $C_1 = -\frac{k}{2}$. The constants k and C are to be determined by boundary conditions. Once pressure field is available, velocity field and other quantities of interest can be obtained accordingly. In view of the situation that $L \gg h$, the order of magnitude analysis and velocity boundary layer approximations suggest viscous effects are primarily due to velocity gradients in y-direction. Such that,

$$\frac{\partial v_y^*}{\partial y^*} \gg \frac{\partial v_y^*}{\partial x^*} \tag{8.72}$$

This approximation implies y- component velocity does not vary with x and hence,

$$v_y^* = f(y^*)$$

Equation 8.68 reduces to:

$$-\frac{\partial p^*}{\partial y^*} + \frac{1}{Re} \frac{\partial^2 v_y^*}{\partial y^{*2}} = 0$$

Perform integrations and apply boundary conditions to result in:

$$v_y^* = \frac{Rek y^{*3}}{6} + Ay^* + B$$

$$k = \frac{3}{Re}; \ A = -1.5; \ B = 0$$

Following the technique of separation of variables, a product function is assumed for the solution of x-component velocity:

$$v_x^* = g(x^*)f(y^*)$$

By observations, x-component velocity exhibits point-symmetry about origin. Solution is conveniently expressed by an odd function in x. Furthermore, the order of magnitude analysis and approximation implies,

$$\frac{\partial^2 v_x^*}{\partial y^{*2}} \gg \frac{\partial^2 v_x^*}{\partial x^{*2}}$$

With these considerations, a first order odd function is selected such that,

$$g(x^*) = x^*$$

$$\frac{\partial^2 v_x^*}{\partial x^{*2}} = 0$$

$$-\frac{\partial p^*}{\partial x^*} + \frac{1}{Re}\frac{\partial^2 v_x^*}{\partial y^{*2}} = 0$$

$$\text{Or } \frac{\partial^2 v_x^*}{\partial y^{*2}} = -Re\,kx^*$$

$$\text{Hence, } v_x^* = -\frac{Re\,kx^* y^{*2}}{2} + y^* f(x^*) + h(x^*)$$

By imposing the boundary conditions, we have

$$f(x^*) = 0, \ h(x^*) = \frac{Re\,kx^*}{2} = 1.5x^*$$

$$C = \frac{1.5}{Re}\left[\left(\frac{L}{h}\right)^2 - 1\right]$$

$$v_x^* = 1.5x^*(1 - y^{*2}) \tag{8.73}$$

$$v_y^* = 0.5y^*(y^{*2} - 3) \tag{8.74}$$

By substitution, the flow fields satisfy continuity equation. Since the curl of velocity does not vanish, velocity equipotential does not exist. From velocity fields, stream function is found by integrating, that is,

$$\psi^* = \frac{x^* y^*(3 - y^{*2})}{2} \tag{8.75}$$

$$\frac{\partial \psi^*}{\partial y^*}\Big|_{y^*=\pm 1} = 0$$

The stream function satisfies biharmonic equation. Figure 8.32 shows the distributions of streamline trajectories symmetric with respect to origin. They are antisymmetric about coordinate axes. Dividing streamline $\psi^* = 0$ passes through origin and splits there to coincide with coordinate axes. It separates

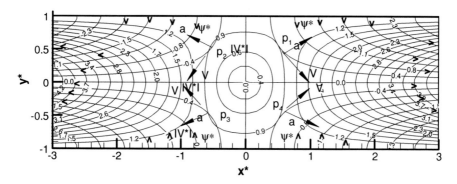

Figure 8.32 Distributions of streamlines and flow field strength on physical plane

the fields into each quadrant and serves as asymptotes for nearby trajectories. Streamlines emanate from two plates perpendicularly and may be represented approximately by a system of hyperbolas.

Flow velocity increases as streamlines squeeze towards channel exit planes. This is reflected by the distributions of flow field strength presented in Figure 8.32. Flow field strength is maximum at points where center line intersects channel exit planes. Kinetic energy density is given by:

$$K = \frac{V^2}{2}$$

$$\text{Or } K^* = \frac{K}{V_0^2} = \frac{v_x^{*2} + v_y^{*2}}{2}$$

Distributions of kinetic energy density are presented in Figure 8.33. The field patterns resemble closely to that of flow field strength depicted in Figure 8.32. Gradient mapping of kinetic energy could be used to visualize the directions along which kinetic energy change most rapidly. Volumetric flow rates per unit depth bounded between two streamlines are given by:

$$Q^* = \frac{\Delta \psi}{V_0 h} = (\psi_1^* - \psi_2^*) \tag{8.76}$$

The pressure field is a system of conic curves given by Equation 8.71, namely,

$$p^* = \frac{3(y^{*2} - x^{*2} - 1 + (L/h)^2)}{2\text{Re}} \tag{8.77}$$

$$\frac{\partial p^*}{\partial y^*}\Big|_{y*=0} = \frac{\partial p^*}{\partial x^*}\Big|_{x*=0} = 0$$

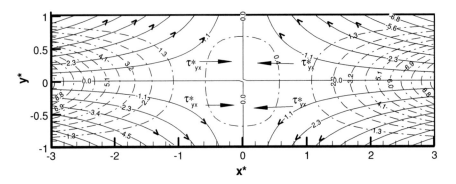

Figure 8.33 Distributions of kinetic energy density contours (dash dot) and stress vector trajectories (arrow)

$$\frac{\partial^2 p^*}{\partial x^{*2}}\Big|_{x*=0} < 0 ; \quad p^*(0, \pm 1) = p^*_{max} = \frac{3(L/h)^2}{2Re} \tag{8.78a}$$

$$\frac{\partial^2 p^*}{\partial y^{*2}}\Big|_{y*=0} > 0 ; \quad p^*(\pm L/h, 0) = p^*_{min} = -\frac{3}{2Re} \tag{8.78b}$$

Pressure contours shown in Figure 8.34 cross both coordinate axes at right angles and are symmetric with respect to these axes as well as origin. It is antisymmetric with respect to lines $y^* = \pm x^*$. The pressure contours consist of a family of hyperbolas and a pair of asymptotes $y^* = \pm x^*$. Maximum pressure occurs at the centers of two plates ($x^* = 0$, $y^* = \pm 1$). Minimum pressure is at the centers of two channel exit planes ($x^* = \pm L/h$, $y^* = 0$) where velocity attains maximum. These extreme pressure values are on the boundary for harmonic function.

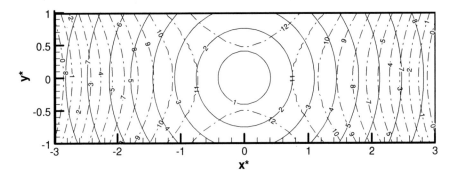

Figure 8.34 Distributions of pressure contours (dash dot) and stress vector magnitude contours (line) on the physical plane

A critical point occurs at origin where velocity vanishes. The gradient of velocity tensor at this point yields,

$$\left| \nabla^* V^* \right| \Big|_{\substack{x*=0 \\ y*=0}} = \begin{bmatrix} \dfrac{\partial v_x^*}{\partial x^*} & \dfrac{\partial v_y^*}{\partial x^*} \\[2mm] \dfrac{\partial v_x^*}{\partial y^*} & \dfrac{\partial v_y^*}{\partial y^*} \end{bmatrix} \Big|_{\substack{x*=0 \\ y*=0}}$$

$$= \begin{bmatrix} 1.5 & 0 \\ 0 & -1.5 \end{bmatrix}$$

Above symmetric matrix yields the following eigenvalues and eigenvectors,

$$\lambda_1 = 1.5; \ \hat{e}_1 = \pm \begin{bmatrix} 1 & 0 \end{bmatrix}$$
$$\lambda_2 = -1.5; \ \hat{e}_2 = \pm \begin{bmatrix} 0 & 1 \end{bmatrix}$$
$$\lambda_1 \lambda_2 < 0$$

The critical point is classified as a saddle point. A pair of principal axes coincides with coordinate axes and defines the orientations of two mutually orthogonal principal planes. The principal axes trajectories crossing at critical point assume the role of dividing streamlines. ($\psi^* = 0$). Upon approaching the critical point along dividing streamline y^*-axis ($\hat{e}_2$, $\lambda_2 < 0$), the vector field undergoes an indeterminate state as it vanishes at the critical point and the eigenvalue changes its sign from negative (in-coming flux) to positive (out-going flux) along dividing streamline (x^*-axis) in opposite directions. Critical point analysis could also be carried out with the employment of Hessian matrix of stream function and its principal state, that is,

$$H_{ij} = \frac{\partial^2 \psi^*}{\partial x_i^* \partial x_j^*} \Big|_{\substack{x*=0 \\ y*=0}} = \begin{bmatrix} 0 & 1.5 \\ 1.5 & 0 \end{bmatrix}$$
$$\lambda_1 = 1.5; \ \hat{e}_2 = \pm \begin{bmatrix} 1 & 1 \end{bmatrix}$$
$$\lambda_2 = -1.5; \ \hat{e}_2 = \pm \begin{bmatrix} 1 & -1 \end{bmatrix}$$
$$\lambda_1 \lambda_2 < 0$$
$$V^* = \nabla^* \psi^* \times \hat{k}$$

This is classified as a saddle point. In the neighborhood of critical point, gradients of stream function align with principal vectors asymptotically along which maximum spatial rates of change of stream function occur. The landscape of streamlines is that of a saddle point with valleys deepening away from the critical point along line $y^* = -x^*$, that is, $\hat{e}_2$ and mountains rising away from it along line $y^* = x^*$, that is, $\hat{e}_1$. The results tally with those of velocity gradient tensor analysis.

Streamline trajectories and pressure contours are then mapped from the physical plane to velocity hodograph in Figure 8.35.

$$p^*(x^*, y^*) = p^*[x^*(v_x^*, v_y^*), y^*(v_x^*, v_y^*)] = p^*(v_x^*, v_y^*)$$
$$\psi^*(x^*, y^*) = \psi^*[x^*(v_x^*, v_y^*), y^*(v_x^*, v_y^*)] = \psi^*(v_x^*, v_y^*)$$

The center plane (y^*-axis) and channel center x^*-axis are mapped to v_y^*-axis and v_x^*-axis in hodograph plane. The critical point at channel center is mapped to origin in velocity hodograph plane. According to hodograph representation and theory, dividing streamlines (Cartesian coordinate axes) are mapped to coincide with hodograph coordinate axes, since they are straight impervious boundaries. It can be seen from Figure 8.35 that maximum velocity ($v_x^* = \pm 4.5$, $v_y^* = 0$) occurs at points of intersections of exit planes and center axis. They are mapped from hodograph plane to points ($\pm 3, 0$) at channel exit planes where minimum pressure occurs. Maximum pressure at ($0, \pm 1$) in the physical plane is mapped to points ($v_x^* = 0$, $v_y^* = \mp 1$). The two horizontal plates $y^* = \pm 1$ corresponding to isoclines with vertical flows in opposite direction are mapped to the negative (upper plate)/positive (lower plate) part of vertical axis on hodograph plane, $v_y^* = \mp 1$. Tangents to the streamline trajectories on the physical plane and its images on hodograph plane are useful to visualize the state of motion since they represent velocity V and acceleration vector a at a point respectively. In Figure 8.35, velocity modulus defines the direction of motion parallel to the tangent (arrow) along a streamline on physical plane. Likewise, acceleration vector points in the direction of velocity change and corresponds to the tangent along the hodograph representation of streamline on hodograph plane. By mapping

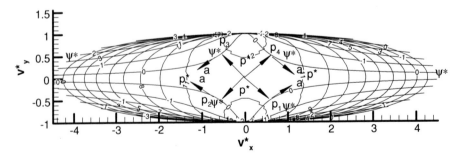

Figure 8.35 Distributions of streamline trajectories and pressure contours on velocity hodograph

tangents (arrow icons) between these two planes and displaying them at the same points on physical plane, we are able to determine the state of fluid motions. For instances, acceleration vectors at four locations (p_1 to p_4) are mapped (or translated like free vectors in parallel) from Figure 8.35 to the corresponding image points in Figure 8.32. By decomposing the acceleration vector into streamline tangential and normal direction in both figures, we observe that the normal components are much greater than the tangential components. The normal components representing centripetal acceleration are predominant with small radius of curvature. In Figure 8.32, they point towards the center of curvature indicating flow direction changes on the physical plane. The tangential components of acceleration are in the same direction of velocity with small magnitude and hence the flow is accelerating slowly.

Mapping of fields may become easier to visualize or to verify with approximation since field patterns/graphs of high order curves may be reduced to lower ones such as lines, circles or conic sections in some local regions. For regions close to the channel center, circles about origin represent lines of constant speed equal to radius $|V^*| = C$ on the hodograph plane are mapped approximately as circles of radius $\frac{|V^*|}{1.5}$ about origin in the physical plane in Figure 8.32. The equation of hodograph at constant speed in the neighborhood of the channel center is:

$$|V^*| = \sqrt{v_x^{*2} + v_y^{*2}} = C$$

With $y^* \to 0$, $x^* \to 0$

then

$$v_x^* \approx 1.5x^*; \quad v_y^* \approx -1.5y^*$$

$$|V^*| \approx 1.5\sqrt{x^{*2} + y^{*2}} = C$$

Above is the equation of position hodograph representation of constant speed contours on physical plane. The same results can be observed in Figure 8.33 where distributions of kinetic energy contours also resemble approximately circles in regions around the channel center. We could trace loci of velocity isoclines on physical plane by mapping lines through hodograph origin at the inclination angles γ equal to the constant flow directions. By applying the same approximation in the region around origin, each isocline is mapped from hodograph plane to another line through origin and reflected about x-axis on physical plane as the inclination angle changes sign.

$$v_y^* = v_x^* \tan \gamma; \quad \to y^* \approx x^* \tan(-\gamma)$$

8.3.4 Evaluation of Rate of Strain Tensor and Viscous Stress Tenor

Rate of strain tensor in two-dimensional fields is given by:

$$\dot{\gamma}_{ij} = \begin{bmatrix} \dot{\varepsilon}_x & \dot{\gamma}_{xy} \\ \dot{\gamma}_{yx} & \dot{\varepsilon}_y \end{bmatrix} = \frac{1}{2}\left[\frac{\partial v_j}{\partial x_i} + \frac{\partial v_i}{\partial x_j}\right]$$

$$\text{Or } \dot{\gamma}_{ij}^* = \frac{1}{2}\left[\frac{\partial v_j^*}{\partial x_i^*} + \frac{\partial v_i^*}{\partial x_j^*}\right]$$

$$\dot{\varepsilon}_x^* = \frac{\partial v_x^*}{\partial x^*} = 1.5(1 - y^{*2})$$

$$\dot{\varepsilon}_y^* = \frac{\partial v_y^*}{\partial y^*} = 1.5(y^{*2} - 1) = -\dot{\varepsilon}_x^*$$

$$\dot{\gamma}_{yx}^* = \frac{1}{2}\left[\frac{\partial v_x^*}{\partial y^*} + \frac{\partial v_y^*}{\partial x^*}\right] = -1.5x^*y^* \tag{8.79}$$

$$\dot{I}_1^* = \dot{\varepsilon}_x^* + \dot{\varepsilon}_y^* = \dot{\varepsilon}_1^* + \dot{\varepsilon}_2^* = 0$$

$$\rightarrow \dot{\varepsilon}_1^* = -\dot{\varepsilon}_2^* \tag{8.80}$$

Viscous stress tensor is evaluated from Newton's law of viscosity.

$$\tau_{ij}^* = -\frac{2}{\text{Re}}\dot{\gamma}_{ij}^*$$

$$\text{Or } \sigma_x^* = 3(y^{*2} - 1)/\text{Re}$$

$$\sigma_y^* = 3(1 - y^{*2})/\text{Re}$$

$$\tau_{yx}^* = -\frac{2}{\text{Re}}\dot{\gamma}_{yx}^* = \frac{3x^*y^*}{\text{Re}}. \tag{8.81}$$

$$\sum \tau_{ii}^* = 0$$

$$\frac{\partial v_y^*}{\partial x^*} = 0$$

Shear stresses are contributed solely by x-component velocity gradient in y-direction. Shear stresses represent viscous fluxes of x-momentum transmitted towards the adjacent decreasing velocity fluid layers along y-direction. The transfer mechanism is set up by fluid layers close to the moving plates with greater shear stress fluxes than layers further away and causes them to remain in motion. Shear stresses resemble a family of hyperbolas symmetric with respect to lines $y^* = \pm x^*$ as well as origin. They are antisymmetric about coordinate axes which also assume the role of asymptotes. In hydrodynamics convention, on a positive y-face in upper section, positive direction of shear stress τ^*_{yx} points towards negative direction of x-axis. On a negative y-face in lower half section, positive direction of shear stress τ^*_{yx} points in the positive direction of x-axis. With shear stress assuming positive in first and third quadrants and negative in second and fourth quadrants, the directions of shear stress are marked with arrows in Figure 8.33. Shear stresses vanish along coordinate axes and assume peak values at channel's four corners.

$$\left| \tau^*_{yx}(\pm L/h, \pm 1) \right|_{\max} = \frac{3L}{h\mathrm{Re}}$$

The force applied to the moving plates is thus transmitted as x-direction momentum from one layer to the next in y-direction. The force acting on the plate per unit depth can be evaluated, that is,

$$F^*_y = \frac{F_y}{\rho h V_0} = - \int_{-L/h}^{L/h} (\sigma^*_y \big|_{y*=\pm 1} - p^* \big|_{y*=\pm 1}) dx^*$$

$$= \frac{2}{\mathrm{Re}} \left(\frac{L}{h} \right)^3 \tag{8.82}$$

8.3.5 Viscous Stress Tensor Visualization

8.3.5.1 Contraction

Viscous stress tensor describes momentum diffusion transport as contiguous action of force transmission in the system. This tensor field could be visualized by stress vector at a point through contraction. They represent external characteristics of momentum in terms of physically measurable quantities, that is, frictional forces Q^*. In Stokes flows, viscous stress force remains operative and is balanced by pressure force below:

$$Q^* = -\nabla^* \cdot [\tau^*] = \frac{2}{\mathrm{Re}} \nabla^* \cdot [\dot{\gamma}^*] = \frac{1}{\mathrm{Re}} \begin{bmatrix} \nabla^{*2} v_x^* \\ \nabla^{*2} v_y^* \end{bmatrix} = \nabla^* p^*$$

$$\because \nabla \cdot Q^* = \nabla^{*2} p^* = 0$$

Frictional force can be visualized by the stress flux function introduced below:

$$Q^*_x = -\frac{3x^*}{\mathrm{Re}} = \frac{\partial \psi^*_m}{\partial y^*} \tag{8.83a}$$

$$Q^*_y = \frac{3y^*}{\mathrm{Re}} = -\frac{\partial \psi^*_m}{\partial x^*} \tag{8.83b}$$

By integration,

$$\psi_m^* = -\frac{3x^*y^*}{\text{Re}}$$

Figure 8.33 presents the distribution of stress vector trajectory resembling a family of hyperbolas. The distribution patterns are symmetric with respect to lines $y^* = \pm x^*$ as well as origin and are antisymmetric about coordinate axes. There is a pair of asymptotes coincide with coordinate axes. Tangents along stress vector trajectories indicate the directions of viscous force. They appear to act against fluid motions in a manner similar to shear stress. The magnitude of stress vector is displayed in Figure 8.34 resembling a system of circles with center at origin corresponding to stress vector critical point. Maximum frictional force magnitudes occur towards channel four corners indicating regions of high shear. Streamline trajectories and pressure contours are then mapped from the physical plane to viscous stress vector hodograph plane in Figure 8.36.

$$\psi^*(x^*, y^*) = \psi^*[x^*(Q_x^*, Q_y^*), y^*(Q_x^*, Q_y^*)] = \psi^*(Q_x^*, Q_y^*)$$
$$p^*(x^*, y^*) = p^*[x^*(Q_x^*, Q_y^*), y^*(Q_x^*, Q_y^*)] = p^*(Q_x^*, Q_y^*)$$
$$|Q^*| = \sqrt{Q_x^{*2} + Q_y^{*2}} = \frac{3\sqrt{y^{*2} + (-x^*)^2}}{\text{Re}} = C \qquad (8.84)$$

The above shows that a circle of radius C about origin in stress vector hodograph plane represents locus of constant frictional force magnitude is mapped to another circle of radius $\frac{\text{Re}C}{3}$ about origin on the physical plane. We could trace loci of frictional force isoclines on physical plane by mapping lines through origin on stress vector hodograph at the inclination angles γ equal to

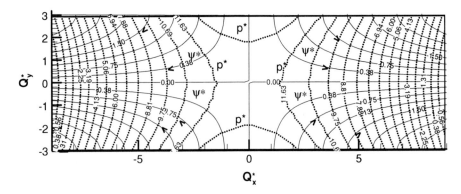

Figure 8.36 Mapping of streamlines (solid) and pressure (dotted) contours from z-plane to viscous stress vector hodograph

the constant force directions. Each isocline is mapped from hodograph plane to another line through origin and reflected about x-axis on physical plane as the inclination angle changes sign. This means each trajectory tangent at the point of intersection between the isoclines and stress vector trajectory on physical plane will have the same direction equal to inclination angle.

$$Q_y^* = Q_x^* \tan \gamma; \rightarrow y^* = x^* \tan(-\gamma)$$

8.3.5.2 Principal Axes Trajectories and Evolving Tensor Ellipses

Visualization of viscous momentum transfer requires the knowledge of principal states of viscous stress tensors. Principal values of viscous stress tensor represent viscous momentum fluxes passing through principal planes driven by momentum concentration (or velocity) gradients. They give rise to extreme normal stresses (principal stresses) and force transmission along respective principal axes trajectories. Principal states are determined as per Equations 6.31 and 6.32.

$$\sigma_{1,2}^* = \pm \frac{3}{\mathrm{Re}} \sqrt{y^{*4} - 2y^{*2} + 1 + x^{*2} y^{*2}}$$

With a pair of equal principal values in opposite sign, it implies contraction/expansion effects are equal and tensor quadric becomes circles. Distributions of principal stress σ_1^* are presented in Figure 8.37 together with a system of orthogonal principal axes trajectories $\hat{e}_1$ and $\hat{e}_2$. Symmetry test shows principal values are symmetric with respect to coordinate axes. The results reflect the effects of symmetry in domain geometry and boundary conditions as well. A pair of principal axes trajectories coincides with coordinate axes while other principal axes trajectories cross them perpendicularly. This could be anticipated since shear stresses vanish along coordinate axes.

Along x-axis : $\sigma_2^*(x^*, 0) = \sigma_x^*(x^*, 0) = \dfrac{-3}{\mathrm{Re}} < 0$ Tension, $\hat{e}_2 = \begin{bmatrix} 1 & 0 \end{bmatrix}$.

Along y-axis : $\sigma_1^*(0, y^*) = \sigma_y^*(0, y^*) = \dfrac{3(1 - y^{*2})}{\mathrm{Re}} \geq 0$ Compression, $\hat{e}_1 = \begin{bmatrix} 0 & 1 \end{bmatrix}$

Viscous momentum transfers from two moving plates towards channel center axis along $\hat{e}_1$ principal axes trajectories under compression $\sigma_1^* > 0$. They cross x^*-axis orthogonally, which corresponds to a principal axis $\hat{e}_2$. Transfer processes are most active near channel corners where principal values have the greatest magnitude numerically. On the other hand, expansion effects, being represented by $\sigma_2^* < 0$ transmit from two exit planes along $\hat{e}_2$ principal axes trajectories towards center vertical plane and cross y^*-axis orthogonally. Thus y^*-axis is one of principal axes $\hat{e}_1$.

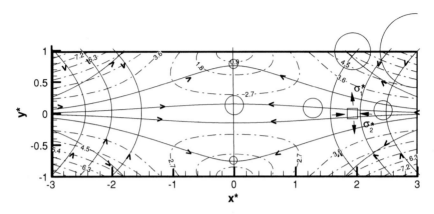

Figure 8.37 Distributions of principal axes trajectories and principal stress σ_1^* (dash dot) with evolving tensor circles. $\hat{e}_1$ principal axes trajectories (up/down directions); $\hat{e}_2$ principal axes trajectories (left/right directions);

Positive principal stresses and negative principal stresses are defined in hydrodynamics conventions with respect to a control volume to represent out-going stress fluxes in compression and incoming stress fluxes in tension as shown in Figure 8.37. The figure also shows evolving tensor circles at selected points along principal axes trajectories. The sizes of circles represent principal values as a measure of momentum diffusive flux strength. There is a degenerate point at the center of each plate at which both eigenvalues assume zero and tensor circles collapse into a point with infinite principal axes. These degenerate points correspond to the points of maximum pressure where pressure is the same in all directions.

It is useful to apply approximation analysis to visualize or to verify principal stress distributions in some local regions. Consider regions along the plate surfaces. We observe that the principal stresses vary linearly with distance from the plate center (degenerate point) along each plate surface. It will assume maximum values at the channel corners.

$$\text{Along plate surfaces: } \sigma_{1,2}^*(x^*, \pm 1) = \pm\frac{3x^*}{\text{Re}}$$

8.3.6 Visualization of Mechanical Energy Transfer/Dissipation

Mechanical energy transfer paths can be visualized by mechanical energy trajectories in terms of energy flux vector written as:

$$\frac{dy^*}{dx^*} = \frac{\dfrac{V^{*2}}{2}v_x^* + (p^*v_x^* + \sigma_x^*v_x^* + \tau_{xy}^*v_y^*)}{\dfrac{V^{*2}}{2}v_y^* + (p^*v_y^* + \sigma_y^*v_y^* + \tau_{xy}^*v_x^*)} \tag{8.85}$$

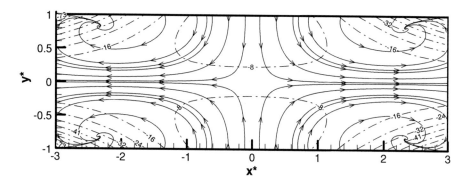

Figure 8.38 Distributions of mechanical energy transfer trajectories (arrows) and viscous dissipation energy contours (dash dot)

Mechanical energy transfer trajectories are shown in Figure 8.38. They are similar to those of streamline patterns in most parts of the channel except at four corner regions. Viscous energy dissipation is evaluated below and is presented in Figure 8.38. Energy dissipation assumes maximum near four corners where maximum shear stress also occurs giving rise to regions of high shear.

$$\Pi^* = -\frac{Re^2}{2}\tau_{ij}^*\tau_{ij}^*$$

8.3.7 Visualization of Vorticity Transport

Viscous flow is rotational in nature with vorticity distributed continuously through the flow fields. Vorticity could be determined from flow fields.

$$\omega^* = \frac{\omega h}{V_0} = \nabla^* \times V^*$$

In two-dimensional flows, the nonzero component rotating about z-axis is given by:

$$\omega_z^* = \left[\frac{\partial v_y^*}{\partial x^*} - \frac{\partial v_x^*}{\partial y^*}\right]$$
$$= 3x^*y^* \tag{8.86}$$
$$= Re\tau_{yx}^* = Re\psi_m^*$$

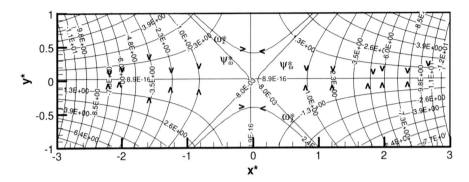

Figure 8.39 Distributions of vorticity and vorticity diffusive flux trajectories (arrows) on the physical plane

Vorticity distribution exhibits a system of hyperbolas and is presented in Figure 8.39. Vorticity is symmetric about origin and has lines of symmetry $y^* = \pm x^*$. It is antisymmetric about coordinate axes implying that vorticity changes its rotation direction from one quadrant to another in accordance with change of sign and that the system is at a steady state. The coordinate axes also serve as asymptotes for the stress vector trajectories and separate vortices into each quadrant. Vorticity intensity is proportional to spatial variables x and y implying that vorticity increases with distance from origin. Vortices generated from two opposite plates are most intensifying and assume maximum strength at channel four corners. They diminish to nil along coordinate axes where shear stress also vanishes. It is noted that distributions of vorticity, stress vector trajectories and viscous stresses are all similar within a constant. This could be expected since vorticity and shear stress are related by vector identity and they track frictional forces through the paths of stress vector trajectories.

Vorticity generated along plate boundaries diffuses inwards. To study vorticity transport phenomena, we take the curl of the momentum equation to yield,

$$\frac{1}{Re}\nabla^* \times \nabla^{*2}V = \frac{1}{Re}\nabla^{*2}\omega_z^* = \nabla^* \times \nabla^* p^* = 0$$

Hence with vanishing convective transport, z-component vorticity is harmonic and attains extreme values at flow boundaries in accordance with the maximum principle. Diffusion transport of vorticity can be visualized in terms of vorticity function in a manner similar to stream function. It is defined below with vorticity diffusive flux vector pointing in the direction of decreasing vorticity:

$$q^* = -\frac{1}{Re}\nabla^*\omega_z^* = -\nabla^*\psi_\omega^* \times \hat{k} \tag{8.87}$$

Perform integrations to yield,

$$\psi_\omega^* = \frac{3(y^{*2} - x^{*2})}{2Re} \tag{8.88}$$

$$\frac{\partial \psi_\omega^*}{\partial x^*}\bigg|_{x*=0} = \frac{\partial \psi_\omega^*}{\partial y^*}\bigg|_{y*=0} = 0$$

Vorticity diffusive flux trajectories cross coordinate axes at right angles serving as lines of symmetry. Figure 8.39 presents vorticity diffusive flux trajectory distributions forming a system of hyperbolas. The trajectories cross vorticity contours orthogonally and are similar to those of pressure contours within a constant. Vorticity generated from moving plate boundaries diffuses towards channel center axis and center vertical plane following the diffusive flux trajectories. At these locations, vorticity diminishes to nil. Diffusion process intensifies towards two exit planes. This is reflected by small curvilinear squares in Figure 8.39. It implies local gradient of vorticity is great and that the region is in a state of high enstrophy density. Dividing diffusive flux trajectory $\psi_\omega^* = 0$ passes through critical point at origin and splits there to coincide with lines $y^* = \pm x^*$. They serve as asymptotes for nearby trajectories and separate the fields into local regions. The gradient mapping of shear stress/vorticity will depict the paths from two opposite plate surfaces towards the center axis along which shear stress/vorticity changes most rapidly. To observe the connection between processes, these paths are then compared with those of principal axes trajectories of viscous stress tensor. The role of viscous effects is to maintain viscous momentum transport processes as well as force balance in the system. In doing so, it establishes viscous stress distribution and makes connection to vorticity distribution in the fields. The origin is a multiple point with coincidence of velocity critical point, stress vector critical point and vorticity diffusive flux critical point.

The gradient of vorticity diffusive flux tensor at the critical point is used to reveal nearby field characteristics, that is,

$$[\nabla^*q^*]\bigg|_{\substack{x*=0 \\ y*=0}} = \begin{bmatrix} \dfrac{\partial q_x^*}{\partial x^*} & \dfrac{\partial q_y^*}{\partial x^*} \\[2mm] \dfrac{\partial q_x^*}{\partial y^*} & \dfrac{\partial q_y^*}{\partial y^*} \end{bmatrix}\bigg|_{\substack{x*=0 \\ y*=0}} = \frac{1}{Re}\begin{bmatrix} 0 & -3 \\ -3 & 0 \end{bmatrix}$$

where

$$q_x^* = -\frac{1}{Re}\frac{\partial \omega_z^*}{\partial x^*} = -\frac{3y^*}{Re}$$

$$q_y^* = -\frac{1}{Re}\frac{\partial \omega_z^*}{\partial y^*} = -\frac{3x^*}{Re}$$

Above symmetric matrix yields the following real eigenvalues and eigenvectors, which indicate a saddle point (Figure 7.2).

$$\lambda_1 = \frac{3}{Re}; \; \hat{e}_1 = \pm\begin{bmatrix} 1 & -1 \end{bmatrix}$$

$$\lambda_2 = -\frac{3}{Re}; \; \hat{e}_2 = \pm\begin{bmatrix} 1 & 1 \end{bmatrix}$$

$$\lambda_1\lambda_2 < 0$$

A pair of principal axes coincides with lines $y^* = \pm x^*$ and defines the orientations of two mutually orthogonal principal planes. Principal axes trajectories crossing at critical point serve as dividing vorticity diffusive flux lines. Similarly, the analysis of Hessian matrix of vorticity function and its principal state at critical point lead to the same findings, that is, a saddle point.

$$H_{ij} = \frac{\partial^2 \psi_\omega^*}{\partial x_i^* \partial x_j^*}\Big|_{\substack{x^*=0 \\ y^*=0}} = \frac{1}{Re}\begin{bmatrix} -3 & 0 \\ 0 & 3 \end{bmatrix}$$

$$\lambda_1 = \frac{3}{Re}; \; \hat{e}_2 = \pm\begin{bmatrix} 0 & -1 \end{bmatrix}$$

$$\lambda_2 = -\frac{3}{Re}; \; \hat{e}_2 = \pm\begin{bmatrix} 1 & 0 \end{bmatrix}$$

$$\lambda_1\lambda_2 < 0$$

Figure 8.40 presents the mapping of vorticity diffusive flux trajectories and streamlines from z-plane to vorticity diffusive flux hodograph. The critical point of vorticity diffusive flux is mapped from origin in z-plane to the origin in vorticity diffusive flux hodograph plane. Dividing vorticity diffusive trajectories $\psi_\omega^* = 0$ coincide with straight lines $y^* = \pm x^*$ in z-plane are mapped to straight lines $q_y^* = \pm q_x^*$ serving as oblique asymptotes in vorticity diffusive flux hodograph plane. The vorticity flux hodograph representation of vorticity diffusive trajectories is conic curves. Velocity critical point and dividing streamlines are mapped to coincide with origin and coordinate axes of vorticity diffusive flux hodograph. By mapping circles of radius C with center at origin from hodograph plane, we could

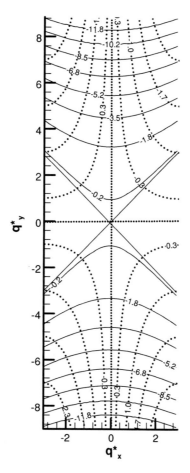

Figure 8.40 Mapping of vorticity diffusive flux trajectories (line) and stream-lines (dotted) from z-plane to vorticity diffusive flux hodograph

visualize lines of constant vorticity diffusive flux magnitude on the physical plane by another circle of radius (ReC)/3 about origin. We could trace loci of vorticity diffusive flux isoclines on physical plane by mapping lines through hodograph origin at the inclination angles equal to the constant vorticity diffusive flux directions. Each isocline is mapped from hodograph plane to another line rotated through a right angle at origin anticlockwise and reflected about x-axis on physical plane.

$$\psi_\omega^*(x^*, y^*) = \psi_\omega^*[x^*(q_x^*, q_y^*), \ y^*(q_x^*, q_y^*)] = \psi_\omega^*(q_x^*, q_y^*) = \frac{Re}{6}(q_x^{*2} - q_y^{*2})$$

$$\psi^*(x^*, y^*) = \psi^*[x^*(q_x^*, q_y^*), y^*(q_x^*, q_y^*)] = \psi^*(q_x^*, q_y^*)$$

$$= \frac{q_x^* q_y^*}{2} \left(\frac{\text{Re}}{3} \right)^2 \left(3 - \left(\frac{q_x^* \text{Re}}{3} \right)^2 \right)$$

$$|q^*| = \sqrt{q_x^{*2} + q_y^{*2}} = \frac{3}{\text{Re}} \sqrt{y^{*2} + x^{*2}} = C$$

$$q_y^* = q_x^* \tan \gamma$$

$$\rightarrow y^* = x^* \cot(\gamma) = x^* \tan[-(\pi/2 + \gamma)]$$

The amount of vorticity in volume diffused per unit depth per unit time bounded between two trajectories is given by the difference of respective flux lines.

$$\frac{Q}{\Delta z} = -\nu \int_1^2 \nabla \omega_z \cdot \hat{b} dl$$

$$\text{Or } Q^* = \frac{Q}{V_0^2 \Delta z} = -\frac{1}{\text{Re}} \int_1^2 \nabla^* \omega_z^* \cdot \hat{b} dl^* = -\int_1^2 \nabla^* \psi_\omega^* \times \hat{k} \cdot \hat{b} dl^* = (\psi_{\omega 2}^* - \psi_{\omega 1}^*)$$

$$(8.89)$$

$\hat{b}$ represents unit normal vector of an arbitrary surface bounded by two trajectories. dl^* is the surface differential arc length. The transfer number is;

$$D = \frac{\text{Re} \Delta \psi_{\omega,\max}^*}{\Delta \omega_{z,\max}^*} = \frac{13.5 - (-1.5)}{9 - (-9)} = 0.83$$

8.3.8 Lamb Vector Visualization and Analysis

Lamb vector fields and its divergences are introduced to visualize the interaction of vorticity and flow fields. They are evaluated and are presented in Figure 8.41.

$$L^* = \frac{Lh}{V_0^2} = \omega^* \times V^*$$

$$|L^*| = \frac{|L|h}{V_0^2} = |\omega^* \times V^*| = |\omega_z^*||V^*| = \sqrt{L_x^{*2} + L_y^{*2}}$$

$$\nabla^* \cdot L^* = \text{Re} V^* \cdot (\nabla^* \cdot [\tau^*]) - \omega^{*2}$$

where

$$L_x^* = 1.5 x^* y^{*2} (3 - y^{*2}); \quad L_y^* = 4.5 x^{*2} y^* (1 - y^{*2})$$

$$\nabla^* \cdot L^* = 1.5 y^{*2} (3 - y^{*2}) + 4.5 x^{*2} (1 - 3 y^{*2})$$

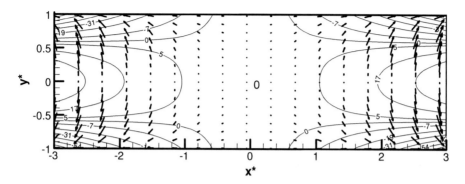

Figure 8.41 Distributions of Lamb vector fields (arrows) and Lamb vector divergence contours.

Lamb vector has a higher order critical point at channel center where both Lamb vector and gradient of Lamb vector tensor vanish. At the multiple point, velocity, vorticity and Lamb vector vanish. The Lamb vector plot in Figure 8.41 shows Lamb vector field strength increases towards two channel exit planes. It is observed that Lamb vectors vanish along coordinate axes and that they are perpendicular to flow velocity (or streamline tangents) everywhere. Lamb vector trajectory represented by an integral curve in terms of an implicit function is obtained by integrating the trajectory equation, that is,

$$\psi_L^* = 3x^{*2} - y^{*2} + 2\ln\left|y^{*2} - 1\right| = C$$

Figure 8.42 presents the distributions of Lamb vector trajectories. The tangent along a trajectory defines the direction of Lamb vector at a point. Both Lamb vector trajectories and Lamb vector divergence contours are symmetric with respect to coordinate axes as well as origin. The zero divergence contours cover the central part of channel where the work done by stress tensor and dissipation by enstrophy are at equal strength and balance each other out. In this region, both straining motion and vortex rotational motion may be present except at critical points. They separate Lamb vector fields into local regions of distinct behaviors. Lamb vector divergences are positive towards channel exits and are marked by hyperbolic contours. It indicates a source of energy available from reversible rate of work done by the stress tensor, which behaves like a release/restoring mechanism in the regions of high shear. Divergences assume negative near four channel corners and along the surfaces of plates. It indicates energy sinks by irreversible rate of work done such as enstrophy acting as the mechanism of dissipation

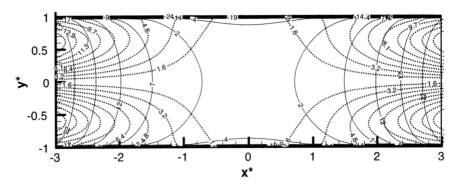

Figure 8.42 Distributions of Lamb vector magnitude contours (dotted) and Lamb vector trajectories (line) on physical plane

in the regions of high vorticity. The results tally with those of vorticity contour distributions. Thus the exit planes are regions of high shear and high vorticity. The strong Lamb vectors fields at the exit planes support the same findings.

Figure 8.42 also depicts the distribution Lamb vector magnitude contours. Lamb vector assumes maximum magnitude at the channel two exit planes $(x^* = \pm 3, y^* = \pm 0.7)$ indicates regions of high vorticity and high velocity. It implies Lamb vector representing combined effects of mechanical energy gradient vector and entropy gradient vector at exit planes are the greatest in accordance with Crocco's theorem. Since streamline curvatures are mild near two exit planes, centripetal accelerations are not significant. This suggests convective accelerations are predominant and hence signifies regions of high shear as well. High vorticity regions penetrate towards the center plane like wedges along the surfaces of two plates accompanied with increasing centripetal acceleration while convective acceleration continues to decrease. This is attributed to the shape of streamlines. Lamb vector magnitude decreases towards channel axis and center plane due to rapidly vanishing vorticity more than velocity strength picking up. One may proceed to apply hodograph representation and mapping techniques to visualize the loci of Lamb vector isoclines or constant Lamb vector magnitudes on Lamb vector hodograph plane/physical plane. Consider approximation analysis in the neighborhood of plate surfaces with y approaching plus/minus unity. The x-component of Lamb vector becomes a linear function of x while the y-component vanishes. The plate surfaces then correspond to Lamb vector isoclines parallel to the positive/negative x-axis. The Lamb vector magnitude is proportional to the distance from the center of plate linearly and assumes maximum at channel exits. Helicity vanishes in two-dimensional fields.

8.4 Example 8.4: Groundwater Flows in an Anisotropic Porous Medium

A flow model is used to simulate groundwater flows in a two-dimensional anisotropic porous basin in Figure 8.43. The two vertical ends of the miniature model basin are considered planes of symmetry as the flow field domain consists of a series of parallel ridges and valleys. The model basin has a longitudinal base length, L coincide with the flat horizontal bottom of an impervious boundary. x-axis of Cartesian coordinate is placed to coincide with the horizontal bottom. Coordinate origin is located at the base center such that the center plane (y-axis) assumes the plane of symmetry and points upwards. As part of boundary conditions, piezometric heads are measured at observation wells located at basin center (y-axis) and basin's two ends ($x = \pm L/2$) at the valleys of the lowest elevation respectively. Water surface elevations are recorded as h_1 at center well and h_2 at basin's end wells with respect to reference datum x-axis. The water table is a free surface of unknown shape along which pressure assumes ambient. It is assumed that there is no source in the domain and fluid is incompressible.

8.4.1 Governing Equation

Two-dimensional steady groundwater movements could be described by continuity equation and an empirical equation known as Darcy's equation:

$$\nabla \cdot V = 0$$

$$V = -k_{ij}\nabla\varphi \tag{8.90}$$

$$\text{where } \varphi = \frac{p}{\rho g} + y$$

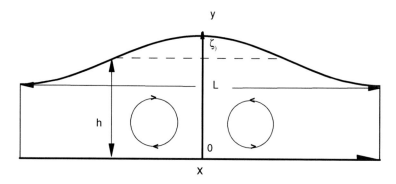

Figure 8.43 A two-dimensional groundwater flow configuration

p is groundwater pressure and ρ is fluid density. φ represents piezometric head composed of hydrostatic pressure and elevation head. Water movement is driven by gradient of piezometric head and depends on hydraulic conductivity. k_{ij} is hydraulic conductivity matrix for an anisotropic linear medium whose principal axes of anisotropy are placed to coincide with coordinate axes, such that

$$k_{ij} = \begin{bmatrix} k_h & 0 \\ 0 & k_v \end{bmatrix}$$

The field is rotational due to nonhomogeneity of medium. Piezometric head contours and streamlines will not cross each other at right angle. By taking divergence of Darcy's equation, the piezometric head and velocity fields are uncoupled and the governing equations reduce to:

$$\frac{\partial}{\partial x} k_h \frac{\partial \varphi}{\partial x} + \frac{\partial}{\partial y} k_v \frac{\partial \varphi}{\partial y} = 0 \qquad (8.91)$$

$$-L/2 \leq x \leq L/2; \; 0 \leq y \leq h_1$$

8.4.2 Boundary Conditions and Model Parameters

Boundary conditions are mixed type. No flow crosses the basin bottom of an impermeable boundary, such that

$$v(x, 0) = V \cdot \hat{b} = v_y = -k_v \frac{\partial \varphi}{\partial y} \Big|_{y=0} = 0$$

$\hat{b}$ is the unit normal of bottom surface parallel to y-axis. Similarly, no flow is allowed to pass through the two basin's end boundaries due to symmetry.

$$v(\pm L/2, y) = V \cdot \hat{b} = v_x = -k_h \frac{\partial \varphi}{\partial x} \Big|_{x=\pm L/2} = 0$$

$\hat{b}$ is the unit normal of the end boundaries parallel to x-axis. The above gives rise to homogenous Neumann conditions and as a result piezometric head will terminate on these boundaries perpendicularly. Such boundaries also serve as streamlines.

Water table is a free boundary of unknown shape across which water may pass through. This is because fluids cannot withstand shear stress and will easily change the surface's shape subject to shearing action. It is an interface separating the unsaturated zone and aquifer. Water table serves as the upper boundary of an aquifer and is defined by the dynamic condition at the interface. When surface tension is negligible, the state of equilibrium yields,

$$p = p_a = C$$
$$\rightarrow \varphi = y \tag{8.92}$$

p_a is the interface pressure equal to atmospheric pressure, which is generally taken as zero reference datum. The water table is assumed by a smooth curve joining the water levels in the observation wells. The following Direchlet conditions apply at the observation wells:

At water table peak point $A: \varphi(0, h_1) = y = h_1 = h + \zeta_0$

At basins two ends : $\varphi(\pm L/2, h_2) = y = h_2 = h - \zeta_0$

where $h = 0.5(h_1 + h_2)$; $\zeta_0 = 0.5(h_1 - h_2)$

The model parameters are:

$h_1 = 1m, \quad h_2 = 0.6m, \quad L = 3m, \quad k_h = 1.e - 8m/s \cdot k_v = 1.e - 9m/s,$
$\mu = 1.e - 3Ns/m^2, \quad \rho = 1000kg/m^3, \quad g = 9.8m/s^2.$

8.4.3 Solution Method and Flow Analysis

Applying the technique of separation of variables in Example 8.3, a solution in terms of a product function is assumed.

$$\varphi = X(x)Y(y) \tag{8.93}$$

Such that,

$$-k_h \frac{X''}{X} = k_v \frac{Y''}{Y} = \lambda = k_h \left(\frac{2\pi}{L}\right)^2$$

λ is an arbitrary constant. For convenience it is set equal to the above value. This leads to a pair of differential equations with the following solutions,

$$X = A\cos\left(\frac{2\pi x}{L}\right) + B\sin\left(\frac{2\pi x}{L}\right)$$

$$Y = C\cosh\left(\sqrt{\frac{k_h}{k_v}}\frac{2\pi y}{L}\right) + D\sinh\left(\sqrt{\frac{k_h}{k_v}}\frac{2\pi y}{L}\right)$$

The unknown coefficients are to be determined by boundary conditions. From Neumann conditions,

$$B = D = 0$$

A suitable solution is given by:

$$\varphi = C_1 \cos\left(\frac{2\pi x}{L}\right) \cosh\left[\sqrt{\frac{k_h}{k_v}} \frac{2\pi y}{L}\right] + C_2 \tag{8.94}$$

The constants are determined from Direchlet boundary conditions. The results are:

$$C_1 = \frac{2\zeta_0}{\cosh[\sqrt{k_h/k_v}\,2\pi\,(h+\zeta_0)/L] + \cosh[\sqrt{k_h/k_v}\,2\pi\,(h-\zeta_0)/L]}$$

$$C_2 = \frac{2\zeta_0 \cosh(\sqrt{k_h/k_v}\,2\pi\,(h-\zeta_0)/L)}{\cosh(\sqrt{k_h/k_v}\,2\pi\,(h+\zeta_0)/L) + \cosh(\sqrt{k_h/k_v}\,2\pi\,(h-\zeta_0)/L} + (h-\zeta_0)$$

$$\frac{\partial\varphi}{\partial y}\Big|_{y=0} = 0; \quad \frac{\partial\varphi}{\partial x}\Big|_{x=0,x=\pm L/2} = 0$$

The center plane is a plane of symmetry for the domain geometry as well as boundary conditions. Distributions of piezometric head contours are shown in Figure 8.44. Piezometric head contours are symmetric with respect to center axis (y-axis). They are concentrated in the basin central part and decrease in the flow directions. Piezometric head approaches uniform in the y-direction as the flow fields become horizontal downstream. This is due to the vanishing

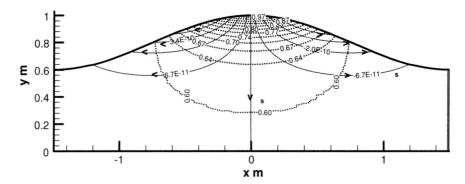

Figure 8.44 Distributions of piezometric head contours (dotted) and streamline trajectories (line) on the physical plane

piezometric gradients in the y-direction and is reflected by the piezometric head distribution, which tends to take the shape of a vertical line. Flow fields driven by piezometric head gradients continue to diminish in the flow directions and eventually vanish downstream where piezometric head assumes constant. The gradient mapping of piezometric head will depict the paths along which piezometric head changes most rapidly. Maximum piezometric head occurs at basin center water table surface point A ($x = 0$, $y = h_1$) equal to elevation. Piezometric head contours intersect coordinate axes and basin's two end boundaries at right angle in accordance with boundary conditions. The drawdown s is evaluated from its definition, that is,

$$s = \varphi(0, h_1) - \varphi$$

From Equation 8.90, the velocity components are obtained accordingly:

$$v_x = C_1 k_h \frac{2\pi}{L} \sin(\frac{2\pi x}{L}) \cosh[\sqrt{k_h/k_v}\, 2\pi y/L] \tag{8.95}$$

$$v_y = -C_1 \sqrt{k_h k_v} \frac{2\pi}{L} \cos(\frac{2\pi x}{L}) \sinh[\sqrt{k_h/k_v}\, 2\pi y/L] \tag{8.96}$$

y-component velocity is symmetric with respect to y-axis and points downwards everywhere. x-component velocity is antisymmetric about y-axis pointing into the positive and negative direction in first and second quadrant respectively. Flow is vertically downwards along the y-axis, which corresponds to a dividing straight streamline. Starting from water table surface upstream, y-component velocity continues to decrease while x-component velocity increases giving rise to horizontal flow. Both velocity components then vanish in the absence of pieozometric gradients downstream. These patterns of variation of velocity components play a part in flow field structure and behavior. Flow field strength characterized by convective momentum and kinetic energy are given by:

$$\rho\,|V| = \rho\sqrt{v_x^2 + v_y^2}$$

$$K = \rho\frac{V^2}{2} \tag{8.97}$$

Distributions of kinetic energy are presented in Figure 8.45. Field strength is the greatest at water table top at point A and decreases rapidly towards the bottom. Distributions of convective momentum depict similar patterns and suggest the same findings. The basin's center (origin) and two bottom corners are stagnation points where velocity vanishes.

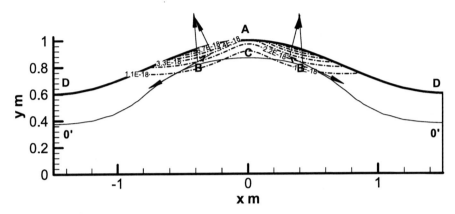

Figure 8.45 Distributions of kinetic energy contours (dotted) and shapes of the water table and phreatic surface

Since velocity field is solenoidal, stream function exists and is found below:

$$\begin{bmatrix} v_x \\ v_y \end{bmatrix} = \begin{bmatrix} -\dfrac{k_h \partial \varphi}{\partial x} \\ -\dfrac{k_v \partial \varphi}{\partial y} \end{bmatrix} = -\nabla \psi \times \hat{k} \tag{8.98}$$

By integration,

$$\psi = -C_1\sqrt{k_h k_v}\,\sin\left(\frac{2\pi x}{L}\right)\sinh\left[\sqrt{k_h/k_v}\,2\pi y/L\right] \tag{8.99}$$

Coordinate axes serve as asymptotes for the streamlines.

$$\underset{x\to 0}{Lim}\,\sinh[\sqrt{k_h/k_y}\,2\pi y/L] = \frac{C}{\sin(\frac{2\pi x}{L})} = \infty$$

$$\underset{y\to 0}{Lim}\,\sin\left(\frac{2\pi x}{L}\right) = \frac{C}{\sinh[\sqrt{k_h/k_y}\,2\pi y/L]} = \infty$$

Distributions of streamline trajectories are presented in Figure 8.44. Streamline distribution patterns are antisymmetric with respect to y-axis. Flow starts from water table top and seeps primarily side-ways towards two basin's ends. Since resistance in horizontal directions is much less than those of vertical, streamlines are nearly parallel in downstream and streamline spacing Δn

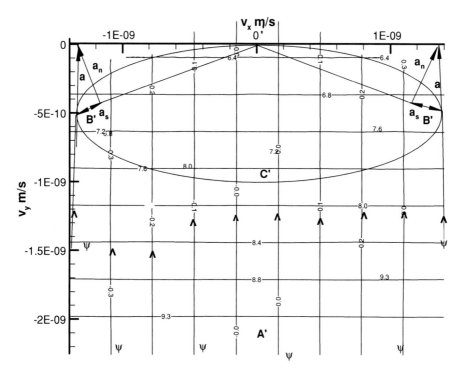

Figure 8.46 Mapping of streamlines (vertical line) and piezometric head (horizontal line) on velocity hodograph

remains practically constant. A region of wider potential spacing Δs corresponds to a higher permeability. The flow field is dominated by x-component velocity and vanishes at points D serving as stagnation points in addition to basin's center and corners. Streamlines and piezometric head contours do not cross at right angle. Maximum y-component downward flow velocity occurs at the peak of water table surface point $A(0, h_1)$. It has a magnitude equal to:

$$v_y \big|_{\max} = v_y(0, h_1) = -2.4e - 9 \, m/s; \quad v_x(0, h_1) = 0$$

Point A is then mapped to image point A' lying on v_y-axis on hodograph plane (Figure 8.46) at which y-component velocity has the same maximum value. This follows the hodograph representation theory for a straight streamline, which coincides with y-axis and passes through point A. In addition, point A also defines the level of water table surface at its peak value h_1 in physical plane (Figure 8.45).

Mapping of streamlines and piezometric head contours from z-plane to velocity hodograph is presented in Figure 8.46, which is constructed with v_x

and v_y assuming horizontal and vertical axes.

$$\psi(x, y) = \psi[x(v_x, v_y), y(v_x, v_y)] = \psi(v_x, v_y)$$
$$\varphi(x, y) = \varphi[x(v_x, v_y), y(v_x, v_y)] = \varphi(v_x, v_y)$$

It is observed that hodograph representation of streamlines is a system of vertical straight lines pointing upwards in hodograph plane. The velocity vectors (modulus) trace hodograph representation of a streamline like a pencil of rays from origin. The velocity vectors show that flow starts primary from downwards flow to become horizontal flows as we follow the path of a streamline. According to hodograph representation and theory, basin's bottom line is mapped to a horizontal line coinciding with v_x-axis since it is a horizontal straight streamline as well as a dividing streamline. Similarly, basin's center plane (line $0A$) is a dividing streamline and two vertical planes passing through points D serve as streamlines of an impervious straight boundary, they are mapped to vertical line $0'A'$ coinciding with v_y-axis inclusively. Water levels at basin's two end wells D are horizontal equipotentials of infinitesimal length and are mapped to vertical v_y-axis in hodograph plane. Conversely, hodograph origin $0'$ corresponds to a critical point and is mapped to points along the basin's two vertical plane on physical plane. The phreatic surface may pass through these points since $0'$ is a point on the ellipse boundary. Likewise, stagnation points at origin 0 and two basin's bottom corners in physical plane are also mapped to coincide with origin $0'$ of hodograph plane.

Determination of phreatic surface location and shape is an essential part of groundwater analysis. Phreatic surface in a water table aquifer is a free surface defined as a boundary of an aquifer where pressure assumes constant. Phreatic surface may coincide with the water table or lie below it. It degenerates into a plane curve in two-dimensional fields separating between saturated and unsaturated zones. Hodograph plane is useful to trace the shape of phreatic surface on the physical plane since its shape is defined on hodograph plane. With reference to Equation 5.26, the boundary condition of phreatic surface shows that the phreatic surface resembles the shape of an ellipse and is presented in Figure 8.46. The ellipse center is located at $(0, -0.5k_v)$ with major axis, $\sqrt{k_h k_v}$ and minor axis, k_v parallel to v_x-axis and v_y-axis respectively. The ellipse is the locus of velocity vector end points that correspond to points of a phreatic surface. By mapping points on ellipse boundary from hodograph plane to physical plane, we obtain the phreatic surface $0'BCB0'$ lying below water table surface. Consider points B' on phreatic surface at the intersection of ellipse boundary and major axis in hodograph plane. They represent maximum x-component velocity and are mapped to image points along the phreatic surface in the physical plane as shown in Figure 8.45. At points B',

$$v_x = \pm 0.5\sqrt{k_h k_v}$$
$$v_y = -0.5k_v$$
$$|V| = \sqrt{v_x^2 + v_y^2} = 1.49e - 9m/s$$

The magnitude of piezometric head gradient is:

$$|\nabla \varphi| = \left|-\frac{V}{k}\right| = \left|-\frac{V}{\sqrt{k_h k_v}}\right| = \frac{1.49e - 9}{3.16e - 9} = 0.47$$

To determine the image points B by graphical means in consideration of kinematic condition, we use the slopes of lines $0'B'$ on hodograph plane to define tangents of free surface at points B equal to velocity $0'B'$ in Figure 8.45. The tangent is constructed by translating line $0'B'$ in parallel until it becomes a tangent at each point B since velocity is the same in direction and magnitude at B and B'. Point C' at the intersection between ellipse boundary and minor axis having only y-component downward flow in hodograph plane is mapped to the image point $C(0, h + 0.3\zeta_0)$ of the same velocity lying directly below water table at point A. It is a point at the intersection of phreatic surface and y-axis in physical plane since vertical downward flow occurs only along y-axis. The piezometric head and water surface elevation are about equal and define part of phreatic surface symmetric with respect to the y-axis.

$$v_x = 0; \; v_y = -k_v = -1e - 9m/s \text{ @ point } C.$$

To investigate the flow fields further, we follow the ellipse boundary $0'B'$ on hodograph plane or the corresponding phreatic surface $0'B$ in physical plane. In this exercise, the x-component velocity continues to increase from zero (points $0'$) to the maximum value (points B' or B) in both planes. The velocity modulus inclination angle on hodograph plane and the corresponding phreatic surface tangent in physical plane also increase from zero steadily following the paths. The water table is sloping downwards from its peak point A in both directions towards basin's two ends. The water table's shape is assumed by a smooth curve as shown in Figure 8.45. It is mapped to an image curve passing through points A' and $0'$ and enclosing the ellipse on hodograph plane. The region bounded by BABC on the physical plane is mapped to the outside of ellipse and bounded by the water table boundary. The region bounded by phreatic surface and basin's boundary is mapped to the inside of ellipse.

Tangents to the hodograph representation of streamline trajectories and vector modulus on hodograph plane are useful to derive qualitative information about the flow fields. They define acceleration vector and velocity vector of a particle at a point respectively. To visualize the action of these two vector quantities, we translate them like free vectors between image points in physical (Figure 8.45) and hodograph planes (Figure 8.46) in a mapping process. Consider points B' in Figure 8.46, the tangents along the hodograph representation of streamline define the acceleration vectors at these points also point upwards in the directions of streamline arrows. When both acceleration vector a and velocity V are displayed at a point, that is, B (or B'), the acceleration vector is decomposed into the streamline tangential and normal components through the law of parallelogram. The tangential component of velocity is represented by modulus $0'B'$ in Figure 8.46 while the normal component vanishes. In Figure 8.45, the tangential component of acceleration vector acts in the opposite direction of motion parallel to modulus $0'B'$ and hence speed decreases. Speed will be constant or increase when this component vanishes or acts in the direction of velocity. In Figure 8.45, the normal component of acceleration always acts towards the center of curvature of trajectory giving rise to centripetal acceleration. It points to the left side and right side of streamlines in the first quadrant and second quadrant respectively. This component is responsible for change in motion direction on the physical plane. By mapping circles or lines about origin from velocity hodograph plane, we could visualize loci of constant speed or isoclines of constant flow direction on physical plane.

The origin is a critical (stagnation) point where velocity vanishes. The gradient of velocity tensor is used to visualize nearby field characteristics.

$$[\nabla V]\Big|_{\substack{x=0 \\ y=0}} = \begin{bmatrix} \dfrac{\partial v_x}{\partial x} & \dfrac{\partial v_y}{\partial x} \\ \dfrac{\partial v_x}{\partial y} & \dfrac{\partial v_y}{\partial y} \end{bmatrix}\Big|_{\substack{x=0 \\ y=0}} = Ak_h \begin{bmatrix} 1 & 0 \\ 0 & -1 \end{bmatrix}$$

$$\text{where } A = \frac{4C_1\pi^2}{L^2}$$

Above symmetric matrix yields the following eigenvalues and eigenvectors,

$$\lambda_1 = Ak_h;\ \hat{e}_1 = \pm\begin{bmatrix} 1 & 0 \end{bmatrix}$$
$$\lambda_2 = -Ak_h;\ \hat{e}_2 = \pm\begin{bmatrix} 0 & 1 \end{bmatrix}$$
$$\lambda_1\lambda_2 < 0$$

The critical point is classified as saddle point. A pair of principal vectors coincides with coordinate axes and defines the orientations of two mutually orthogonal principal planes. The principal axes trajectories crossing at critical point assume the roles of dividing streamlines and serve as asymptotes for nearby streamlines. Upon approaching the critical point along dividing streamline y-axis ($\hat{e}_2, \lambda_2 < 0$), the vector field undergoes an indeterminate state as it vanishes at the critical point and eigenvalue changes its sign from negative (in-coming flux) to positive (out-going flux) along dividing stream-line (x-axis) in opposite directions.

Volumetric flow rate per unit depth bounded between two streamlines is given by:

$$\frac{Q}{\Delta z} = -\int_1^2 \nabla\psi \times \hat{k} \cdot \hat{b} dl = \psi_2 - \psi_1$$

$\hat{b}$ represents unit normal vector of an arbitrary surface bounded by the two streamlines. dl is the surface differential arc length.

8.4.4 Model Transformation by Affine Mapping

The degree of anisotropy can be assessed in Figure 8.44 by the angle of intersection between two crossing curves against a true right angle or by comparison with an irrotational field in an equivalent model plane. The latter is accomplished by curvilinear transformation of the original domain into a model ($x^* y^*$) plane where the two systems of curves become orthogonal. The following affine mapping functions are used:

$$\begin{bmatrix} x^* \\ y^* \end{bmatrix} = \begin{bmatrix} \sqrt{\frac{k_h}{k_v}} & 0 \\ 0 & 1 \end{bmatrix} \begin{bmatrix} x \\ y \end{bmatrix}$$

This yields the following Lame coefficients and Jacobian determinant of co-ordinate transformation:

$$h_1 = \frac{\partial x}{\partial x^*} = \sqrt{\frac{k_v}{k_h}}; \; h_2 = \frac{\partial y}{\partial y^*} = 1; \; J = \frac{\partial(x, y)}{\partial(x^*, y^*)} = h_1 h_2 = \sqrt{\frac{k_v}{k_h}}$$

Hence,

$$\varphi(x, y) = \varphi(x(x^*, y^*), y(x^*, y^*)) = \varphi^*(x^*, y^*)$$
$$\psi(x, y) = \psi(x^*(x^*, y^*), y^*(x^*, y^*)) = \psi^*(x^*, y^*)$$

The anisotropic dilation transformation results in a distorted domain with its horizontal dimension compressed and the size modified by a factor equal to Jacobian determinant. Furthermore, Equation 8.91 becomes Laplacian type for an isotropic and homogeneous medium. By substitution, the average hydraulic conductivity k for the model domain can be determined accordingly.

$$k\nabla^{*2}\varphi^* = 0$$
$$\text{where } k = \sqrt{k_h k_v} \tag{8.100}$$

It is equal to geometric mean. Figure 8.47 depicts the distributions of piezometric head contours which are independent of hydraulic conductivity in the

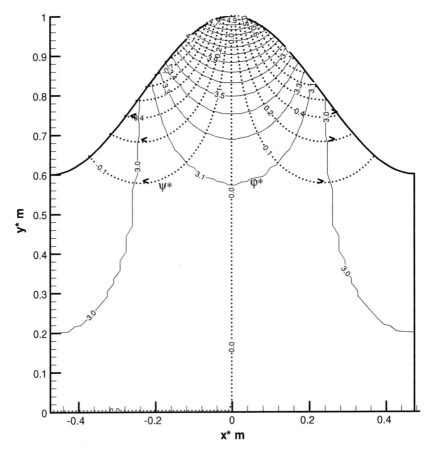

Figure 8.47 Curvilinear mapping of streamlines (dotted) and piezometric head (line) from original domain to model plane

model plane. The transfer number for the model configuration is:

$$D = \frac{\Delta \psi^*_{max}}{\Delta \varphi^*_{max}} = \frac{1.72 - (-1.72)}{5 - 2.9} = 1.64 \tag{8.101}$$

Graphical construction of field maps for an anisotropic medium is difficult because equipotentials and flux lines do not cross at right angles and cells are not orthogonal curvilinear squares. The construction work could be done with convenience and advantage using mapping techniques. The method includes transformation of boundary conditions in addition to governing equations and domain geometry. With the employment of a model plane transformed from the original domain, orthogonal curvilinear square field maps can be constructed in the usual manner and results are then transformed back to the original by an inverse mapping procedure. The coordinate of every crossing point of a field map in the model plane is scaled accordingly by the same mapping relation to yield a corresponding point in original configuration. By tracing smooth curves through these points, a field map is produced in the original domain.

Equipotentials are useful to assess the effects of anisotropy since its existence implies the field is irrotational and diffusion transport process is optimal. Figure 8.48 presents the distributions of piezometric head contours for the given anisotropic medium and equipotential contours in an assumed homogenous medium. Both systems of contours terminate on x-axis and two vertical ends perpendicularly. Comparison of two distributions shows that piezometric head finds it difficult to penetrate downwards because of high resistance in these directions with $k_h > k_v$. Anisotropic effects are more pronounced at the basin bottom than those in the water table top as the two patterns deviate more severely.

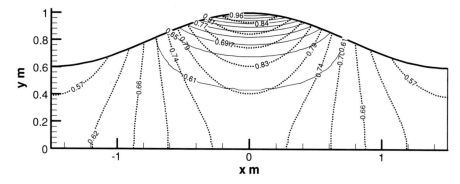

Figure 8.48 Comparison of equipotentials (dotted) and piezometric head distributions (line)

8.4.5 Visualization of Piezometric Head Diffusion

Diffusion of piezometric head in porous media could be studied using acoustic pressure wave equation. By neglecting wave propagation term, the wave phenomena degenerate into diffusion giving rise to a general diffusion equation in groundwater mechanics,

$$\nabla \cdot (\alpha_{ij} \cdot \nabla \varphi) = \frac{\partial \varphi}{\partial t}$$

The diffusion coefficient matrix for a porous medium is given by (Bear, 1988):

$$\alpha_{ij} = \begin{bmatrix} \alpha_h & 0 \\ 0 & \alpha_v \end{bmatrix} = \begin{bmatrix} k_h & 0 \\ 0 & k_v \end{bmatrix} \frac{\rho g h}{n_e} \tag{8.102}$$

n_e is a scalar quantity representing effective porosity. For steady state, piezometric head diffusive flux function is introduced written in terms of diffusive flux vector, which points towards the decreasing direction.

$$\begin{bmatrix} -\alpha_h \dfrac{\partial \varphi}{\partial x} \\ -\alpha_v \dfrac{\partial \varphi}{\partial y} \end{bmatrix} = -\nabla \psi_\varphi \times \hat{k}$$

It is observed that piezometric head diffusive flux function and stream function are similar within a proportional constant in Darcy's model.

$$\psi_\varphi = \frac{\rho g h}{n_e} \psi$$

A critical point also occurs at origin where piezometric head diffusive flux vector vanishes as well. The principal state analysis of the Hessian matrix of piezometric diffusive flux function gives the same result as those of stream function. It is classified as a saddle point.

Energy diffused per unit depth per unit time bounded between two piezometric head diffusive flux lines is proportional to the streamline difference,

$$\frac{Q}{\Delta z} = \int_1^2 \left(\frac{k_{ij}\rho g h}{n_e} \cdot \nabla \varphi \right) \cdot \hat{b} dl = \frac{\rho g h}{n_e} \left[\int_1^2 k_h \frac{\partial \varphi}{\partial x} dy - \int_1^2 k_v \frac{\partial \varphi}{\partial y} dx \right] = \frac{\rho g h}{n_e} (\psi_2 - \psi_1)$$

$\hat{b}$ represents unit normal vector of an arbitrary surface bounded by the two diffusive flux lines. dl is the surface differential arc length.

8.4.6 Viscous Stress Tensor Evaluation

Viscous stress tensor is evaluated from velocity fields. The shear stress component is given by:

$$
\tau_{yx} = -\mu \left[\frac{\partial v_x}{\partial y} + \frac{\partial v_y}{\partial x} \right]
$$

$$
= -\frac{\mu k C_1}{L^2} \left(\frac{k_h}{k_v} + 1 \right) (2\pi)^2 \sin \left(\frac{2\pi x}{L} \right) \sinh \left[\sqrt{\frac{k_h}{k_v}} 2\pi y/L \right] \quad (8.103)
$$

Distributions of shear stress contours are similar to those of streamlines within a proportional constant and are presented in Figure 8.49. Shear stress is anti-symmetric about center plane.

$$
\text{Since } L > h + \zeta
$$

$$
\rightarrow \frac{\partial v_x}{\partial y} \gg \frac{\partial v_y}{\partial x}
$$

Shear stress is primarily contributed by x-component velocity gradient in y-direction. Consider the negative y-face coincide with basin bottom boundary. Positive direction of shear stress points to positive x-axis direction following hydrodynamics convention. With $\tau_{yx} < 0$ in first quadrant, the direction of shear stress is marked with an arrow pointing in negative x-axis direction. With $\tau_{yx} > 0$ in second quadrant, the direction of shear stress is also marked with an arrow pointing towards positive x-axis direction. They appear to act against fluid motions. Shear stress vanishes at basin's two end boundaries and coordinates axes, implying that these boundaries/axes are principal axes.

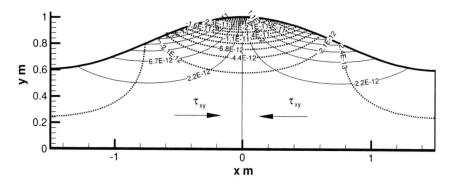

Figure 8.49 Distributions of stress component contours: τ_{yx} (line) and σ_x (dotted)

Normal stress components are:

$$\sigma_x = -\frac{2\mu k_h C_1}{L^2}(2\pi)^2 \cos\left(\frac{2\pi x}{L}\right) \cosh\left[\sqrt{\frac{k_h}{k_v}}2\pi y/L\right] \tag{8.104a}$$

$$\sigma_y = \frac{2\mu k_h C_1}{L^2}(2\pi)^2 \cos\left(\frac{2\pi x}{L}\right) \cosh\left[\sqrt{\frac{k_h}{k_v}}2\pi y/L\right] \tag{8.104b}$$

Hence, $\Lambda_1 = \sigma_x + \sigma_y = \sigma_1 + \sigma_2 = 0$

$$\rightarrow \sigma_1 = -\sigma_2$$
$$\rightarrow \sigma_1 > 0;\ \sigma_2 < 0$$
$$\frac{\partial \sigma_x}{\partial y}\Big|_{y=0} = -\frac{\partial \sigma_y}{\partial y}\Big|_{y=0} = 0$$
$$\frac{\partial \sigma_y}{\partial x}\Big|_{x=0, x=\pm L/2} = -\frac{\partial \sigma_x}{\partial x}\Big|_{x=0, x=\pm L/2} = 0$$

Figure 8.49 presents the distributions of normal stress contours symmetric with respect to center plane. They cross coordinate axes and two basin's end boundaries orthogonally and vanish at $x = \pm L/4$.

8.4.7 Viscous Stress Tensor Visualization

Principal states of viscous stress tensors describe the ways in which viscous momentum transfers in fluid systems. Principal values of viscous stress tensor represent viscous momentum fluxes acting on principal planes driven by momentum concentration (or velocity) gradients. They give rise to extreme normal stresses (principal stresses) and contiguous action of force transmissions along respective principal axes trajectories. Principal states of tensor are determined as per Equations 6.31 and 6.32. With a pair of equal principal values in opposite sign, it implies that contraction/stretching effects are equal everywhere and the tensor quadric becomes circles. Distributions of principal stress σ_1 are presented in Figure 8.50 together with a system of principal axes trajectories $\hat{e}_1$ and $\hat{e}_2$. Both distributions of principal axes $\hat{e}_1$ and $\hat{e}_2$ axes are symmetric with respect to y-axis. It is observed that y-coordinate is a principal axes $\hat{e}_1$ along which shear stress components vanish and normal stresses assume principal stresses.

$$\text{Along } y-\text{axis parallel to } \hat{e}_1 = \begin{bmatrix} 0 & 1 \end{bmatrix}: \sigma_1(0, y) = \sigma_y(0, y)$$

$$= \frac{2\mu k_h C_1}{L^2}(2\pi)^2 \cosh\left[\sqrt{\frac{k_h}{k_v}}2\pi y/L\right] > 0, \text{ Compression}$$

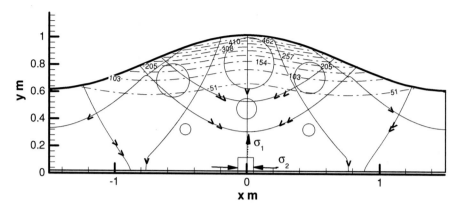

Figure 8.50 Distributions of σ_1^* principal eigenvalues (dash dot) and evolving stress tensor circles along $\hat{e}_1$ principal axes trajectories. $\hat{e}_2$ principal axes trajectories (double arrow); $\hat{e}_1$ principal axes trajectories (single arrows)

Viscous momentum transfers from the water table surface downwards along $\hat{e}_1$ principal axes trajectories under compression. $\hat{e}_1$ principal axes intersect x-axis and two basin's end boundaries orthogonally, which correspond to principal axes $\hat{e}_2$. On the other hand, expansion effects, being represented by $\sigma_2 < 0$, transmit from the water table surfaces along $\hat{e}_2$ principal axes trajectories towards center plane. They cross y-axis orthogonally which corresponds to a principal axis $\hat{e}_1$. Thus x-axis is one of the principal axes $\hat{e}_2$.

$$\text{Along } x - \text{axis parallel to } \hat{e}_2 = \begin{bmatrix} 1 & 0 \end{bmatrix}: \ \sigma_2(x,0) = \sigma_x(x,0)$$

$$= -\frac{2\mu k_h C_1}{L^2}(2\pi)^2 \cos\left(\frac{2\pi x}{L}\right) < 0, \text{ Tension}$$

There are two degenerate points along basin's bottom at $(\pm L/4, 0)$ where stress tensor vanishes. The tensor circle degenerates into a point where the normal stresses have been found to vanish earlier on. Indeed, all curves passing through any degenerate points are principal axes including x-axis. In hydrodynamics convention, the incoming and out-going directions of momentum fluxes are defined with respect to a control volume as tensile stresses (negative principal values) and compressive stresses (positive principal values) respectively as shown in Figure 8.50. The figure also shows evolving tensor circles at selected points along principal axes trajectories. The sizes of circles represent principal values as a measure of viscous momentum flux strength and transfer process activity. Viscous momentum transfer is localized in the upper region and is most active near water table surfaces where principal values have the greatest magnitude numerically.

8.4.8 Visualization of Mechanical Energy Transfer and Dissipation

Mechanical energy transfer trajectories are used to describe energy transfer paths in terms of energy flux vector:

$$\frac{dy}{dx} = \frac{v_y \dfrac{V^2}{2} + g\varphi v_y + \dfrac{1}{\rho}(\sigma_y v_y + \tau_{xy} v_x)}{v_x \dfrac{V^2}{2} + g\varphi v_x + \dfrac{1}{\rho}(\sigma_x v_x + \tau_{xy} v_y)} \tag{8.105}$$

Mechanical energy transfer trajectories are depicted in Figure 8.51 which shows energy is transferred from the water table top in the central region towards the bottom. It then changes its direction sideways in a manner similar to streamlines.

Energy dissipation is evaluated by the viscous dissipation function, that is,

$$\Pi = -\frac{1}{2\rho\mu}[\tau]:[\tau] \tag{8.106}$$

Distribution of energy dissipation is presented in Figure 8.51. Dissipation process took place along the water table surface where maximum dissipation occurs. It is active only in the upper basin region.

8.4.9 Visualization of Vorticity and Transport

Vorticity could be evaluated from velocity fields, that is,

$$\omega = \nabla \times V$$

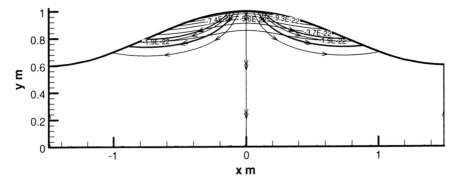

Figure 8.51 Distributions of mechanical energy transfer trajectories (arrow) and viscous dissipation energy contours (line)

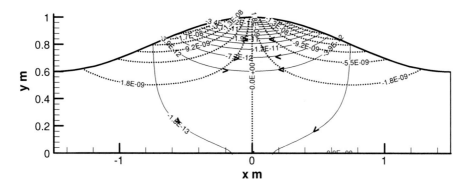

Figure 8.52 Distributions of vorticity (dotted) and vorticity diffusive flux trajectories (arrows)

where $\omega_x = \omega_y = 0$

$$\omega_z = \frac{\partial v_y}{\partial x} - \frac{\partial v_x}{\partial y}$$

$$= kC_1 \left(\frac{2\pi}{L}\right)^2 \left(1 - \frac{k_h}{k_v}\right) \sin\left(\frac{2\pi x}{L}\right) \sinh\left[\sqrt{k_h/k_v}\, 2\pi y/L\right] \quad (8.107)$$

Vorticity distributions are shown in Figure 8.52 antisymmetric with respect to center plane. With opposite sign between two adjacent quadrants, it implies vorticity is rotating in opposite direction in a steady state flow field. A pair of vorticity systems is developed as illustrated in Figure 8.43. They are separated by coordinate axes along which vorticity vanishes together with shear stress. Distributions of vorticity, streamlines, piezometric head diffusive trajectories and shear stresses are all similar within a constant. The gradient mapping of shear stress/vorticity will depict the paths from water table surface towards the center/bottom axis along which shear stress/vorticity changes most rapidly. To observe the connection between processes, these paths are then compared with those of principal axes trajectories of viscous stress tensor through which viscous momentum transfer. This is because the role of viscous effects is to maintain viscous momentum transport processes as well as force balance in the system. In doing so, it tracks the viscous stress distribution and makes the connection with vorticity distribution in the fields. Vortices are generated along the water table surface and diffuse downwards. To visualize vorticity diffusion transport, we call upon the vorticity function defined by Equation 6.46 in terms of vorticity diffusive flux and present results in Figure 8.52. Distributions of vorticity diffusive flux trajectories are symmetric with respect to center axis. The trajectories cross vorticity contours at right angle. Both contours squeeze towards water table peak indicating a region of high enstrophy density. Vorticity transports towards basin's bottom and center plane following the trajectories. Tangents along a vorticity diffusive flux trajectory define the transport directions. Two dividing vorticity diffusive flux trajectories pass through critical point (origin) and separate the fields into three local regions, that is, right, central and left. Vorticity generation and diffusion processes are most intensifying at the water table surface in the central region within which vorticity diffuses towards center plane. Vorticity transports towards the bottom in the right and left regions. The trajectories could be used to evaluate the amount of vorticity in volume transported per unit width per unit time bounded in-between two trajectories similar to stream function and streamlines.

The origin is a multiple point with coincidence of velocity critical point and vorticity diffusive flux critical point. The gradient of vorticity diffusive flux vector is used to visualize the topology of vorticity diffusive flux in the neighborhood.

$$[\nabla q]\Big|_{\substack{x=0\\y=0}}$$

$$= B \begin{bmatrix} \sin\left(\frac{2\pi x}{L}\right)\sinh\left(\sqrt{k_h/k_v}\,\frac{2\pi y}{L}\right) & -\sqrt{k_h/k_v}\cos\left(\frac{2\pi x}{L}\right)\cosh\left(\sqrt{k_h/k_v}\,\frac{2\pi y}{L}\right) \\ -\sqrt{k_h/k_v}\cos\left(\frac{2\pi x}{L}\right)\cosh\left(\sqrt{k_h/k_v}\,\frac{2\pi y}{L}\right) & -\frac{k_h}{k_v}\sin\left(\frac{2\pi x}{L}\right)\sinh\left(\sqrt{k_h/k_v}\,\frac{2\pi y}{L}\right) \end{bmatrix}$$

$$= B\sqrt{\frac{k_h}{k_v}}\begin{bmatrix} 0 & -1 \\ -1 & 0 \end{bmatrix}$$

where $B = vkC_1\left(1 - \frac{k_h}{k_v}\right)\left(\frac{2\pi}{L}\right)^4$

Vorticity diffusive flux: $q = -v\nabla\omega_z$.

The symmetric matrix yields the following eigenvalues and eigenvectors indicating that the critical point is a saddle point.

$$\lambda_1 = B\sqrt{\frac{k_h}{k_v}}; \ \hat{e}_1 = \pm\begin{bmatrix} 1 & -1 \end{bmatrix}$$

$$\lambda_2 = -B\sqrt{\frac{k_h}{k_v}}; \ \hat{e}_2 = \pm\begin{bmatrix} 1 & 1 \end{bmatrix}$$

$$\lambda_1\lambda_2 < 0$$

References

Aggarwai, S.K. & Manhapra, A. 1989. Use of heatlines for unsteady buoyancy-driven flow in a cylindrical enclosure. *Journal of Heat Transfer ASME*, **105**, pp. 576–578.

Bear, J. 1988. *Dynamics of fluids in porous media*. Dover Publications, Inc.

Bejan, A. 1995. *Convection heat transfer*. John Wiley & Sons, Inc.

Belfiore, L.A. 2003. *Transport phenomena for chemical reactor design*. John Wiley & Sons, Inc.

Bello-Ochende, F.L. 1986. Analysis of heat transfer by free convection in tilted rectangular cavities using the energy-analogue of the stream function. *International journal of Mechanical Engineering education*, **15** (2), pp. 90–89.

Bello-Ochende, F.L. 1988. A heat function formulation for thermal convection in a square cavity. *International communication heat and mass transfer*, **15**, pp. 193–202.

Bird, R.B. Stewart, W.E. & Lightfoot, E.N. 1960. *Transport phenomena*. John Wiley & Sons, Inc.

Blake, K.R. Bejan, A. & Poulikakos, D. 1993. Natural convection near 4C in a water saturated porous layer heated from below. *International journal of heat and mass transfer*, **36** (16), pp. 3957–3966.

Brodkey, R. 1995. *The phenomena of fluid motions*. Brodkey Publishing.

Cabral, B. & Leedom, L.C. 1993. Imaging vector fields using line integral convolution. In: *Proceedings ACM Siggraph 93, International Conference on Computer Graphics and Interactive Techniques*, Anaheim, CA, 1993, pp. 263–270. ACM.

Carslaw, H.S. & Jaeger, J.C. 1988. *Conduction of heat in solids*. Oxford Science Publications.

Cedergren, H.R. 1989. *Seepage, drainage, and flow nets*. John Wiley & Sons, Inc.

Chen, L.D. & Roquemore, W.M. 1990. Planar visualization of propane jets and propane jet diffusion flames. In: *Fluid Measurements and Instrumentation forum*, 1990. ASME Fluids Engineering Division.

Chou, P.C. & Pagano, N.J. 1967. *Elasticity, tensor, dyadic and engineering approach*. D. Van Nostrand Company, Inc.

Cook, D.M. 1975. *The theory of the electromagnetic field*. Prentice-Hall.

Visualization of Fields and Applications in Engineering, First Edition. Stephen Tou.
© 2011 John Wiley & Sons, Ltd. Published 2011 by John Wiley & Sons, Ltd.

Costa, V.A.F. 2003. Unified streamline, heatline and massline methods for the visualization of two-dimensional heat and mass transfer in anisotropic media. *International Heat and Mass Transfer*, **46** (8), pp. 1309–1320.

Courant, R. & John F. 1974, Reprint of 1989. *Introduction to Calculus and Analysis Volume II/1,2.* Springer.

Delmarcelle, T. & Hesselink, L. 1992. Visualization of second order tensor fields and matrix data. *Proc. IEEE visualization'92*, pp. 316–323. CS Press, Los Alamitos, CA.

Delmarcelle, T. & Hesselink, L. 1993. Visualizating second order tensor fields with hyperstreamlines. *IEEE computer graphics and applications*, **13** (4), pp. 25–33. Special issue on scientific visualization.

Delmarcelle, T. & Hesselink, L. 1994. The topology of second order tensor fields. *Proc. IEEE visualizaion'94*, CS Press, Los Alamitos, CA.

Dickinson, R.R. 1989. A unified approach to the design of visualization software for the analysis of field problems. *Proc. SPIE*, 1083, pp. 173–180.

Dickinson, R.R. 1991. Interactive analysis of the topology of 4D vector fields. *IBM Journal of Research and development*, **35**, pp. 59–66.

Fayyad, U. Grinstein, G.G. & Wierse A. 2002. *Information Visualization in Data Mining and Knowledge Discovery*. Morgan Kaufmann Publisher, Academic Press.

Fung, Y.C. 1965. *Foundations of solid mechanics*. Prentice-Hall.

Gallagher, R.S. ed., 1995. *Computer visualization, graphics technique for scientific and engineering analysis*. CRC Press Inc.

Graebel, W.P. 2001, *Engineering fluid mechanics*. Taylor & Francis Publishers.

Greenkorn, R.A. 1983. *Flow phenomena in porous media*. Marcel Dekker, Inc.

Haber, R.B. 1990. Visualization technique for engineering mechanics. *Computing systems in Engineering*, **1** (1), pp. 37–50.

Helman, J. & Hesselink, L. 1989a. Representation and display of vector field topology in fluid flow dataset. *IEEE computer*, **22** (8), pp. 27–369.

Helman, J. & Hesselink, L. 1989b. Automated analysis of fluid flow topology. *Proc. SPIE*, **1083**, pp. 144–152.

Helman, J. & Hesselink, L. 1990. Surface representation of 2-3 dimensional fluid flow topology. *Proc. IEEE, visualization'90*, pp. 6–13. CS Press, Los Alamitos, CA.

Helman, J. & Hesselink, L. 1991. Visualization of vector field topology in fluid flows. *Proc. IEEE, computer graphics and applications*, **11** (3), pp. 36–46. Special issue on visualization.

Jaeger, J.C. 1969. *Elasticity, fracture and flow with engineering and geological applications*. J.W. ArrowSmith.

Kellogg, O.D. 1953. *Foundations of potential theory*. Dover Publications, Inc.

Kerlick, G.D. 1990. Moving iconic objects in scientific visualization. *Proc. IEEE visualization'90*, pp. 124–130. CS Press, Los Alamitos, CA.

Kimura, S. & Bejan, A. 1983. The 'heatline' visualization of convective heat transfer. *Journal of Heat Transfer*, ASME, **105**, pp. 916–919.

Kimura, S. & Bejan, A. 1985. Natural convection in a stably heated corner filled with porous medium. *Journal of Heat Transfer*, ASME, **107**, pp. 293–298.

Konopinski, E.J. 1981. *Electromagnetic fields and relativistic particles*. McGraw-Hill.

Kraus, J.D. 1992. *Electromagnetics*. 4th ed. McGraw-Hill.

Kreyszig, E. 2005. *Advanced Engineering Mathematics*. 9th ed. John Wiley & Sons, Inc.

Landau, L.D. & Lifshitz, E.M. 1975. *The Classical Theory of Fields*. Vol. 2 (4th ed.). In: *Course of Theoretical Physics*. Butterworth-Heinemann.

Landau, L.D. & Lifshitz, E.M. 1987. *Fluid mechanics*, Vol 6 (1st ed.). In: *Course of Theoretical Physics*. Butterworth-Heinemann.

Love, A.E.H. 1944. *A treatise on the mathematical theory of elasticity*. Dover Edition.

Milne-Thomson, M.I. 1962. *Theoretical hydrodynamics*. McMillan company.

Moon, P. & Spencer, D.E. 1961. *Field theory handbook*. Springer-Verlag.

Morega, A. & Bejan, A. 1993. Heatline visualization of forced convection laminar boundary layers. *International Journal of Heat and Mass Transfer*, **36** (16), pp. 3957–3966.

Morega, A. 1988. The heat function approach to the thermo-magnetic convection of electroconductive melts. Revue Roumaine des Sciences Techniques, Série Électrotechnique et Énergétique, 33, pp. 359–368.

Mukhopadhyay, A. Qin, X. Puri, I.K. Aggarwai, S.K. 2002. On extension of heatline and massline concepts to reacting flows through use of conserved scalars. *Journal of Heat Transfer*, **124** (4), pp. 791–199.

Mukhopadhyay, A. Qin, X. Puri, I.K. Aggarwai, S.K. 2003. Visualization of scalar transport in non-reacting and reacting jets through a unified heatline and massline formulation. *Numerical Heat Transfer*, Part A applications, **44**, pp. 683–704.

Needham, T. 1997. *Visual Complex Analysis*. Clarendon Press.

Pagendarm, H.G. 1993. Scientific visualization in computational fluid dynamics. In: Power, H. Murphy, T.K.S. Hernandez, S. & Conner, J.J., ed. *Visualization and Intelligent Design in Engineering and Architecture*. WIT Press.

Pagendarm, H.G. 1999. Visualization environments supporting human communications. *Future Generation Computer Systems*, **15** (1), pp. 109–117.

Panton, R.L. 1984. *Incompressible flow*. John Wiley & Sons, Inc.

Pytel, A. and Kiusalaas, J. 2010, *Engineering Mechanics, Dynamics*. (third ed.) Cengage Learning.

Phillips, O.M. 1991. *Flow and reactions in permeable rocks*. Cambridge University Press.

Pickett, R.M. & Grinstein, G.G. 1988. Iconographic displays for visualization of multidimensional data. *Proceedings of the 1988 IEEE International Conference on Systems, Man, and Cybernetics*, pp. 514–519.

Polya, G. & Latta, G. 1974. *Complex variables*. John Wiley & Sons, Inc.

Pollack, G.L. & Stump, D.R. 2002. *Electromagnetism*. Addison Wesley.

Pozrikidis, C. 1997. *Introduction to theoretical and computational fluid dynamics*. Oxford University Press.

Prandtl, L. & Tietjens, O.G. 1957. *Fundamentals of hydro and aeromechanics*. Dover Publications, Inc.

Ramo, S. Whinnery, J.R. & Duzer, T.V. 1984. *Fields and waves in communication electronics*. John Wiley & Sons, Inc.

Saffman, P.G. 1992. *Vortex dynamics*. Cambridge University Press.

Schinzinger, R. & Laura, P.A.A. 1991. *Conformal mapping: methods and applications*. New York: Elsevier.

Sigfridsson, A.E. Ebbers, T. Heiberg, E. & Wigstorm, L. 2002. Tensor field visualization using adaptive filtering of noise fields combined with glyph rendering. *Proceedings of IEEE Visualization Conference*, pp. 372–378.

Sokolnikoff, I.S. 1983. *Mathematical theory of elasticity*. Robert E. Krieger Publishing Co.

Trevisan, O.V. & Bejan, A. 1987. Combined heat and mass transfer by natural convection in a vertical enclosure. *Journal of Heat Transfer*, ASME, **109**, pp. 104–112.

Wijk, J.J. van., 1990. A raster graphics approach to flow visualization. *Proceedings Eurographics'90*, North-Holland, Amsterdam, 1990, pp. 251–259, Elsevier Scientific Publishers.

Woods, R.J. 1903. *Strength of materials and elasticity of structural members*. Edward Arnold, London.

Index

Visualization of Fields and Applications in Engineering, First Edition. Stephen Tou.
© 2011 John Wiley & Sons, Ltd. Published 2011 by John Wiley & Sons, Ltd.

Breinigsville, PA USA
07 April 2011
259263BV00001B/1/P